Inference in Statistical Modelling and Machine Learning

Statistical modelling and machine learning offer a vast toolbox of inference methods with which to model the world, discover patterns and reach beyond the data to make predictions when the truth is not certain.

This concise book provides a clear introduction to those tools and to the core ideas – probabilistic model, likelihood, prior, posterior, overfitting, underfitting and cross-validation – that unify them. Toy and real examples illustrate diverse applications ranging from biomedical data to treasure hunts, while the accompanying datasets and computational notebooks in R and Python encourage hands-on learning. Instructors can benefit from online lecture slides and solutions to all the exercises.

Requiring only first-year university-level knowledge of calculus, probability and linear algebra, the book equips students in statistics, data science and machine learning, as well as those in quantitative applied and social science programmes, with the tools and conceptual foundations to explore more advanced techniques.

JAMES BURRIDGE is Professor of Probability and Statistical Physics at the University of Portsmouth, where he teaches probability, stochastic processes and statistical learning. He models language, birdsong, rocks, tessellations and games and develops commercial applications of machine learning in green technology.

NICK TOSH is Lecturer in Philosophy at the University of Galway, where he teaches critical thinking, logic and the philosophy of science. Until 2024, he coordinated Galway's Arts with Data Science BA.

Inference in Statistical Modelling and Machine Learning

A Concise Introduction

James Burridge
University of Portsmouth

Nick Tosh
University of Galway

Shaftesbury Road, Cambridge CB2 8EA, United Kingdom

One Liberty Plaza, 20th Floor, New York, NY 10006, USA

477 Williamstown Road, Port Melbourne, VIC 3207, Australia

314–321, 3rd Floor, Plot 3, Splendor Forum, Jasola District Centre,
New Delhi – 110025, India

Cambridge University Press is part of Cambridge University Press & Assessment,
a department of the University of Cambridge.

We share the University's mission to contribute to society through the pursuit of
education, learning and research at the highest international levels of excellence.

www.cambridge.org
Information on this title: www.cambridge.org/9781009630726
DOI: 10.1017/9781009630696

When citing this work, please include a reference to the DOI 10.1017/9781009630696

First published 2026

Cover image: Jan Toorop, *Twee gestileerde vrouwelijke figuren met klok in de hand*.
Rijksmuseum, Amsterdam. Public domain.

A catalogue record for this publication is available from the British Library

A Cataloging-in-Publication data record for this book is available from the Library of Congress

ISBN 978-1-009-63068-9 Hardback
ISBN 978-1-009-63072-6 Paperback

Additional resources for this publication at www.cambridge.org/9781009630689.

Contents

Preface ix
Acknowledgements xi
Notation xii

1 Orientation 1
1.1 Risky Inference 1
1.2 Supervised Learning 2
1.3 Unsupervised Learning 3
1.4 Reasoning with Probability: Frequentist and Bayesian Approaches 6
1.5 Deep Learning 7
1.6 Chapter Summary 7

2 Supervised Learning Warm-Up 8
2.1 The Queen's New Weapon 8
2.2 Best-Fit Curves 8
2.3 Cross-Validation 13
2.4 Exploiting Background Knowledge 14
2.5 Background Knowledge Is Essential 16
2.6 Some New Data 17
2.7 Quantifying Uncertainty in the Final Prediction Function 19
2.8 Chapter Summary 21
2.9 Further Exercises 22

3 Unsupervised Learning Warm-Up 24
3.1 Treasure Hunt 24
3.2 Approximating Distributions 25
3.3 Mixture Models 27
3.4 Cross-Validation Again 33
3.5 Chapter Summary 34
3.6 Further Exercises 34

4 Interlude: Probability, Likelihood and Bayes 36
4.1 Alien Invasion Detector 36
4.2 The Principle of Maximum Likelihood 37
4.3 A Preview of Bayesian Inference 39
4.4 Chapter Summary 40

5 Probabilistic Modelling 41
5.1 Overview: From Supervised Learning to Probabilistic Modelling 41
5.2 A Closer Look 45
5.3 Maximum Likelihood Can Lead to Overfitting 51
5.4 The Unsupervised Case 53
5.5 Chapter Summary 55
5.6 Further Exercises 55

6 Frequentist and Bayesian Uncertainty 57
6.1 Tasks and Datasets 57
6.2 Models, Densities, Likelihoods – and Uncertainty 58
6.3 Lizard Sex Ratios 61
6.4 Frequentist Uncertainty: The Basics 64
6.5 Bayesian Uncertainty: The Basics 73
6.6 Frequentist Tools 81
6.7 Bayesian Tools 88
6.8 Connections between the Sampling and Posterior Distributions 91
6.9 Chapter Summary 93
6.10 Further Exercises 93

7 Frequentist Linear Regression 96
7.1 Predicting Body Fat 96
7.2 Data Exploration 98
7.3 Model Interpretation 99
7.4 Measuring Performance and Comparing Models 100
7.5 Fitting Single-Predictor Models 102
7.6 Multiple-Predictor Models 109
7.7 Best Subset Selection 116
7.8 The Bias–Variance Trade-Off 118
7.9 Chapter Summary 121
7.10 Further Exercises 122

8 Directed Graphical Models 124
8.1 Directed Acyclic Graph Models 124
8.2 Conditional Independence 127
8.3 Chapter Summary 130
8.4 Further Exercises 131

9 Bayesian Linear Regression, Priors, and Regularisation 133
9.1 Bayesian Linear Regression 133
9.2 Application to Body Fat Data 142
9.3 Gaussian Priors and Ridge Regression 147
9.4 Laplace Priors and the Lasso 150
9.5 Chapter Summary 151
9.6 Further Exercises 152

10 Bayesian Methods 153
10.1 The Components of a Bayesian Model 153

10.2 Modelling Rainfall Statistics 154
10.3 Posterior Predictive Checks 157
10.4 A New Model 160
10.5 Markov Chain Monte Carlo 163
10.6 How to Choose Priors 173
10.7 MCMC for the New Model 174
10.8 Model Comparison 178
10.9 The Queen's Archers 180
10.10 Archery Training and Informative Priors 181
10.11 Accuracy Estimation and Hierarchical Models 186
10.12 Chapter Summary 191
10.13 Further Exercises 191

11 Classification 192
11.1 The Concept of Classification 192
11.2 Probabilistic Binary Classification: Preliminaries 193
11.3 Binary Logistic Regression 195
11.4 Probabilistic Multiclass Classification: Preliminaries 208
11.5 Multinomial Logistic Regression 211
11.6 k-Nearest Neighbours 215
11.7 Multinomial Logistic Regression versus k-NN in Higher Dimensional Spaces 218
11.8 Chapter Summary 220
11.9 Further Exercises 220

12 Unsupervised Learning: A Deeper Dive 222
12.1 Clustering 222
12.2 Density Estimation 229
12.3 Dimensionality Reduction 241
12.4 Chapter Summary 253
12.5 Further Exercises 254

13 Neural Networks and Deep Learning 255
13.1 Introduction 255
13.2 Single-Neuron Networks 258
13.3 Single-Layer Networks 260
13.4 Hidden Layers and the Universal Approximation Theorem 262
13.5 Deep Neural Networks 264
13.6 Training Neural Networks 266
13.7 Classifying Human Speech Sounds with MLPs 272
13.8 Convolutional Neural Networks 275
13.9 Other Architectures 284
13.10 Backpropagation 285
13.11 Vanishing and Exploding Gradients 289
13.12 Chapter Summary 291
13.13 Further Exercises 291

14 Expanding the Toolkit 293
14.1 More Prediction Function Families 293

14.2 Differential Equations and Stochastic Models 295
14.3 Generative Language Models 296
14.4 Chapter Summary 298

Appendix A **Probability Theory** 299

Appendix B **Linear Algebra** 301

Appendix C **Jensen's and Gibbs' Inequalities** 304

References 306
Index 307

Preface

If you know that all men are mortal and that Socrates is a man, you can infer that Socrates is mortal. If you know that a circle has radius 1, you can infer that its circumference is 2π. These are safe inferences: there is no possibility of their conclusions being false when their premises are true.

Other inferences are risky. Given sufficient information about the current state of the atmosphere, you can infer that it will rain in London tomorrow; but even if your data is solid and your analysis competent, you could be wrong. Given sufficient information about the current state of life on Earth, you can infer that the theory of evolution is true; but again, you could be wrong. It is logically possible that powerful aliens created the planet last week, complete with all apparent traces of a deeper past.

This is a book about risky inference. We are interested in inferences that 'go beyond' the current data to make predictions about future data, or to draw conclusions about the processes that generated the data we have. Either way, they involve a kind of guesswork.

It is notoriously difficult to pin down detailed general rules for doing risky inference well. Philosophers of science once hoped to do so; they have largely given up. Researchers in statistics and machine learning have taken a more practical approach. By tackling a wide variety of real-world inference problems, they have over the years built up a remarkable toolbox of mathematical and computational techniques for learning from data. The sheer number of these techniques is intimidating. Luckily, though, a few core ideas recur again and again: probabilistic model, likelihood, prior, posterior, overfitting, underfitting, regularisation and cross-validation. Anyone who achieves fluency with these core ideas has learned how to learn about inference. When they need to master a specific technique from the inference toolbox, they can consult a standard reference – and expect to understand it.

There is no shortage of books covering the core ideas. But they generally cover a good deal else besides, and weigh in at 600 pages or more. We think the core ideas can be conveyed and motivated in 300 pages. That is what we have attempted to do here. Our book is aimed at readers coming to the fields of inference, statistical modelling, and machine learning for the first time. Prerequisites are therefore minimal: calculus, linear algebra and probability at first-university-course level, and basic Python or R for those keen to dig into the supporting notebooks. (There is no code in the main text.)

After a brief orientation (Chapter 1), we begin with an informal introduction to overfitting, underfitting and cross-validation in supervised (Chapter 2) and unsupervised (Chapter 3) contexts. Then in Chapters 4–6 we introduce probabilistic models, likelihood functions and the core ideas of Bayesian and frequentist inference. This mathematical apparatus helps us think in a more principled way about the material covered in the introductory chapters. It also sets the stage for new ideas and methods covered in Chapters 7–12: multiple predictor models, graphical models, regularisation, Bayesian methods, classification, Gaussian mixture models and techniques for dimensionality reduction and clustering.

Chapter 13 is an introduction to neural networks and deep learning. Once again we emphasise core ideas. Overfitting and underfitting are ever-present concerns. Priors are encoded in network architectures. Probabilistic models are implicit in loss functions. We want to give readers the confidence to explore the deep learning literature and to experiment with deep learning methods. We can do this in one chapter thanks to the conceptual foundations laid in earlier chapters.

There are many tempting directions for further study, and in Chapter 14 we touch *very* lightly on some of them. Splines, generalised additive models and local models make brief appearances here; the question of how differential equations and stochastic processes can be incorporated into statistical models gets a whole page; and we close with a high-level summary of the probabilistic modelling strategy implicit in large language models.

Short, straightforward exercises are sprinkled throughout the book. Readers should tackle these as they encounter them, for they are integral to the main flow of ideas. Most chapters end with a 'further exercises' section, containing a small number of more challenging problems. Doing these is recommended but not essential: if you skip one, you will still be able to follow the rest of the book.

Acknowledgements

Burridge thanks Jesse, Anna, Samia, Jane and Jerry. Tosh is grateful to his partner Julianne, to his family and to all the students who studied Arts with Data Science at the University of Galway between 2018 and 2024. Both authors are grateful to Natalie Tomlinson and Anna Scriven at Cambridge University Press for their expert guidance throughout the writing and publication process.

Notation

The table gives a list of symbols which are used particularly frequently within the book, along with brief descriptions of their meanings. By default we assume that vectors are column vectors. Where this is important, we use transpose notation to make it explicit. For example, $\boldsymbol{x} = (x_1, x_2)^T$ denotes the column vector obtained by transposing the row vector (x_1, x_2) (see Appendix B for details).

Symbol	Meaning
X	Random variable
$\boldsymbol{X}$	Random vector
$\boldsymbol{x}$	Vector
$\underline{X}$	Vector of i.i.d. random variables representing separate observations
$\underline{x}$	An observation of $\underline{X}$
$\underline{\boldsymbol{A}}$	Matrix
$\mathcal{D}$	Random dataset
D	Dataset (often conceived as an observation of $\mathcal{D}$)
$\mathcal{L}$	Likelihood function
ℓ	Log-likelihood function
l	Loss function
$\mathfrak{g}$	The *true* form of a function
$\mathcal{N}(\mu, \sigma^2)$	Univariate normal distribution with mean μ and variance σ^2
$\mathcal{N}(x\|\mu, \sigma^2)$	Probability density of the univariate normal distribution
$\mathcal{N}(\boldsymbol{\mu}, \underline{\boldsymbol{\Sigma}})$	Multivariate normal distribution with mean $\boldsymbol{\mu}$ and covariance matrix $\underline{\boldsymbol{\Sigma}}$
$\mathcal{N}(\boldsymbol{x}\|\boldsymbol{\mu}, \underline{\boldsymbol{\Sigma}})$	Multivariate normal density
$\mathcal{E}$	Zero mean noise random variable
$\mathbb{E}$	The expectation operator
Var	The variance operator
Cov	The covariance operator
Pr	A probability measure
MSE	Mean squared error
RMSE	Root mean squared error
$\|\|\boldsymbol{x}\|\|$	Magnitude of a vector
$\boldsymbol{e}_k$	Unit vector in kth coordinate direction
$\mathbb{H}$	Cross-entropy
$\mathbb{KL}$	Kullback–Leibler divergence

1

Orientation

1.1 Risky Inference

Inference is the process of arriving at a conclusion based on evidence or data. Sometimes it is possible to draw conclusions which are certain to be true. Here is a well-known example. You are in a strange land where each person either always lies or always tells the truth. One day you meet a couple, Hilda and Godiva. You ask Hilda whether at least one of them is a liar; Hilda says 'yes'. What kind of people are Hilda and Godiva? Hilda's answer is your evidence, and there is only one possible conclusion. Godiva is a liar and Hilda is a truth teller.

■ ***Exercise*** 1.1 To explain this inference, we can list all the possible combinations of liar (L) and truth teller (T). Writing Hilda's type first and Godiva's second, these are $\{TT, TL, LT, LL\}$. Why is TL the only combination consistent with the evidence?

Now suppose that liars and truth tellers occasionally act against type. That is, a truth teller might spontaneously lie once in a while, and a liar will speak the truth from time to time. What can you infer from Hilda's answer now? 'Godiva is a liar and Hilda is a truth teller' might still be a good guess, but (on our new supposition) *all* the hypotheses listed in Exercise 1.1 are consistent with the evidence. For example, Hilda and Godiva might both be natural truth tellers, but on the evening you met them, Hilda decided to lie. Whatever your conclusion, you can no longer be sure it is correct. You are dealing with a *risky inference* problem. To get a better grip on it, more information – some *observational data* – would be helpful. How often do liars and truth tellers behave against type? Are liars and truth tellers equally common in the population, or does one type predominate? Do liars tend to couple up with liars and truth tellers with truth tellers, or do opposites attract? If you knew the answers to those questions, you might be able to attach *probabilities* to the four possible conclusions. But you still won't know for sure which one is true.

■ ***Exercise*** 1.2 Suppose that people act against type 20% of the time, and that a priori (i.e. before Hilda speaks) you attach equal probability to each of the four combinations of liar and truth teller. Show that after Hilda speaks, the probability you should attach to the conclusion 'Godiva is a liar and Hilda is a truth teller' is $\frac{4}{7}$.

Risky inference problems are very common, and this is a book about how to tackle them. The methods we present will not always involve probabilistic reasoning, but they often will.

As our title suggests, we engage with two closely related disciplines: statistical modelling and machine learning. Both are centrally concerned with problems of risky inference, and they tackle these problems by learning from data. Historically the goal of statistical modelling was to infer probabilistic models from data, and to do so in a mathematically rigorous way. The goal of machine learning was to enable computers (machines) to learn from data without explicit programming. The boundary between the two has blurred. Statistical modelling now relies heavily on computational tools, and probabilistic methods are fundamental to machine learning, providing theoretical and practical tools for handling uncertainty and generalisation. Some differences persist. Machine learning researchers remain more willing than statisticians to use 'black box' models (e.g. large neural networks) whose inner workings are difficult for humans to interpret. In contrast, statistical modelling places greater emphasis on understanding data-generating processes, favouring models that are relatively simple and human-interpretable. This book does not dwell on disciplinary distinctions. With the partial exception of Chapter 13, our focus is on ideas and techniques that are central to both fields.

1.2 Supervised Learning

Here are some examples of risky inference tasks:

- given a series of clinical measurements from a single individual, including blood sugar levels, weight, height, blood pressure and so on, predict whether that person has diabetes;
- given a photograph of a plant, predict the plant's species;
- given an audio recording of a person speaking, predict the person's age.

Although these tasks look superficially dissimilar, they share a common structure. There is some variable, y, whose value you wish to predict. (In simple cases y will just be a number or a categorical label.) In order to make your prediction, you have some case-specific information, x. This might be one or more numbers or measurements, an image, an audio file, some text, or some combination of these things. We often refer to x as the *predictor*, and y as the *response*. The aim is to find a *prediction function*, $\hat{g}$, which makes a prediction, $\hat{y}$, of y, given x:

$$\hat{y} = \hat{g}(x).$$

We follow standard practice here and use hat symbols to indicate predictions or estimates. The prediction function gets a hat because we think of it as an estimate of an unknown ideal function g that makes the best possible predictions of y given x.

Now, suppose that a dataset consisting of many previous examples of (x, y) pairs is available. If $\hat{g}$ is a good prediction function, we would expect it to perform well on these past examples. That is, we would expect $\hat{g}(x)$ to lie close to y for most of the (x, y) pairs in the dataset. Seeking a function that has this property is a powerful and useful strategy, but we will see later that it must be applied cautiously. It is almost always possible to construct a 'wiggly' prediction function which matches previous examples very well.[1] Unfortunately, such functions can often be poor at predicting y for values of x which were not in the original dataset. This pitfall is called *overfitting* (see Figure 1.1), and we will need to find ways to

[1] One situation in which this is not possible is where the dataset contains many pairs of examples of the form (x, y) and (x, y'), with y and y' taking wildly dissimilar values.

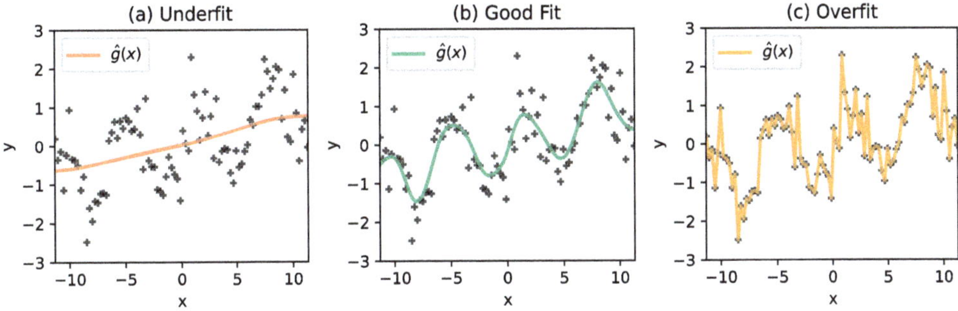

Figure 1.1 Three different candidates for prediction function $\hat{g}$, fitted to the same set of predictor–response (x, y) pairs. Intuitively, it seems clear that in graph (a), $\hat{g}$ is too simple to capture the systematic relationship between x and y. The prediction function in graph (b) seems to do a good job, while in graph (c), the prediction function appears to have fit to noise in the data.

avoid it. Broadly speaking, the trick will be to constrain – or at least to encourage – $\hat{g}$ to take a reasonably simple form. (Of course, what counts as 'reasonably simple' will often depend on the context.)

The process of searching for an optimal $\hat{g}$ is known as *supervised learning*, because for each predictor value we have a response y available to help us decide how well we have done. Supervised learning tasks are called *regression* tasks when the response is a continuously varying quantity, and *classification* tasks when the response is a categorical label.

■ ***Exercise*** 1.3 For each of the risky inference tasks listed at the start of this section, state whether it is best thought of as a regression or a classification task.

■ ***Exercise*** 1.4 You have reason to believe that the distance travelled by a cannonball, y, is directly proportional to the volume of gunpowder used to fire it, x. You perform three experiments, obtaining the (x, y) pairs $(1, 5)$, $(2, 11)$ and $(3, 14)$. How much gunpowder would you use if you wanted the cannonball to travel a distance $y = 7$? This is an example of risky inference: there isn't a provably correct answer. How would you justify yours?

We will study a regression task not too different from that of Exercise 1.4 in Chapter 2, and we will see many further examples of regression in Chapters 5, 7, 9 and 13. In Chapters 11 and 13, we will also study classification.

1.3 Unsupervised Learning

Supervised learning is not the only kind of inference we will be interested in. Consider the following example. As a bank customer, you make purchases of all sorts of sizes in different geographical locations or online stores. It is in your bank's interest to detect if someone has

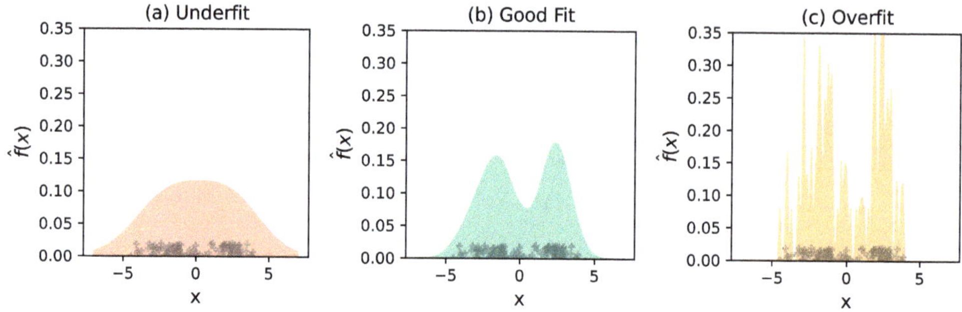

Figure 1.2 Three different candidates for a density function $\hat{f}$, fitted to the same set of data $x_1, x_2, \ldots, x_{50}$ (plotted as crosses at the base of each plot with some vertical jitter to aid visualisation). The data forms two visible clusters on either side of the origin. In graph (a), $\hat{f}$ is too simple to capture the two clusters. The density function in graph (b) seems to do a good job at capturing the clusters, while in graph (c), the density function has fit to noise in the data.

stolen your card or otherwise obtained access to your account. At first sight, this kind of fraud detection might look like a supervised problem: the predictors are the transaction sizes and locations, and the response is either 'fraudulent' or 'genuine'. In reality, fraud is quite rare, and there are lots of different ways in which it is committed, so it would be difficult to collect enough examples to learn an effective prediction function. Let's think about this another way. We do have plenty of examples of legitimate transactions for every cardholder. To keep things simple, we'll imagine that we know only the sizes of these transactions, which we could write as a long list $x_1, x_2, x_3, \ldots$. Suppose we could find a function $\hat{f}$ which measured how *typical* a transaction size was, with $\hat{f}(x)$ taking large values for typical values of x and small values for unusual ones. We could use this to detect *anomalies*: transaction sizes which stand out from the others. If x can be regarded as a continuous variable, it is natural to constrain $\hat{f}$ to be a probability density. The task of learning $\hat{f}$ from the data will then be an example of *density estimation*. This is an *unsupervised* task because we aren't given 'true' values of the density to compare our estimates to. Despite this, it is still possible to devise methods for learning good densities. Our density should describe the existing observations well, but not too well. For example, it would clearly be foolish to assume that in order to be typical, a new transaction would need to be exactly the same size as one of the previous transactions. That assumption would amount to *overfitting* – that is, matching previous observations too closely, and not allowing for the possibility of future deviations. (For stylised examples, see Figure 1.2.)

More generally, unsupervised learning involves the search for structure or patterns within a set of observations. Density estimation is one way to discover structure. However, there are other approaches which can often yield more practically useful and interpretable results. As an example, consider the problem of learning the sets of distinct sounds, or *phones*, that humans use to communicate. Each human language uses a subset of the possible sounds that the vocal tract is physically capable of making, and as a child learns to speak and understand, they must learn what their language's set is. How might we, as scientists, discover how many phones a given language uses? An experienced speaker of standard English knows instinctively that the vowel sounds in 'had', 'head', 'hit', 'heed', 'hot', 'hood', 'hud' and 'heard' are all different, but are these *all* the possible vowels? How would we go about counting how many there are in the language as a whole? One approach would be to measure frequency spectra, represented as high-dimensional vectors, emitted from the vocal tract when uttering a wide range of vowels. We could then attempt to identify groups of similar spectra. Identifying groups in this way is an inference problem known as *clustering*.

A typical clustering method is to assign each observation a group label and then iteratively relabel so that observations with the same label are close to each other. Unfortunately, if we tried to cluster vowel spectra using the above method, we would soon run into 'the curse of dimensionality'. In low dimensions, simple measures of distance between observations are usually meaningful. For example, we might be interested in identifying clusters of disease outbreaks based on a set of geographical coordinates of individual cases. In this case, the Euclidean distance between cases makes sense as a measure. With high-dimensional observations such as vowel spectra, standard distance measures may cease to be useful, as Exercise 1.5 illustrates.

■ ***Exercise*** 1.5 Consider a three-dimensional vector $\boldsymbol{U} = (U_1, U_2, U_3)$, where each coordinate U_i is selected uniformly and independently at random from $[0, 1]$. The vector $\boldsymbol{U}$ represents the location of a random point inside a unit cube. By adding extra coordinates, so that we have d dimensions, we can represent a random point inside a d-dimensional hypercube,

$$\boldsymbol{U} = (U_1, U_2, \ldots, U_d).$$

For a given value of d, imagine that you have randomly selected many such points. Consider any pair of them. The square of the difference in their ith coordinates is a random variable with some mean μ and standard deviation σ.[a] (Of course, neither μ nor σ depends on d.)

(a) Explain why the square of the distance between two such points has mean μd and standard deviation $\sigma\sqrt{d}$. (Hint: recall that the variance of the sum of independent random variables is the sum of their variances.)

(b) Explain why the ratio of the largest interpoint distance to the smallest interpoint distance tends to one as the dimension d increases (but the number of points is held fixed).

[a] The interested reader may wish to show that $\mu = 1/6$ and $\sigma^2 = 7/180$.

The above exercise reveals that in high dimensions, the contrast between narrowly and distantly separated observations begins to disappear, making it difficult to find groups. One way to tackle this problem is to look for ways of describing our observations using a smaller number of 'factors of variation'. Vowel sounds, for example, can often be described in terms of a small number of resonant frequencies. In general, however, we want automatic methods for boiling high-dimensional data down to a small number of features which describe its essence as efficiently as possible. Such *dimensionality reduction* techniques, in common with other unsupervised methods, are about finding patterns and structure in data. They have broad applicability, from visualising and explaining observations to improving predictions made using them.

We will play with a simple density estimation task in Chapter 3 and consider density estimation problems using frequentist and Bayesian methods in Chapters 6 and 10. We take a closer look at unsupervised learning in general in Chapter 12.

1.4 Reasoning with Probability: Frequentist and Bayesian Approaches

Risky inference involves guesswork. Given a number of alternative explanations of our observations, or alternative predictions, we want to find methods to choose between them and to quantify how certain we are about our guesses. The mathematical tool which allows us to quantify uncertainty and chance is probability. It is invaluable for doing inference. However, defining exactly what we mean when we make statements like 'the probability that the next coin toss will yield heads is 50%' or 'the probability that the defendant is innocent is only 1%' is difficult. There are two alternative ways to do it: *frequentist* and *Bayesian*. People sometimes talk about *frequentist inference* and *Bayesian inference* as opposed schools of thought. We prefer to see them as two conceptual toolkits, both of which are useful; but it is true that there are two distinct concepts of probability in play.

Frequentist inference relies on a definition of probability as long-run relative frequency. When construed frequentistically, the statement 'my coin lands heads with probability 0.5 when tossed' *means* that the proportion of heads in a sequence of N tosses of my coin would converge to 0.5 as N tends to infinity. Bayesian inference, by contrast, uses probabilities to represent 'degrees of belief' in *anything* that we are not sure about, be that the value of a constant, a prediction, or a scientific hypothesis. The starting point of Bayesian inference is a precise probabilistic statement of the prior beliefs we are bringing to the inference problem. Inference then consists of updating our beliefs in light of observations. Some people feel that the Bayesian interpretation of probability yields a more unified and satisfying approach to inference. It allows us to address a wider range of questions in a way that is closer to how we naturally think about the process of learning from observations. On the downside, we lose the objectivity of the frequentist definition of probability: degrees of belief are inherently subjective. However, we will see that the conclusions reached using the two approaches are often similar. Box 1.1 summarises the two interpretations of probability.

Chapters 4–6 develop some conceptual and mathematical machinery for thinking about inference probabilistically. Chapter 6 introduces core ideas of frequentist and Bayesian inference, and contrasts them. We will draw on both throughout the remainder of the book. Bayesian statistical modelling is covered in detail in Chapter 10.

★ Box 1.1 Frequentist and Bayesian probability

According to the **frequentist** view, the probability of an event A is the long run relative frequency with which it occurs over a repeated sequence of trials. If $N(A)$ is the number of times that A occurs out of N trials, then $\mathbf{Pr}(A) = \lim_{N\to\infty} N(A)/N$. According to the **Bayesian** view, $\mathbf{Pr}(A)$ is our degree of belief that A will occur or has occurred.

1.5 Deep Learning

Neural networks – architectures for building complex functions loosely inspired by the brain – are currently driving a revolution in artificial intelligence. The received view in the 1990s was that neural networks were difficult to train. Precisely because they were such supremely flexible function approximators, it seemed inevitable that large networks would overfit to their training data. Now, however, we routinely see neural networks with hundreds of billions of tunable parameters trained successfully on just a few tens of gigabytes of data. Exactly why contemporary architectures work so well remains to some extent an open question, but depth appears to be an important factor. Networks built out of many layers of neurons, with the ith layer's output serving as the $(i + 1)$th layer's input, are often easier to train than shallower networks with fewer layers. 'Deep learning' has therefore become the favoured catch-all label for designing, training and applying large neural networks.

This is our topic in Chapter 13. It is, of course, a *big* topic. We aim merely to introduce deep learning and to show readers who have worked through Chapters 1–12 that they are ready to explore it in detail. Much of the terrain will be familiar: training methods are partly inspired by probabilistic inference ideas; constraining network complexity is essential to avoid overfitting; and neural networks often learn most effectively when their architectures reflect prior knowledge about the domains in which they will be applied.

Inference has been the engine of science for generations, but there has never been a more exciting time to study it as a subject in its own right.

1.6 Chapter Summary

The purpose of this chapter has been to give an overview of the kinds of inference problems we are going to solve, and of the different ways that inference is done. Let us summarise the main points, with key words in bold.

- We are interested in **risky inference**: drawing conclusions, learning, and making predictions when we cannot be certain about the truth.
- Learning how to predict a *response* from one or more *predictors* based on previous predictor-response pairs is called **supervised learning**. Learning to predict a continuous response is called **regression**, whereas learning to predict a categorical label is **classification**.
- Inference problems in which there is no response variable (and we simply seek patterns and structure in observations) are called **unsupervised learning** problems. These include **density estimation**, **clustering** and **dimensionality reduction**.
- **Overfitting** occurs where we model data in excessive detail and fail to distinguish systematic patterns from noise.
- The **Bayesian** and **frequentist** interpretations of probability are associated with two distinct approaches to inference.
- Large **neural networks** can be used for supervised and unsupervised inference tasks. Modern architectures are remarkably effective, confounding the naive expectation that models with billions of parameters will inevitably overfit to their training data.

2

Supervised Learning Warm-Up

2.1 The Queen's New Weapon

As the Queen's mathematician, you may be called upon to apply your mind to any problem of interest to her. A war is brewing, and the Queen's inventors have devised a new weapon whose mechanism of action remains a closely kept secret. The inventors are willing to divulge that the measured power of the weapon, y, appears to depend on a quantity x which they can control. Each day, the inventors perform a test using a different value of x, with the aim of finding how it affects y. Each evening they hand you an envelope bearing the royal seal, which contains the results (x_k, y_k) of that day's test. Here k denotes the test number. They are interested to see if you can come up with a way of predicting y from x in future.

Notice that this is a supervised learning task. More specifically, it is a regression task.

2.2 Best-Fit Curves

So far, you have received 40 envelopes, and the collected results – which we'll call D for dataset – are shown in Figure 2.1a. It seems that larger x values are generally associated with larger y values, but the trend perhaps starts to break down at the largest x values. One way

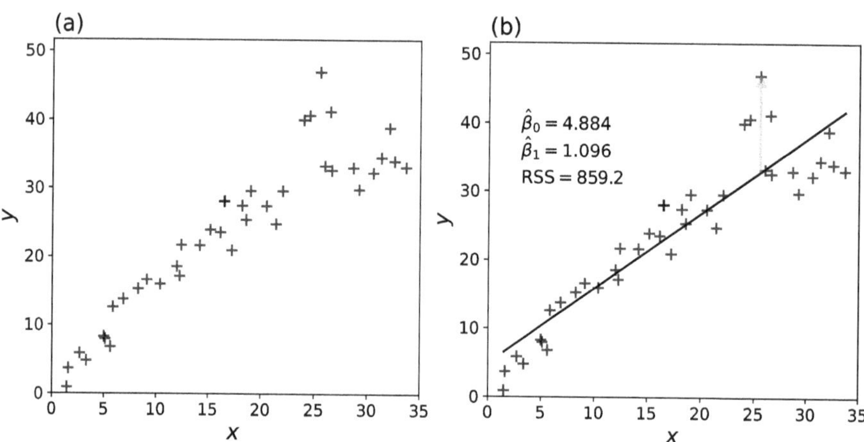

Figure 2.1 (a) Raw data D provided by the Queen's inventors. (b) A straight prediction line fitted to D by minimising the residual sum of squares. The grey arrow shows the prediction error for one datapoint.

to capture the fact that y seems to depend on x is to fit a curve of some sort to the data. But what kinds of curves should you consider? And how exactly should you measure goodness of fit?

If you consider only straight lines and measure goodness of fit using the so-called residual sum of squares (RSS), then you will find that the best fit is given by the line in Figure 2.1b. Let's unpack this. The purpose of the line is to make predictions. Given the next day's x value, x_{41}, you could use the line to read off a predicted y value, $\hat{y}_{41}$. The line, then, is a graphical representation of a prediction function, $\hat{g}_{\text{LIN}}$, that maps x values to $\hat{y}$ values. That is, given a *predictor* value x, we predict a *response*

$$\hat{y} = \hat{g}_{\text{LIN}}(x).$$

The grey arrow in Figure 2.1b is an example of a *residual*, that is, a prediction error. If we square each residual and sum the results, we get the RSS:

$$\text{RSS}(\hat{g}_{\text{LIN}}; D) = \sum_{(x,y)\in D} (y - \hat{g}_{\text{LIN}}(x))^2 .$$

The RSS is an example of a *loss function*. That means it is a way of measuring how well a prediction function $\hat{g}$ approximates a dataset D, with smaller values representing better approximations. The RSS is not the only way to do this, but it is computationally convenient, and there are sometimes principled reasons for preferring it to other measures. We'll take up that story in Chapter 5. For now, to keep things simple, let's just assume that RSS is the measure you care about. If you fit a prediction function by minimising RSS, you are doing *least-squares regression*, also sometimes called *ordinary least squares*.

The line in Figure 2.1b represents $\hat{g}_{\text{LIN}}$, the linear prediction function that achieves the smallest possible RSS. Here it is important to emphasise that 'smallest possible' means smallest possible *for a linear function*. That is,

$$\hat{g}_{\text{LIN}} = \underset{g\in \text{LIN}}{\arg\min}\, \text{RSS}(g; D),$$

where LIN is the family of all straight-line functions of the form $g(x) = \beta_0 + \beta_1 x$. If you are unfamiliar with arg min notation, see Box 2.1.

★ Box 2.1 arg min **and** arg max **notation**

The expression

$$\underset{a\in A}{\arg\min}\, \phi(a) \tag{2.1}$$

denotes the element of set A at which the function ϕ defined on A achieves its minimum in A. So, for example,

$$\underset{x\in\mathbb{R}}{\arg\min}\, (x^2 - 2) = 0.$$

In this book, expressions of the form (2.1) presuppose that ϕ has a unique minimum in

A; they are inappropriate if the presupposition is false. The expression

$$\arg\max_{a \in A} \phi(a)$$

denotes the element of set A at which ϕ achieves its *maximum* in A.

■ ***Exercise*** 2.1 Here is a toy example where we can minimise the RSS by hand. Suppose you want to fit the constant prediction function $g_{\text{CONST}}(x) = c$ to a dataset $D = \{(x_i, y_i)\}_{i=1}^n$. The RSS depends only on c and the data, and can be written

$$\text{RSS}(c; D) = \sum_{i=1}^{n} (y_i - c)^2.$$

Show that the RSS is minimised when c is the mean of the y values in D. That is,

$$\arg\min_{c \in \mathbb{R}} \text{RSS}(c; D) = \frac{1}{n} \sum_{i=1}^{n} y_i.$$

Hint: view the RSS as a function of c and find its stationary point.

■ ***Exercise*** 2.2 Let LINO be the set of all possible linear functions which pass through the origin. Functions in the LINO family have the form

$$g(x) = \beta x,$$

where β is a parameter that distinguishes between members of the family. Given a dataset $D = \{(x_i, y_i)\}_{i=1}^n$, the RSS when a member of LINO is used as the prediction function is

$$\text{RSS}(\beta; D) = \sum_{i=1}^{n} (y_i - \beta x_i)^2.$$

(a) Find an expression for the parameter value $\hat{\beta}$ which minimises the RSS.
(b) Consider the following dataset of predictor–response pairs:

i	1	2	3	4
x_i	-2	-1	1	2
y_i	-1.5	-0.5	0.6	1.1

(2.2)

Calculate $\hat{\beta}$ using this data.
(c) Plot both the data and the curve $y = \hat{\beta} x$ on the same pair of axes.

Looking at Figure 2.1, you might worry that the linear prediction strategy is too simple. Doesn't y start to *decrease* with x once x gets very large? Perhaps you should consider quadratic prediction functions, that is, functions of the form $g(x) = \beta_0 + \beta_1 x + \beta_2 x^2$. That

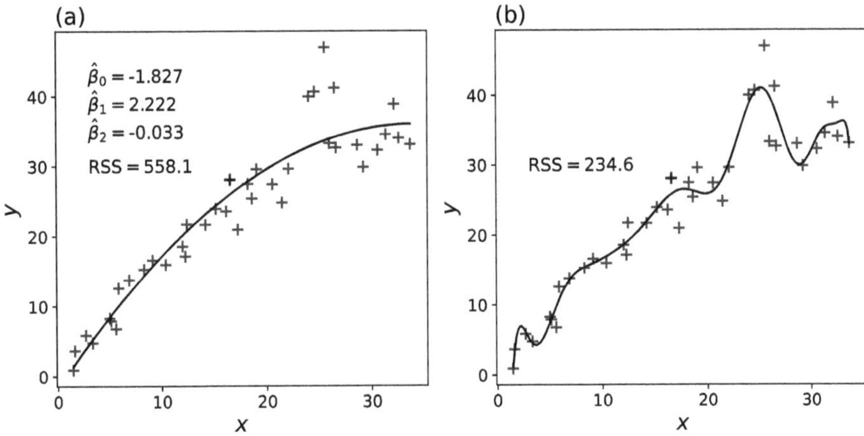

Figure 2.2 (a) Quadratic and (b) degree-15 polynomial prediction curves fitted to our experimental data by minimising the residual sum of squares. For the degree-15 fit, the 16 $\hat{\beta}_i$ estimates ($i = 0, \ldots, 15$) are not shown.

would at least allow for a bit of curvature. Figure 2.2(a) shows the best quadratic fit,

$$\hat{g}_{\text{QUAD}} = \underset{g \in \text{QUAD}}{\arg\min}\, \text{RSS}(g; D),$$

where QUAD is the family of all quadratic functions of x.

Comparing Figures 2.2a and 2.1b, you will see that $\hat{g}_{\text{QUAD}}$ achieves a lower RSS than $\hat{g}_{\text{LIN}}$. In other words, the best quadratic fit to D approximates D more closely than the best linear fit. Does this mean you should prefer $\hat{g}_{\text{QUAD}}$ to $\hat{g}_{\text{LIN}}$ when predicting the result of tomorrow's test? Not necessarily. To see that a simple comparison of RSS scores does not settle the issue, consider another function family – the family of degree-15 polynomial functions of x, that is, functions of the form $g(x) = \sum_{i=0}^{15} \beta_i x^i$. We'll call this family P15. High-degree polynomials can be very 'wiggly'. It is therefore possible to find functions in P15 that snake their way through the datapoints in D, passing most of them quite closely. Figure 2.2b shows $\hat{g}_{\text{P15}}$, the best fit available in the family. It achieves a much lower RSS on D than $\hat{g}_{\text{LIN}}$ or even $\hat{g}_{\text{QUAD}}$. But suppose you were told that tomorrow's x value would be $x_{41} = 29$. Would you trust $\hat{g}_{\text{P15}}$'s prediction for *tomorrow*'s y value over $\hat{g}_{\text{QUAD}}$'s prediction? Probably not! The sceptical intuition here is strong but justifying it requires a bit of work. Let us go slowly.

To obtain $\hat{g}_{\text{P15}}$, we exploited the flexibility of the P15 family and found a wiggly curve that is a close fit to the data. But suppose the data had been a little different. For example, suppose the Queen's inventors had tested the same set of x values but observed slightly different y values. Then we'd still have achieved a close fit with the P15 family, but we'd have done it using a different wiggly curve. Figure 2.3 demonstrates this. In each panel, the original dataset D is plotted (black crosses) along with a 'perturbed' dataset D' (blue circles), which we generated by nudging each point in D up or down a little at random. Dataset D' looks rather like dataset D, but the best degree-15 fit to D' looks noticeably different to the best degree-15 fit to D (Figure 2.3a). By contrast, the degree-1 (straight line) fits to D and to D' are so similar that it is difficult to distinguish them (Figure 2.3b). Now, we made up the

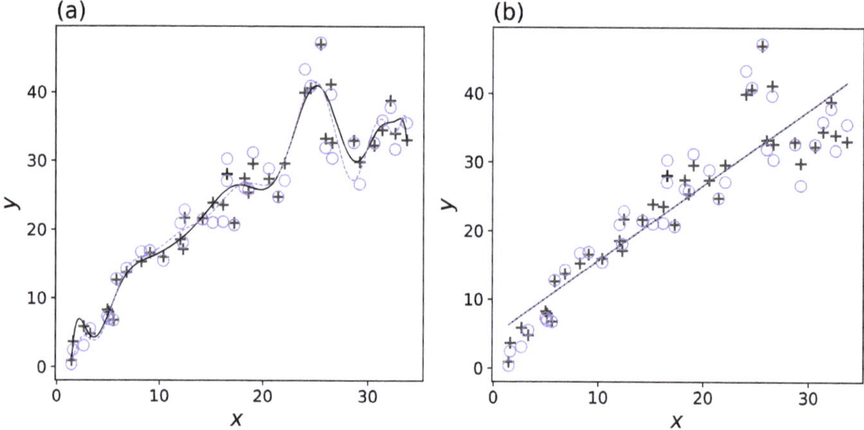

Figure 2.3 Degree-15 (a) and linear (b) fits to original dataset D (black crosses) and to perturbed dataset D' (blue circles). Prediction curves fitted to D are plotted in solid black and those fitted to D' are plotted in dashed blue.

perturbations in D', but if the inventors actually repeated all their experiments, you probably would not expect them to get *exactly* the same results each time. Rather, you would expect the y value obtained from any particular x value to bounce around a bit from trial to trial in an inherently unpredictable ('noisy') way. Using a curve-fitting procedure that is very sensitive to the precise values of y recorded in the dataset is therefore dangerous. The curve obtained from such a procedure may turn out to be driven more by noise than by any systematic relationship between predictor and response.

This is a reason to be wary of $\hat{g}_{\text{P15}}$ and other very 'wiggly' fits, but it isn't a knock-down argument against them. Remember, you know almost nothing about the Queen's new weapon or the inventors' experiments. All you have is a bunch of (x, y) pairs, plus a *tiny* bit of contextual information: you know that x is controlled by the inventors, and that y measures 'power' (whatever that means). So when you look at the data in Figure 2.1, how are you supposed to tell what is systematic trend and what is noise? Perhaps the systematic relationship between x and y is very simple – just a linear trend – and all the apparent wiggliness is noise. If so you should use the best-fit line $\hat{g}_{\text{LIN}}$ to predict tomorrow's y value. Or perhaps there is hardly any noise in the data, and the systematic relationship between x and y is genuinely very complicated. If that's right then you should use a more wiggly prediction function to predict tomorrow's y value. (How much more wiggly, though? Would $\hat{g}_{\text{P15}}$ be overkill?) Or perhaps there is a complicated systematic relationship between x and y, but you have no hope of spotting it because there is *also* a lot of noise. Then a simple prediction function would probably be safer. It may seem, given your ignorance, that you are in no position to pick a good strategy. Perhaps your best bet is to ask the Queen's inventors for more information and to refuse to predict tomorrow's y value until they get back to you. But in fact there is something you can do now.

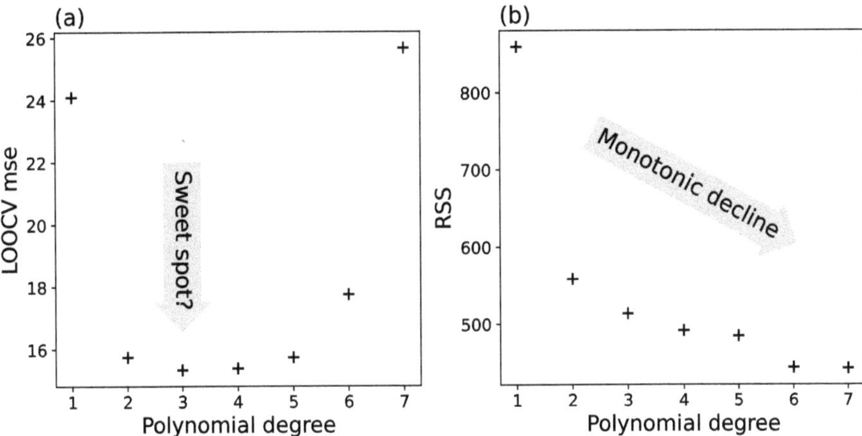

Figure 2.4 LOOCV mean squared error (a) and residual sum of squares (b) for polynomials of degrees 1 through 7 fitted to experimental data D. LOOCV MSE for degree 0 (not plotted) is 139.5. RSS for degree 0 (not plotted) is 5,305.

2.3 Cross-Validation

Here is the key insight: the task of predicting a y value from an x value and a dataset of 40 (x, y) pairs – which is what you have in dataset D – is not so very different from the task of predicting a y value from an x value and a dataset of 39 (x, y) pairs. But you can estimate how good any given curve-fitting strategy is at the latter task right now! Just fit your curve to 39 of the 40 available datapoints and see how well it predicts y from x for the 40th. In fact you can do this 40 times, letting each datapoint take its turn as the prediction target. Then you can use the average of the 40 squared prediction errors as your measure of the curve-fitting strategy's predictive power. Good strategies should yield small mean squared errors (MSEs).

This trick – fitting your curve to most (but not all) of the available datapoints and then testing it against the others – is known as *cross-validation*. If you leave out just one datapoint each time you fit your curve, you're doing *leave-one-out cross-validation* (LOOCV). We'll see another form of cross-validation (k-fold cross-validation) in Chapter 7. Figure 2.4a shows the LOOCV MSE for various polynomial curve-fitting strategies. The strategies are distinguished by degree. Fitting a straight line is the degree-1 strategy, fitting a quadratic is the degree-2 strategy, and so on. (The degree-zero strategy is fitting a constant, that is, a straight line of zero slope.) Notice that the LOOCV MSE initially declines with degree. A sloping line fitted to 39 of the datapoints of D tends to do a better job of predicting the fortieth than a horizontal (degree-0) line; a quadratic curve does better still, and a cubic better again. A quartic – degree 4 – does nearly as well as the cubic, but increasing the degree further appears to hurt performance.

We can now justify our intuition that fitting a degree-15 polynomial to D was not a sensible way to predict tomorrow's y value from tomorrow's x value. The LOOCV MSE for the degree-15 curve-fitting strategy turns out to be a catastrophic 1230. In other words, fitting a degree-15 polynomial to 39 of our datapoints is a *very* poor way to predict the 40th.

Of course, if we actually used the degree-15 strategy to predict tomorrow's y value we'd be fitting our curve to 40 datapoints, not 39; maybe the extra datapoint would help a bit. But the lesson is nevertheless clear: if you want to predict tomorrow's y value, don't use a degree-15 polynomial. You'll almost certainly do better if you use a cubic.

It's instructive to compare Figure 2.4a to Figure 2.4b, which plots RSS against degree for polynomials fitted to the whole dataset (no cross-validation). The latter plot shows how well different best-fit curves approximate *the specific datapoints they were fitted to*. Not surprisingly, curves drawn from more flexible function families achieve tighter fits. By contrast, Figure 2.4a shows how well different curve-fitting strategies perform when used to predict y from x for datapoints that were *not* used to fit the curves in the first place. Here the relationship between flexibility of function family and performance is more interesting. If the function family is not flexible enough, predictive performance is poor, because the best-fit curve can't capture the systematic trend in the data. We refer to this as *underfitting*. But if the function family is too flexible, the best-fit curve will end up chasing noise. It will approximate the datapoints it is fitted to too well, resorting to wild wiggles and swerves in order to do so; and these wiggles and swerves will make it worse at predicting y from x for *other* datapoints. We refer to this as *overfitting*.

Cross-validation is a powerful technique for finding sweet spots between underfitting and overfitting, and we will make heavy use of it throughout this book. However, it is not magic. If you think you've just learned that fitting a cubic polynomial to dataset D is the best way to tackle the inventors' prediction task, then you need to reconsider. You have seen some evidence that cubics do better than polynomials of degree n for values of n other than 3.[1] But that's just polynomials! There are plenty of other function families you could try. Maybe what you *really* need is a Fourier series, or a sum of exponentials, or so on – the possibilities are literally endless.

2.4 Exploiting Background Knowledge

A powerful new weapon cannot stay secret for long. Drinkers in the local tavern report seeing the Queen's inventors blasting stone balls out of an iron tube. The word is that the blast comes from a combustible powder made from a mixture of two substances, A and B. The inventors have been trying to determine what fraction of A to include in the mixture to create the biggest explosion.

Your relation to the dataset has just changed significantly. Variables x and y are no longer bloodless abstractions. If the rumours are right, then x measures the fraction of A in the inventors' powder, and y the power of the resulting explosion. (Perhaps y is the distance travelled by the projectile.) Looking back at the numbers, it seems likely that x is a percentage. Since the explosion is driven by a chemical reaction between A and B, y should be zero when x is 0 (no A) and also when x is 100 (no B). Moreover, there should be an optimal A percentage x_{opt} that ensures a complete burn of the mixture. When x is below x_{opt}, the supply of A is the limiting factor determining the size of the explosion, so y ought to be proportional to x. But once x exceeds x_{opt}, the situation flips: now there is a surplus of A, and the supply of

[1] The evidence is not very strong: setting n to 2, 4 and 5 worked nearly as well, and of course there are many other values we haven't tried yet.

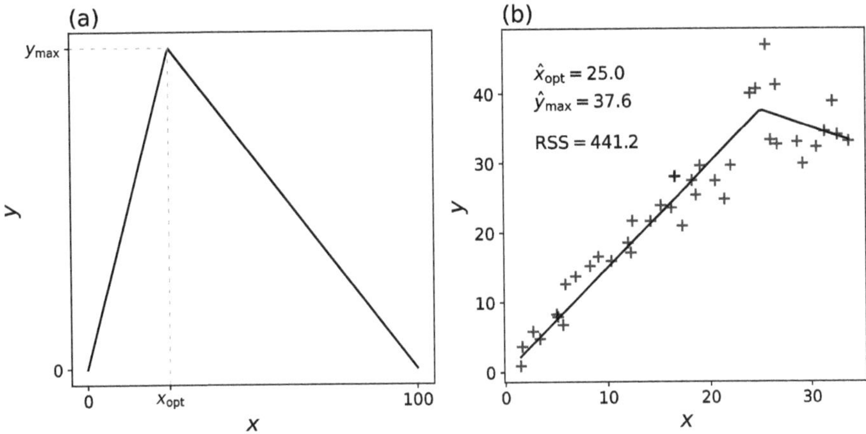

Figure 2.5 (a) One example of a function in the piecewise linear family AB-MIX. (b) A piecewise linear prediction function fitted to experimental data D by minimising the residual sum of squares.

B is the limiting factor. For these larger x values, y should *decrease* linearly with x, because increasing x reduces the supply of B.

Your prediction function $\hat{g}(x)$ ought therefore to be *piecewise linear*, rising from 0 at $x = 0$ to some maximum value y_{max} at $x = x_{opt}$ and then falling back to 0 again at $x = 100$. Figure 2.5a shows an example. We have a new function family to play with here; we'll call it AB-MIX to remind ourselves of the tavern rumours that inspired it. To specify a particular member of AB-MIX, you specify two values: an optimum A percentage x_{opt} and an associated maximum power y_{max}. The former must lie in the interval $(0, 100)$; the latter just needs to be non-negative.

It is a straightforward computation to find the function in AB-MIX that gives the best fit to dataset D:

$$\hat{g}_{\text{AB-MIX}} = \mathop{\arg\min}_{g \in \text{AB-MIX}} \text{RSS}(g; D).$$

It is plotted in Figure 2.5b.

If you were given tomorrow's x value, you could use that function to predict tomorrow's y value: $\hat{y}_{41} = \hat{g}_{\text{AB-MIX}}(x_{41})$. In light of the tavern rumours, this seems like a promising strategy. Would it actually work? Well, you don't have a crystal ball, so you can't be sure; but you can compute the LOOCV MSE. It turns out to be 12.9. That's the best cross-validation performance we've seen so far. The piecewise linear strategy is even better than the cubic strategy (LOOCV MSE 15.3) at predicting y from x for one datapoint of D given the other 39. It is reasonable to expect it will also do better at predicting y_{41} from x_{41} given all of D. However, it is very important to get clear on *why* it is reasonable to expect this. Please ponder that question for a few minutes – and then read on!

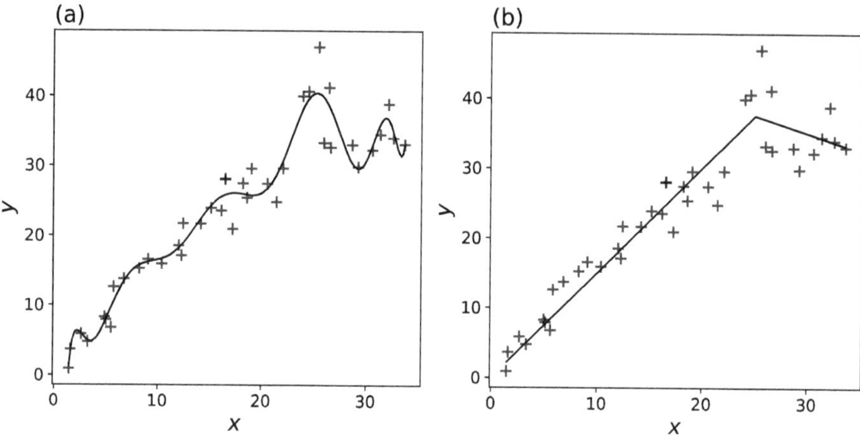

Figure 2.6 Degree-13 polynomial (a) and piecewise linear function (b) fitted to experimental data D by minimising the residual sum of squares.

2.5 Background Knowledge Is Essential

Wasn't there something a little *unscientific* about appealing to tavern rumours? We had a perfectly good dataset, after all. Shouldn't we have stuck to that, and left the tavern drinkers out of it? Perhaps you had a worry along those lines as you read and reflected on Section 2.4. There is a superficially tempting way of dispelling the worry, and it goes like this.

> Relax! The tavern rumours were just our *inspiration for trying* the piecewise linear strategy. After we tried it, our *reason for trusting* it was that it performed well in cross-validation – period. Thus, our justification for using $\hat{g}_{\text{AB-MIX}}$ to predict tomorrow's y value rests entirely on mathematical facts about the dataset.

Is that right? The idea that cross-validation performance is the only relevant criterion when justifying one prediction strategy over another is simple and appealing. Cross-validation scores are well defined, objective, and easy to verify. All competent scientists will agree that your piecewise linear strategy achieves an LOOCV MSE of 12.9 on dataset D, irrespective of the taverns they happen to frequent or the rumours they happen to trust. But should cross-validation scores trump all other considerations? The answer is no.

Let's briefly revisit our polynomial fits to D. In Figure 2.4a, we plotted LOOCV MSE for polynomial families of degrees 1 through 7. Increasing the degree beyond 3 appeared to hurt performance. We should now confess that appearances were deceptive. Check a wider range of degrees, and you will find that degree 13 does surprisingly well. In fact, it achieves a LOOCV MSE on D of 11.5, outperforming not only the cubic family but also the piecewise linear strategy. If cross-validation scores were trumps, then using a degree-13 fit to D to predict tomorrow's y value would be more reasonable than using a piecewise linear fit.

But take a look at Figure 2.6, which plots degree-13 and piecewise linear fits to D side by side. Would you really want to use the wiggly curve to predict tomorrow's y value? We hope your gut says no. The degree-13 strategy's strong cross-validation performance on D was most likely a fluke – a trick of the noise. After all, we *found* the degree-13 strategy by trying out lots of different polynomial families and picking the one with the best CV score on

D. That's clearly a procedure that favours families that happen to do flukily well on D. But flukes, by definition, are unlikely to be repeated. You should expect our 'winning' polynomial family to do less well on future data. (This general phenomenon is known as *winner's curse*.)

The piecewise linear strategy is a better bet than the degree-13 strategy for predicting future y values from future x values, even though the degree-13 strategy achieved a slightly better cross-validation score on presently available data. The crucial point is that the piecewise linear strategy was based on a well-informed guess about what the Queen's inventors are up to. The case for using it does not rest entirely on its strong cross-validation performance. By contrast, the degree-13 strategy is supported *only* by its cross-validation score on D. In the absence of any other reason to take it seriously, you should attribute its narrow victory in the cross-validation Olympics to fluke.

If you're still uncomfortable with the idea of overruling a cross-validation 'win', consider this: even before running any computational experiments, we could have stated with certainty that *somewhere out there in mathematical space* there exist function families that cross-validate well on D but make bad predictions $\hat{y}(x)$ for values of x not appearing in D. Indeed, given D, it is quite easy to write down a silly function family that will achieve *zero* cross-validation error on D, while making ludicrous out-of-sample predictions – say, that $\hat{y}(x) = -1000$ for every x not in D.

■ ***Exercise* 2.3** Write down such a function family. (Hint: there are many ways to do this, but the simplest approach is to define a 'family' that contains just one function.)

Consequently, you can't respond to the inventors' challenge by conducting a completely open-minded trawl through the universe of function families, with cross-validation performance your only criterion for acceptance. That approach is obviously a non-starter in practice (life is short and computational resources are finite), but it is also a non-starter in principle.

So it turns out that your justification for using $\hat{g}_{\text{AB-MIX}}$ to predict tomorrow's y value does *not* rest entirely on mathematical facts about the dataset. It is also important that AB-MIX is a plausible family of hypotheses about y's systematic dependence on x. In particular, it is more plausible than the family of degree-13 polynomials. This judgement is not driven by D, but by your background knowledge. You have the tavern rumours, and you have your general scientific education: you know that systematic relationships between variables of interest rarely look like degree-13 polynomials. To use the technical term favoured in statistics and machine learning, your judgement that AB-MIX is a plausible family is driven by your *priors*.

2.6 Some New Data

The Queen's inventors have made a special effort to get you more data today, and this evening's envelope contains 15 new (x, y) pairs. The new data is plotted in Figure 2.7a. This is a good opportunity for you to test your chosen prediction strategy. In fact, let's call the new batch of 15 datapoints the *test set*, D_{test}. We'll continue to use the symbol 'D' for the original set of 40 datapoints, which we'll now also refer to as the *training set*.

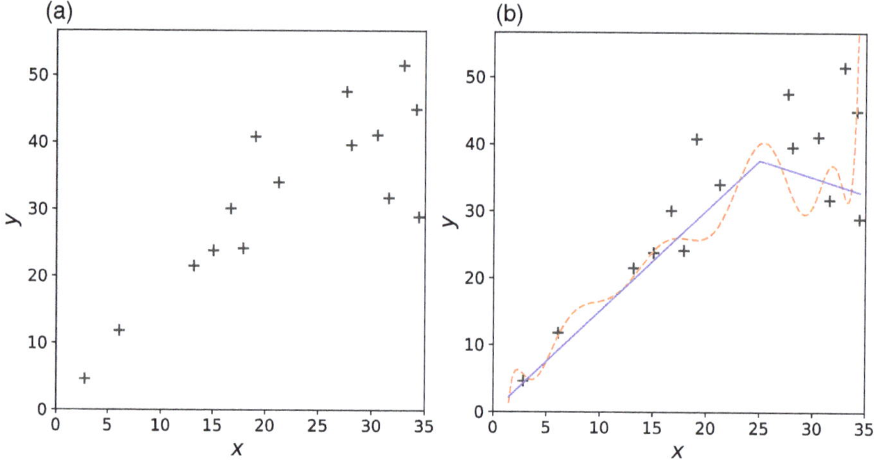

Figure 2.7 (a) New experimental data D_{test}. (b) The new data D_{test} plotted alongside degree-13 polynomial (dashed orange) and piecewise linear (solid blue) fits to *original* data D.

The prediction function you picked after analysing the training set D was

$$\hat{g}_{\text{AB-MIX}} = \underset{g \in \text{AB-MIX}}{\arg\min} \ \text{RSS}(g; D).$$

You preferred this to the degree-13 polynomial fit,

$$\hat{g}_{\text{P13}} = \underset{g \in \text{P13}}{\arg\min} \ \text{RSS}(g; D),$$

even though the degree-13 polynomial strategy did slightly better in cross-validation on D. So, which of these prediction functions, $\hat{g}_{\text{AB-MIX}}$ or $\hat{g}_{\text{P13}}$, does in fact do better at predicting the y values in D_{test} from the corresponding x values? We can answer that question by computing each prediction function's MSE over D_{test},

$$\text{MSE}(\hat{g}, D_{\text{test}}) = \frac{1}{|D_{\text{test}}|} \sum_{(x,y) \in D_{\text{test}}} (y - \hat{g}(x))^2 .$$

The result, not surprisingly, is that $\hat{g}_{\text{AB-MIX}}$ does much better than $\hat{g}_{\text{P13}}$. The MSEs on D_{test} are 29.7 and 168.0, respectively. Figure 2.7b plots D_{test} again, this time with $\hat{g}_{\text{AB-MIX}}$ and $\hat{g}_{\text{P13}}$ shown too. It is important to remember that these functions were fitted to D, not to D_{test}. A degree-13 polynomial fitted to the data in the Figure 2.7b would not have missed the rightmost point so disastrously. Indeed, the plot illustrates a general problem with fitting high-degree polynomials to sets of (x, y) pairs: the function you end up with tends to shoot abruptly up or down for values of x that lie outside the range encountered in the training set. As it happens, the largest x value in D_{test} exceeded the largest x value in D. That was bad news for $\hat{g}_{\text{P13}}$, but no problem for $\hat{g}_{\text{AB-MIX}}$. This is not the only reason to be wary of high-degree polynomial fits, however: even if the rightmost point is dropped from the test set, $\hat{g}_{\text{AB-MIX}}$ still has a smaller MSE than $\hat{g}_{\text{P13}}$.

The MSE of $\hat{g}_{\text{P13}}$ on D_{test} is a more reliable indicator of its likely future predictive power than its LOOCV MSE on D. That is because we selected degree 13 by trying many different degrees and picking the one that worked best (in the LOOCV MSE sense) on D. As noted earlier, that makes it likely that degree 13 did *flukily* well on D. Thus, its performance on D is not a good guide to its performance on future data. The general principle here is *don't use the same data for optimising a prediction strategy and for estimating that strategy's future performance*. Notice that we have actually applied this principle twice in the course of our supervised learning warm-up: first when computing cross-validation scores to help select a good prediction strategy, and second just now, when estimating the likely future performance of the strategy we selected. The two stages here are potentially confusing, so let's recap.

As we noted earlier, it would have been a serious error to infer that a high-degree polynomial fit to D would perform well in future simply because it achieved low RSS on D. This would, in fact, have been a violation of the principle we just articulated, because we would have been using the same data to optimise a prediction function (picking polynomial coefficients to minimise RSS) and to estimate that function's future performance (treating low RSS on D as evidence that errors would likely be low in future). With cross-validation, though, there is no violation: the datapoints used as prediction targets are not also used to shape the predictions. That's the first application of the principle. The second comes after we have made our final choice of prediction function. Cross-validation scores helped us make that choice, even if they didn't dictate it; and we used all the data in D to compute those scores. Thus, we used all the data in D to optimise our prediction strategy. Thus, to get an unbiased estimate of our strategy's likely future performance, we needed some new data.

Of course, we can't always rely on getting new data just when we need it, but we can do something nearly as good. At the very start of our project, just after we receive our dataset, we can pick a random sample of the datapoints, call that sample the test set, and then *lock it away*. We make no use of the test set whatsoever until we want to estimate the likely future performance of the inference procedure we have selected.

2.7 Quantifying Uncertainty in the Final Prediction Function

Now that you have the additional 15 datapoints in D_{test}, it would be silly not to use them to fine-tune your prediction function. Merging datasets D (40 points) and D_{test} (15 points) together yields a dataset D_{final} of 55 points. This final dataset is plotted in Figure 2.8a, along with the best piecewise linear fit to it. The new estimates of x_{opt} and y_{max} are $\hat{x}_{opt} = 25.5$ and $\hat{y}_{max} = 39.8$.

Congratulating yourself on a job well done, you report these estimates back to the Queen's inventors. But they are not satisfied. 'Presumably you don't think the optimal percentage of substance A is *exactly* 25.5%', they reply. 'So when you say your best estimate is that it's 25.5%, what are you really telling us? Are you saying the true value of x_{opt} probably lies between 25.45 and 25.55? If not, what interval would you propose?' This is a difficult question, and it is not immediately clear how to answer it. Given the data you have, D_{final}, it really is true that $\hat{x}_{opt} = 25.5$ is the RSS-minimising estimate of x_{opt}. There is no significant uncertainty about *this*; you could provide second and third decimal places if the inventors were interested. The problem, of course, is that the data you happen to have on the Queen's new weapon is not the ultimate authority on the topic: it's just the data you happen to have. If the inventors repeated their experiments and produced a new set of 55 datapoints – call it

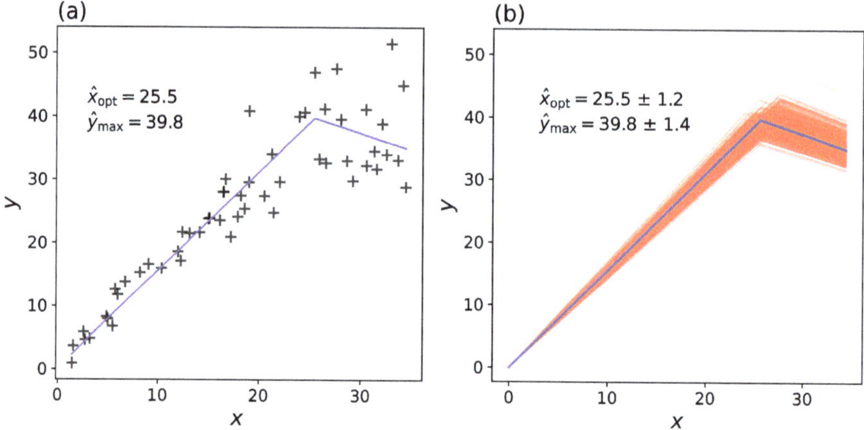

Figure 2.8 (a) Dataset D_{final} and the best piecewise linear fit to it. (b) Piecewise linear fits (thin orange lines) to 500 bootstrap replicas of D_{final}. The fit to D_{final} is represented by the thicker blue line for reference.

D_{rep} – it wouldn't be identical to D_{final}. Since D_{rep} wouldn't be identical to D_{final}, it's unlikely that the best estimate of x_{opt} based on D_{rep} would be precisely equal to the best estimate of x_{opt} based on D_{final}. It would, however, be no less reasonable an estimate. So, one way to conceptualise the uncertainty in your estimate of x_{opt} is to think about how much that estimate would bounce around if you repeated the data generation and analysis procedure from scratch (i.e. not retaining previous data), again and again and again. If through some clever magic you learned that the standard deviation of these individual estimates would be, say, 0.9, then you could report '±0.9' as a rough measure of the uncertainty in your estimate of x_{opt}. Now you don't, in fact, have a magical way of learning this standard deviation, but the situation is not entirely hopeless. There are ways you can estimate it using just the data you do have.

The method we'll describe here is called *bootstrapping*. (The term is an allusion to pulling oneself up by one's own bootstraps.) Bootstrapping is appropriate only if the datapoints in the actual dataset can reasonably be regarded as a random sample from some population of possible datapoints. Let's assume that we can think of our dataset that way. Unfortunately we're not in a position to list all the points that are in the population of possible datapoints, but we can list some of them: the ones we actually observed. We can take the set of actually observed datapoints as a crude substitute for the population of all possible datapoints, and we can simulate the process of drawing random samples of size 55 from the true population of all possible datapoints by drawing samples of size 55 from the set of 55 actually observed datapoints, *with replacement*. Replacement is essential: after a datapoint has been drawn, it must remain available as a possible pick for the next draw. If you draw a sample of size 55 from a set of 55 points *without* replacement, you will just reproduce the original set. But if you draw with replacement, you may (indeed you almost certainly will) draw some of the points more than once and miss out on others altogether. This process is called *resampling*. The new dataset it generates is called a *replica dataset* or *bootstrap replicate*. Why is resampling a good way of simulating reruns of the process that generated the actual dataset? Well, it may not be: your actual dataset may not be sufficiently representative of the range of possible datapoints. But the hope is that the actual dataset gives a rough indication of the sorts of datapoints that can be produced by the data-generating process. In reruns of the process, you might, just by chance, see more examples of one sort and fewer examples of another. Resampling approximates this variation.

Figure 2.8b shows 500 bootstrap prediction functions (thin orange lines), each one fitted to a different bootstrap replicate of the original dataset. The fit to the original dataset is also shown (thick blue line). The distribution of bootstrap prediction functions – the spread of orange in the plot – offers you one way of quantifying your uncertainty in the prediction function, and hence in your estimates of x_{opt} and y_{max}. Each bootstrap prediction function corresponds to a bootstrap replica estimate of x_{opt}, and a bootstrap replica estimate of y_{max}. It turns out that the standard deviation of the 500 replica x_{opt} estimates is approximately 1.2. So, you might tell the inventors that your estimate of x_{opt} is $\hat{x}_{\text{opt}} = 25.5 \pm 1.2$.

Notice that bootstrapping is a way of measuring uncertainty in the prediction function, not a way of measuring uncertainty in predictions of y given x. To see the difference, imagine you had 55,000 (as opposed to 55) datapoints, all produced by the same data-generating process, and all with x values lying in the range plotted in Figure 2.8. Your uncertainty about the best piecewise linear prediction function would now be dramatically less. Bootstrap prediction functions – that is, prediction functions fitted to bootstrap replicates of the enormous actual dataset – would all look much the same. But the data would still look noisy. Cross-validated MSE would be only slightly less than with the original dataset of 55 points.

We'll revisit bootstrapping and explore other ways of measuring uncertainty in Chapter 6.

2.8 Chapter Summary

We have taken our first look at a supervised learning problem. We explored methods for constructing prediction functions and comparing their performance. Let us summarise the key points.

- Given a set of predictor–response pairs, we can learn the parameters of a **prediction function** by minimising a **loss**. One example of a loss is the sum of the squared errors between the actual and predicted responses. This sum is known as the **residual sum of squares**.
- We can usually reduce the loss by using more **flexible** prediction function families. But if we use a function family that is too flexible, we will end up fitting to noise in the data, and failing to capture the systematic component of the predictor–response relationship. This is known as **overfitting**.
- Overfitting leads to poor prediction performance on unseen data, that is, on data that was not used to learn the parameters of the prediction function.
- We use **cross-validation** to estimate what the performance of a prediction method (e.g. 'fit a fifth-order polynomial by minimising RSS') would be on unseen data.
- However, when cross-validation is used to compare a large number of prediction methods, the winner's score is not a reliable guide to its performance on unseen data. The winner will probably perform less impressively in future.

- When deciding which prediction methods to compare, we must rely on **prior knowledge** to pick plausible candidates.
- Bootstrapping offers one way to quantify uncertainty in prediction functions.

2.9 Further Exercises

■ ***Exercise* 2.4** A solar equinox occurs when day and night are of equal duration, and a solar solstice occurs when the day is at its shortest or longest. One equinox occurs on 20th March, and in the northern hemisphere this marks the halfway point between the winter and summer solstices.

Functions in the DAY family all have the form

$$g(x) = 12 + \beta \sin\left(\frac{2\pi x}{365}\right),$$

where β is a free parameter.

(a) Let x be the number of days since the March equinox, and y be the number of hours of daylight (at a given location). Suppose we want to predict y from x. Given a dataset $D = \{(x_i, y_i)\}_{i=1}^n$, write down an expression for the RSS when a member of DAY is used as the prediction function.

(b) Find an expression for the parameter value $\hat{\beta}$ which minimises the RSS.

(c) The following dataset was collected in London:

k	1	2	3
x_k	13	26	41
y_k	13	14	15

(2.3)

Estimate the number of hours of daylight in London at the summer solstice on 21st June ($x = 93$). For reference, the exact answer is 16 hours and 38 minutes.

■ ***Exercise* 2.5** A rare species of jerboa lives in the sands surrounding a desert research station, where you work as an ecologist. Your research project requires you to venture into the desert to observe the jerboa during the night hours when they leave their burrows to feed. You would like do so at times when you are more likely than not to see one. One night you camp out in the desert. Starting from 6 p.m. you divide the night into 8-minute intervals (100 of them) and record whether or not you see a jerboa in each interval. Your results are marked on the grid in Figure 2.9, labelled 'Train'. A grey square indicates that you saw a jerboa in that interval.

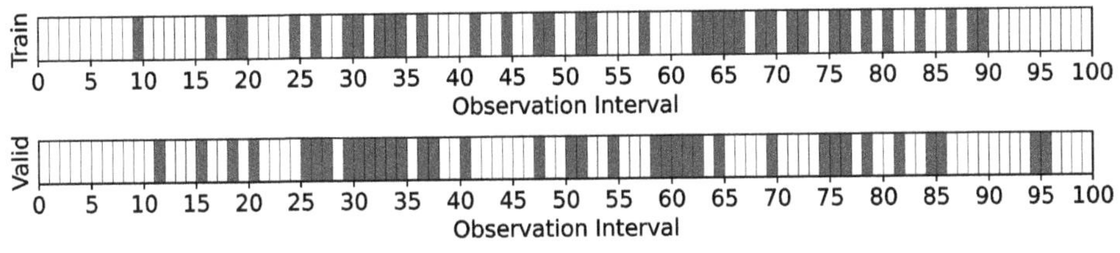

Figure 2.9 Jerboa observations: training and validation data

Let x be an interval number, and $y(x) \in \{0, 1\}$ be the *indicator variable* for observing a jerboa in that interval. Here $y(x) = 1$ if you see one, and $y(x) = 0$ if not. Consider the following supervised learning problem: given the interval number x, predict y. We will construct a prediction function by defining k disjoint 'jerboa' intervals $J_1, J_2, \ldots, J_k$ and let

$$\hat{y}(x) = g(x) = \begin{cases} 1 & \text{if } x \in J_i \text{ for some } i \in \{1, \ldots, k\}, \\ 0 & \text{otherwise.} \end{cases}$$

We also define the RSS $= \sum_{x=1}^{100} (y(x) - g(x))^2$. To measure the performance of g on *unseen* data, you collect a second set of data labelled as 'Valid'.

(a) Consider the $k = 1$ prediction function with $J_1 = [15, 85]$. Calculate the RSS for the training and the validation data.
(b) Construct your own prediction functions for $k \in \{2, 3, 5, 10\}$ based on the principle of minimising the training RSS (you will need to choose your jerboa intervals 'by eye'). Calculate the RSS scores of your prediction functions using the validation data.
(c) Plot graphs of training and validation RSS vs k. How does training RSS behave with increasing k? What is the optimal choice of k?
(d) Are your results consistent with the hypothesis that the jerboa species is crepuscular, that is, most active during twilight?
(e) In fact *most* species of jerboa are crepuscular. Given this background knowledge, is there a case for changing the prediction function family you used for the supervised learning problem in this exercise?

3

Unsupervised Learning Warm-Up

3.1 Treasure Hunt

At times of danger to the realm, the Queen's predecessors buried hoards of treasure at random locations near the capital. The Queen now wishes to draw on these reserves. The locations of buried hoards were recorded on slips of paper by the officials sent to bury them. To preserve secrecy, each slip was posted through a narrow slit in a sealed strong box in the Royal archives. To their dismay, upon breaking the box open the Queen's ministers have discovered that it is infested with bookworms. Only 90 readable slips remain, a small fraction of the original total. We will assume that the bookworms chose which slips to eat 'uniformly at random' from the population. Figure 3.1 is a map of the locations recorded on the surviving slips. The set of 90 points constitutes this chapter's dataset, D.

The only way to find the missing hoards is a laborious search by digging. To help allocate scarce digging labour more efficiently, the Queen's ministers would like you to analyse D and provide them with a 'hoard score' function, f. The hoard score $f(x, y)$ should be a non-negative number proportional to the estimated chance of finding a new hoard when digging at point (x, y).[1] How are you supposed to learn this function from the data? Well, the rough

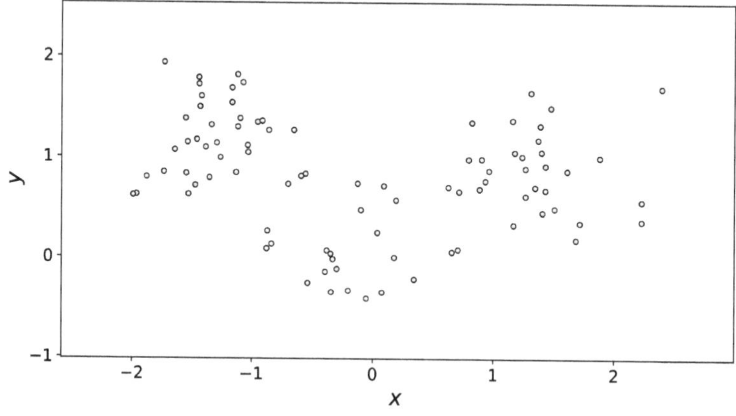

Figure 3.1 Map of known hoards. The coordinate system is a state secret, but you can assume it is orthonormal. This set of (x, y) pairs constitutes dataset D.

[1] 'Digging at point (x, y)' means digging a hole of some standard size – say, a spade-width or two – centred on (x, y).

idea is that f should be high in regions of the plane where there are many known hoards and low in regions where there are few known hoards. How 'wiggly' the function should be is up to you.

Notice that this is an unsupervised learning task. You should *not* think of the datapoints in Figure 3.1 as (feature, label) pairs, because you are not hoping to learn to predict y values from x values, or vice versa. The two coordinate components are conceptually on a par as far as you are concerned; you are interested in a two-dimensional pattern.

Since the Queen's ministers will use the hoard score function to compare possible digging locations, the function's overall scale doesn't much matter. But you will need to choose *some* scale. You can do that by fixing the integral of f over the whole plane,

$$\int_{-\infty}^{\infty}\int_{-\infty}^{\infty} f(x, y)\, dxdy = c,$$

where c is an arbitrary scale constant. If you happened to know the total number of hoards, n_{total}, then you could set $c = n_{\text{total}}$ and think of f as a hoard density function. If instead you set $c = 1$, you can think of f as a probability density function. Although we won't be emphasising probabilistic ideas before Chapter 4, we will adopt the latter convention ($c = 1$) now. Thus, 'density' will refer, by default, to probability density. Readers who don't want to wait until Chapter 4 to start thinking probabilistically can try Exercise 3.1.

■ ***Exercise* 3.1** When $c = 1$ then f is the probability density function of a pair of random variables (X, Y). What is the 'trial' that generates realisations of this pair?

Having chosen a scale, how do you go about finding a good hoard score function? In this chapter, we will pursue one simple approach just far enough to illustrate underfitting, overfitting and cross-validation in an unsupervised context.

3.2 Approximating Distributions

Before we start the search for a two-dimensional hoard score function, let's work through a simpler one-dimensional problem. Figure 3.2 shows a cluster of hoards along a line, plotted as red dots with coordinates $x_1, x_2, \ldots, x_n$. To aid visualisation, we have added some vertical 'jitter' to each point. Suppose we want to describe the distribution of points using a density which smooths out some of the detail. One option is to divide the line into equal width intervals or 'bins' and to count the number of points in each bin. We can then define a function, called a *histogram*, whose height in each interval is proportional to the count, scaled so that the total area under the histogram is 1. An example is shown in Figure 3.2, using a bin width of 1/2. An alternative is to fit a *parametric* density function to the data. In the case of a single clump of points like ours, a sensible choice is the Gaussian 'blob'

$$f(x; \mu, \sigma) = \frac{1}{\sigma\sqrt{2\pi}} \exp\left(-\frac{(x-\mu)^2}{2\sigma^2}\right),$$

where μ is the centre of the blob and σ measures its width. We use a semicolon to separate the parameters of f from its argument, x. The factor outside the exponential ensures that the area under the curve is 1.

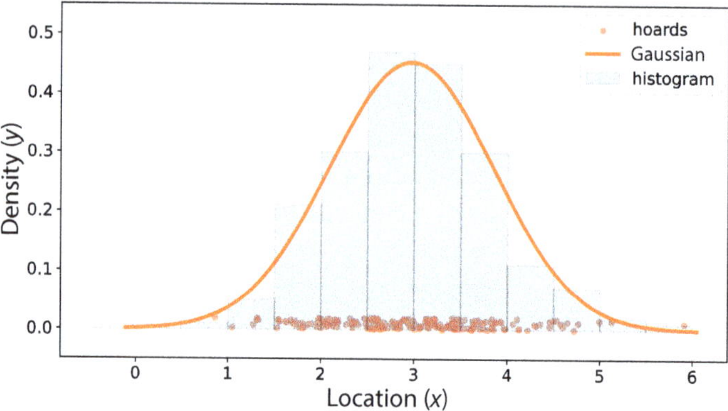

Figure 3.2 A one-dimensional distribution of hoards, plotted as red dots with some vertical jitter to aid visualisation. The blue shading shows a histogram of the data with bin width 0.5. The orange curve shows a Gaussian distribution, fitted to the data.

■ ***Exercise*** 3.2 Show that

$$\int_{-\infty}^{\infty} \frac{1}{\sigma\sqrt{2\pi}} \exp\left(-\frac{(x-\mu)^2}{2\sigma^2}\right) dx = 1.$$

How should we choose μ and σ so that our density is a good fit to the data? We want the density to be large where the datapoints (hoards) are and small where they aren't. That is a bit vague, though. Really what we need is a mathematical measure of fit. A natural first thought is to use the sum of the values of the density function at each of the datapoints. The bigger that sum, the more concentrated the density is near the hoards. Another option is to use the product of the same values. Let's start by considering the sum,

$$L_{\text{sum}}(\boldsymbol{\theta}) = \sum_{i=1}^{n} f(x_i; \mu, \sigma).$$

Here we have wrapped up both parameters into a single *parameter vector* $\boldsymbol{\theta} = (\mu, \sigma)$. If $L_{\text{sum}}(\boldsymbol{\theta})$ is our measure of fit, then we should pick a $\boldsymbol{\theta}$ that makes it as large as possible. This leads to disaster. It is possible to make L_{sum} as big as we like by placing the centre of our Gaussian blob on one of the hoards and then letting $\sigma \to 0$. Our fit measure L_{sum} shoots off to infinity. We end up with a density that is infinitely concentrated, giving an infinite L_{sum} score, but a terrible description of our data. Let's try the product instead.

$$L_{\text{prod}}(\boldsymbol{\theta}) = \prod_{i=1}^{n} f(x_i; \mu, \sigma).$$

Here we don't have the same problem. If we concentrate the density more and more tightly on one datapoint, so that the densities at all the other points shrink towards zero, the overall product of densities *also* shrinks towards zero. That sounds more promising! Let's give up on L_{sum} and try to maximise L_{prod}.

■ ***Exercise*** 3.3 What is the probabilistic interpretation of $L_{\text{prod}}(\boldsymbol{\theta})$?

Now, in practice there is a small hitch: products of many densities tend to be either unmanageably large or unmanageably small. We can fix that by working with the logarithm of L_{prod} instead, which we'll simply call 'ℓ':

$$\ell(\boldsymbol{\theta}) = \log L_{\text{prod}}(\boldsymbol{\theta}).$$

The logarithm is an increasing function, so this transformation does not affect the locations of any maxima in parameter space; but it does helpfully turn the product of densities into a sum:

$$\begin{aligned}\ell(\boldsymbol{\theta}) &= \log \prod_{i=1}^{n} f(x_i; \mu, \sigma) \\ &= \sum_{i=1}^{n} \log f(x_i; \mu, \sigma) \\ &= -\frac{1}{2\sigma^2} \sum_{i=1}^{n} (x_i - \mu)^2 - n \log \sigma - \frac{n}{2} \log(2\pi).\end{aligned}$$

The first term in the last line looks rather like the negative of our RSS from Chapter 2. We can view it as a penalty paid for deviations of the data from the centre of the blob. To maximise this term, we should choose μ so that the squared deviations of the datapoints from it are as small as possible. The effect of increasing σ is to reduce this penalty, but this is balanced by the negative $\log \sigma$ term, which increases as σ shrinks.

It is possible to obtain exact expressions for the values of μ and σ which maximise $\ell(\boldsymbol{\theta})$. They are the sample mean and mean squared deviation from the sample mean:

$$\begin{aligned}\hat{\mu} &= \frac{1}{n} \sum_{i=1}^{n} x_i, \\ \hat{\sigma}^2 &= \frac{1}{n} \sum_{i=1}^{n} (x_i - \hat{\mu})^2.\end{aligned}$$

Figure 3.2 shows the density with these parameters.

■ ***Exercise*** 3.4 The maximum of $\ell(\boldsymbol{\theta})$ can be found by looking for its stationary point, which solves

$$\frac{\partial \ell}{\partial \mu} = 0, \quad \frac{\partial \ell}{\partial \sigma} = 0.$$

By computing these derivatives, show that $\ell(\boldsymbol{\theta})$ is maximised when μ and σ^2 are the sample mean and the mean squared deviation from the sample mean.

3.3 Mixture Models

It is time to return to the Queen's treasure hoards. We are now working in two dimensions, where the density of a circularly symmetric Gaussian blob centred at the coordinate vector

$\boldsymbol{\mu} = (\mu_x, \mu_y)$ is

$$f(x, y) = \frac{1}{2\pi\sigma^2} \exp\left(-\frac{(x - \mu_x)^2 + (y - \mu_y)^2}{2\sigma^2}\right).$$

If we use vector notation, with $\boldsymbol{x} = (x, y)$, we can write this as

$$f(\boldsymbol{x}) = \frac{1}{2\pi\sigma^2} \exp\left(-\frac{\|\boldsymbol{x} - \boldsymbol{\mu}\|^2}{2\sigma^2}\right).$$

As in the one-dimensional case, the prefactor ensures that the density integrates to 1.

After staring at Figure 3.1 for a few seconds, it is clear that a single blob is not going to be sufficient to describe the distribution of hoard finds. You might tentatively judge that there are three main clusters of points – a 'top-left' cluster, a 'top-right' cluster, and a 'bottom-middle' cluster. The latter two look like roughly circularly symmetric blobs. The top-left cluster perhaps shows hints of a more interesting structure, but that could be a trick of the noise. Let's keep things simple for now and try a weighted sum of three Gaussian blobs. A function in this family has the form

$$f(\boldsymbol{x}; \boldsymbol{\theta}) = \sum_{i=1}^{3} \frac{w_i}{2\pi\sigma_i^2} \exp\left(-\frac{\|\boldsymbol{x} - \boldsymbol{\mu}_i\|^2}{2\sigma_i^2}\right). \tag{3.1}$$

Vectors $\boldsymbol{\mu}_1$, $\boldsymbol{\mu}_2$ and $\boldsymbol{\mu}_3$ are the blob centres; scalars $\sigma_1, \sigma_2, \sigma_3 > 0$ are their spatial sizes and scalars $w_1, w_2, w_3 \geq 0$ are the weights in the weighted sum which satisfy

$$w_1 + w_2 + w_3 = 1,$$

making f a density. We can think of w_i as the proportion of hoards that are associated with blob i. As with the one-dimensional case, to condense our notation we bundle up all the parameters – here $\boldsymbol{\mu}_1$, $\boldsymbol{\mu}_2$, $\boldsymbol{\mu}_3$, σ_1, σ_2, σ_3, w_1 and w_2 – into a single vector, $\boldsymbol{\theta}$. We don't include w_3 in the parameter vector because it is fixed by w_1 and w_2 via the requirement that $w_1 + w_2 + w_3 = 1$.

■ ***Exercise* 3.5** Show that $\boldsymbol{\theta}$ is an 11-dimensional vector.

We now need to adjust the parameter vector $\boldsymbol{\theta}$ to find a hoard score function that fits data D well. Figure 3.3 shows some initial experimentation. The hoard score function plotted in panel (d) looks like it might fit the data better than the functions plotted in panels (a)–(c). Inspired by our attempts to fit a single blob to a cluster of points in one dimension, let us try using the same log-product measure of fit,

$$\ell(\boldsymbol{\theta}) = \sum_{\boldsymbol{x} \in D} \log f(\boldsymbol{x}; \boldsymbol{\theta}). \tag{3.2}$$

Our task is to choose $\boldsymbol{\theta}$ so as to make $\ell(\boldsymbol{\theta})$ as large as possible. The multiple-blob version of this problem turns out to be considerably more difficult than the single-blob version we solved earlier. First, the optimal parameter values cannot be found exactly. This means that numerical optimisation methods must be used. However, there are many choices of parameter vector $\boldsymbol{\theta}$ that are *locally* optimal, and it may be difficult to find the best one. Second, when multiple blobs are involved, provided one or more of them are large enough to cover all the datapoints, the log-product measure can be made arbitrarily large by collapsing the remaining blobs onto

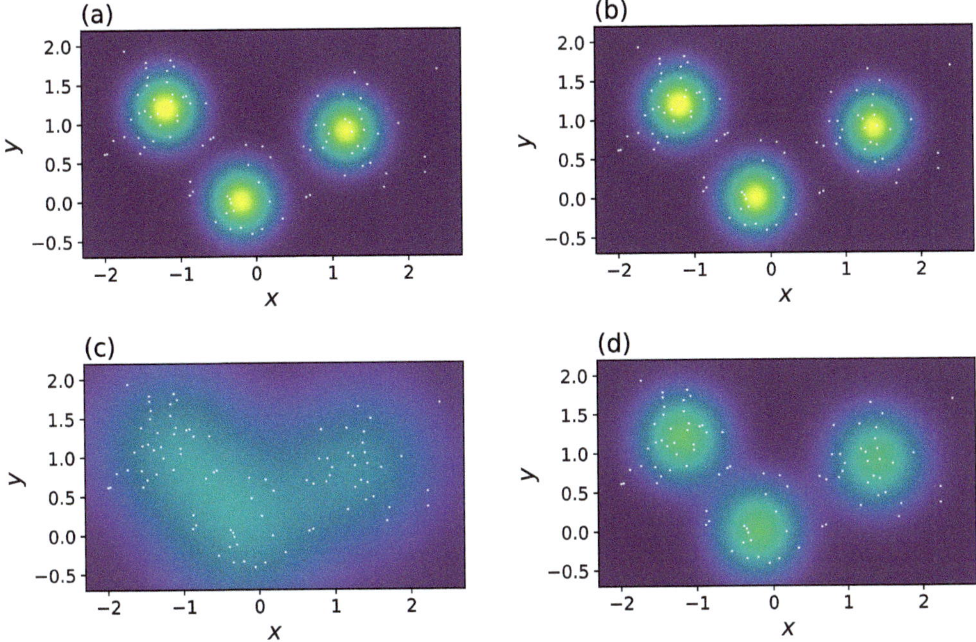

Figure 3.3 Dataset D plotted four times, with overlaid heat maps of some trial-and-error 'fits' of the hoard score function family defined in equation (3.1). Panel (a) is a first attempt. Nudging the top-right Gaussian blob to the right a bit yields panel (b). The blobs in (a) and (b) look too concentrated. Greatly increasing all the σ_i values yields panel (c); here the blobs look too diffuse. Panel (d) is a compromise. But is it the best we can do?

single datapoints. Intuitively, these collapsed densities are not good fits to the data, despite their sky-high ℓ scores. To avoid finding bad solutions, we need to instruct our optimisation algorithm to avoid certain regions of parameter space – those in which any of the blob sizes get too small. Figure 3.4 provides a caricature of this situation.

Fortunately, provided we start with some reasonable initial parameter vector guess $\boldsymbol{\theta}_0$ and then look for a local maximum of $\ell(\boldsymbol{\theta})$ in the vicinity of that initial guess, things tend to work pretty well. We have a good chance of finding a $\hat{\boldsymbol{\theta}}$ such that

$$\ell(\hat{\boldsymbol{\theta}}) \approx \max_{\boldsymbol{\theta} \in \mathcal{S}} \ell(\boldsymbol{\theta}),$$

where $\mathcal{S}$ is what's left of parameter space after we impose a constraint of the form $\sigma_i \geq \sigma_{\min}$ for $i = 1, 2, 3$ to exclude the 'bad' regions where blobs get infinitely concentrated on single hoards.[2] A statistician would describe what we are doing here as *fitting a Gaussian mixture model*. There are standard algorithms for the job. Figure 3.5 shows the fit obtained by

[2] Why did we not write '$\hat{\theta} \approx \arg\max_{\theta \in \mathcal{S}} \ell(\theta; D)$'? The answer is that a unique arg max is impossible in this case. Suppose $\hat{\theta} = (\hat{\mu}_1, \hat{\mu}_2, \hat{\mu}_3, \hat{\sigma}_1, \hat{\sigma}_2, \hat{\sigma}_3, \hat{w}_1, \hat{w}_2)$ maximises $\ell(\theta; D)$. Then so does $\hat{\theta}' = (\hat{\mu}_2, \hat{\mu}_1, \hat{\mu}_3, \hat{\sigma}_2, \hat{\sigma}_1, \hat{\sigma}_3, \hat{w}_2, \hat{w}_1)$. The order in which we describe the three blobs in our parameter vector is a purely notational matter.

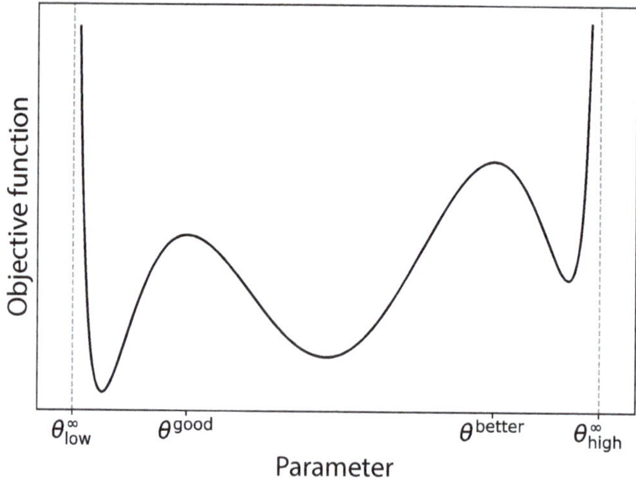

Figure 3.4 An objective function with local maxima at θ^{good} and θ^{better}, and 'blow-ups' near $\theta^{\infty}_{\text{low}}$ and $\theta^{\infty}_{\text{high}}$. If the objective function ceases to be a useful measure near $\theta^{\infty}_{\text{low}}$ and $\theta^{\infty}_{\text{high}}$, then the blow-ups are pathological, and optimisation algorithms will need to be told to avoid these regions.

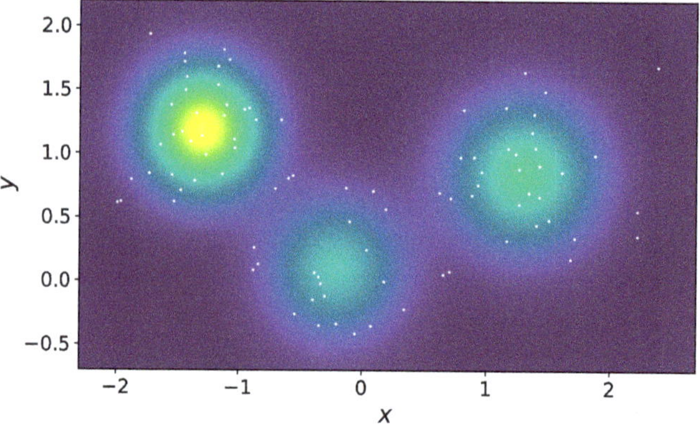

Figure 3.5 Dataset D plotted with overlaid heat map of an algorithmically optimised fit of the hoard score function family defined in equation (3.1).

applying one of them.[3] You'll notice it looks quite similar to Figure 3.3d. The top-left blob is brighter now because the tuning algorithm has picked up on the fact that the top-left cluster contains more points that the other two; the algorithm has duly increased the weight of the corresponding blob.

[3] The algorithmic details are not our focus here, but for the record, we used K-means to generate an initial parameter-vector guess $\theta^{(0)}$, and then expectation–maximisation to iteratively generate improved guesses $\theta^{(1)}, \theta^{(2)}, \ldots$, continuing until $\theta^{(t+1)}$ differed little from $\theta^{(t)}$. We discuss both algorithms in Chapter 12.

The algorithmic fitting procedure seems preferable to the eyeball-driven trial-and-error tinkering illustrated in Figure 3.3. However, we are still relying on our intuitive visual judgement that there are three clusters in the data, and hence that there should be three Gaussian blobs in our hoard score function. Might appearances be deceptive? Maybe we should be fitting a hoard score function with 2, or 4, or 10 Gaussian blobs! Well, we can try. The procedure in each case is essentially the same as the procedure described earlier – but we'd better tweak our notation to keep track of how many blobs we're using. Let's write '$\boldsymbol{\theta}_k$' for a parameter vector specifying a k-blob hoard score function, and '$\mathcal{S}_k$' for the region of k-blob parameter space we're willing to consider. As noted earlier, we will exclude 'bad' regions of parameter space in which any of the σ_i ($i = 1, 2, ..., k$) becomes very small.

■ ***Exercise*** 3.6 Recall that $\boldsymbol{\theta}_3$ is a vector with 11 components. Generalise this result. How many components does $\boldsymbol{\theta}_k$ have?

Our previous formulae now generalise in straightforward ways. To fit a k-blob hoard score function to our data, we need to find a $\hat{\boldsymbol{\theta}}_k$ such that

$$\ell_k(\hat{\boldsymbol{\theta}}_k) \approx \max_{\boldsymbol{\theta}_k \in \mathcal{S}_k} \ell_k(\boldsymbol{\theta}_k),$$

where

$$\ell_k(\boldsymbol{\theta}_k) = \sum_{\mathbf{x} \in D} \log f_k(\mathbf{x}; \boldsymbol{\theta}_k)$$

and

$$f_k(\boldsymbol{x}; \boldsymbol{\theta}_k) = \sum_{i=1}^{k} \frac{w_i}{2\pi\sigma_i^2} \exp\left(-\frac{|\boldsymbol{x} - \boldsymbol{\mu}_i|^2}{2\sigma_i^2}\right). \tag{3.3}$$

Figure 3.6 shows algorithmic fits to D for $k = 2$, $k = 3$, $k = 4$ and $k = 10$. You probably still have the intuition that $k = 3$ is the best choice, but it would be nice to be able to appeal an objective measure to support your intuition. What should that measure be?

You might be tempted to say 'just use the ℓ measure again! For each k, find $\ell_k(\hat{\boldsymbol{\theta}}_k)$. Larger values signal better fits!' But take a look at Figure 3.7a. If you interpret it in the way just suggested, it's telling you that $k = 3$ is far from the best choice. The $k = 10$ fit that looked so unnatural in Figure 3.6 does better on this measure; $k = 16$ does better still. A plot covering a wider range of k values would show that $k = 90$ does 'best'. The $k = 90$ fit, $f_{90}(-; \hat{\boldsymbol{\theta}}_{90})$, puts a tiny Gaussian blob on top of each of the 90 datapoints, with $\sigma_i = \sigma_{\min}$ for each i. (The notation '$f_{90}(-; \hat{\boldsymbol{\theta}}_{90})$' is shorthand for the function $\boldsymbol{x} \mapsto f_{90}(\boldsymbol{x}; \hat{\boldsymbol{\theta}}_{90})$.) There is clearly a sense in which this fit describes dataset D very well. Indeed, a heat map of $f_{90}(-; \hat{\boldsymbol{\theta}}_{90})$ would look a lot like a scatter plot of the raw data. But describing the data wasn't your assigned task. The Queen's ministers want to know where to send their diggers to look for new hoards – for hoards whose locations are presently unknown. If they take their cue from $f_{90}(-; \hat{\boldsymbol{\theta}}_{90})$, they'll dig at the 90 *known* hoard locations and ignore the rest of the plane. That doesn't sound like a great strategy. The lesson here is that picking k to maximise $\ell_k(\hat{\boldsymbol{\theta}}_k)$ is not the right way to decide how many Gaussian blobs to use. You want to pick k such that $f_k(-; \hat{\boldsymbol{\theta}}_k)$ assigns high values to locations at which *as yet unknown* hoards lie. Now obviously there is no foolproof way to do this. The unknown hoards could in principle be anywhere. But assuming that the

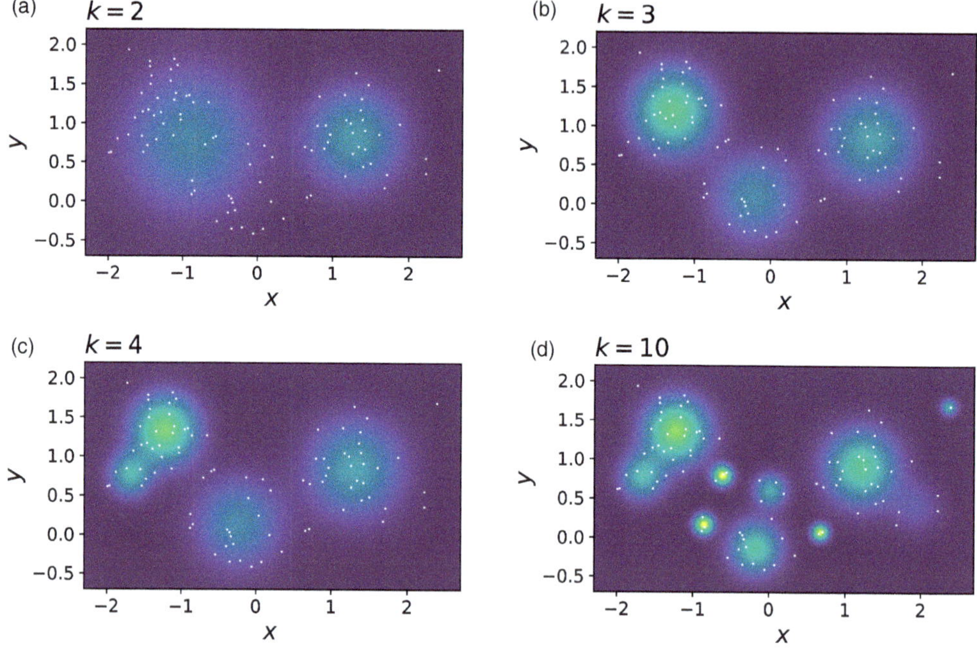

Figure 3.6 (a–d) Dataset D plotted four times, with overlaid heat maps of algorithmically optimised fits of k-blob hoard score functions. The $k = 3$ panel looks slightly different from Figure 3.5 because the colour scheme here is adapted to fit the range of hoard score values across the four panels.

90 known hoards can be regarded as a fair (i.e. random) sample of all the hoards, there is a sensible way to make a data-driven choice of k. You can use cross-validation!

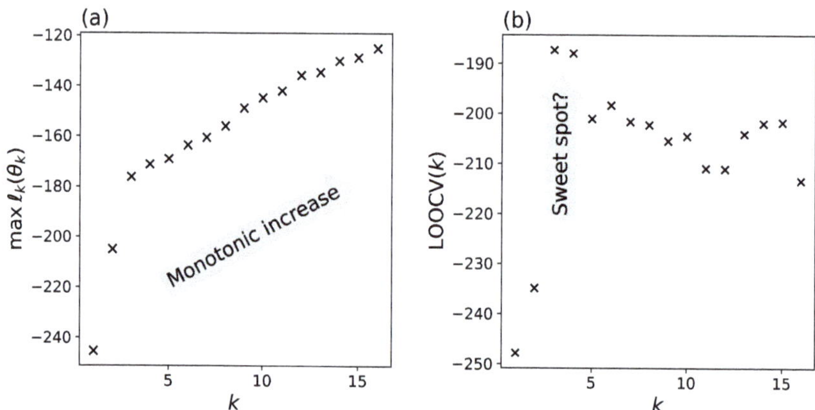

Figure 3.7 (a) Best achievable logarithm of k-blob hoard score product plotted against k. (b) Leave-one-out cross-validation score plotted against k.

■ ***Exercise*** 3.7 Before reading on, try to describe an LOOCV scheme for assessing candidate values of k. Then compare your answer to the scheme described in Section 3.4.

3.4 Cross-Validation Again

There are 90 points in dataset D. Pick an arbitrary ordering and label the points $\boldsymbol{x}_1, ..., \boldsymbol{x}_{90}$. If you drop the ith point, you're left with an 89-point dataset we'll call D_{-i}. (Careful! D_{-i} is not the ith point of D; it's the whole of D *except* for the ith point.) You can fit a k-blob hoard score function to dataset D_{-i}, resulting a parameter vector which we write $\hat{\boldsymbol{\theta}}_{k,-i}$, and a corresponding hoard score function $f_k(-;\hat{\boldsymbol{\theta}}_{k,-i})$. The only novelty here is notational: we need to keep track of which datapoint has been excluded, hence the annoying '$-i$' subscripts. Now, you didn't make any use of datapoint $\mathbf{x}_i$ when finding $\hat{\boldsymbol{\theta}}_{k,-i}$. Thus, from the perspective of hoard score function $f_k(-;\hat{\boldsymbol{\theta}}_{k,-i})$, datapoint $\mathbf{x}_i$ is rather like the location of an as-yet-undiscovered hoard. The hoard score is a measure of how effective the k-blob strategy was at anticipating the location of the ith hoard based on the locations of all the others. To assess the usefulness of the k-blob strategy for the Queen's ministers' purposes, a natural thing to do is to compute $f_k(\boldsymbol{x}_i;\hat{\boldsymbol{\theta}}_{k,-i})$ for each value of i, and then aggregate the results. Let's continue our previous practice of aggregating by product. As usual with products of many factors, working with the logarithm of the product (yielding a sum of logarithms) is more convenient than working with the product directly. Thus, we define the LOOCV score as follows:

$$\text{LOOCV}(k) = \sum_{i=1}^{90} \log f_k(\boldsymbol{x}_i;\hat{\boldsymbol{\theta}}_{k,-i}).$$

Figure 3.7b plots LOOCV(k) against k. The cross-validation scores support the intuitive verdict: the three-blob strategy is best. Its margin of victory over the second-best four-blob strategy is narrow, but the general trend of LOOCV(k) decreasing with k for $k \geq 3$ is clear.

We encountered a similar phenomenon in Chapter 2. To see the connection, notice that increasing the value of k increases the *flexibility* of the function family you are attempting to fit to the data. Any hoard score function you can construct with three blobs you can also construct with four (just make one of the four blob weights zero), but the converse doesn't hold: with four blobs to play with, you can create patterns that cannot be matched with three, as illustrated in Figure 3.6. In Chapter 2, higher-degree and hence more flexible polynomial families allowed for more 'wiggly' fits to the Queen's inventors' dataset. Wiggliness was helpful for fitting any specific set of datapoints. After all, if you know where all the datapoints are, you know where to put the wiggles. That was the lesson of Figure 2.4b. But beyond a certain point – indeed, beyond degree 3 – extra wiggliness was *not* helpful for predicting y from x for 'hold-out' datapoints that were not used to optimise the fit. That was the lesson of Figure 2.4a. Something very similar is going on here. Higher k values allow for more wiggly fits, where the value of the fitted function changes rapidly on short spatial scales in order to accommodate the data. Thus, for a given dataset, increasing k will allow you to obtain closer and closer fits *to that dataset*, until you 'max out' at a fit that puts a tiny blob on top of every datapoint. But beyond a certain point – indeed, for $k > 3$ – extra wiggliness is *not* helpful for anticipating the likeliest locations of hold-out datapoints.

Figure 3.7b suggests that you will underfit D if you use fewer than three blobs, and that you will overfit D if you use many more than three blobs. These are the same concepts of underfitting and overfitting that we encountered in Chapter 2. Whether the task is supervised or unsupervised, a well-chosen function family must be flexible enough but not *too* flexible.

3.5 Chapter Summary

In this chapter, we learned a distribution function from a set of datapoints in two dimensions. Although this was an unsupervised problem with no response variable to predict, the workflow nevertheless resembled that of Chapter 2.

- Within a function family, we found the best-fitting distribution by optimising a quantitative measure of fit: the logarithm of the product of the distribution function values at each datapoint.
- This procedure led to **overfitting** when the function family was too **flexible**, and **underfitting** when it was not flexible enough.
- We used **cross-validation** to find a sweet spot – a function family that was just flexible enough.

3.6 Further Exercises

■ ***Exercise*** 3.8 King Adwald the Fearful, on hearing rumours of an approaching band of raiders, asked his servant to hide a hoard of two thousand silver coins at a certain secret location in the river Nadder. Fortunately for the local population, the rumours of the raiding party were unfounded. Unfortunately for Adwald, his servant failed to properly secure the bag containing the coins, and the river has carried each coin to a different location downstream. As the court mathematician, you are in charge of the search process. So far, ten coins have been returned by loyal locals, who found them by chance at the following distances (measured to the nearest foot) downstream from the hiding place.

k	1	2	3	4	5	6	7	8	9	10
x_k	11	14	23	51	62	74	79	103	148	512

(3.4)

This is your dataset, D. Suppose that your approach to the search problem is to use D to estimate a density function $f(x)$ which is large where coins are likely to be found, and small elsewhere. This can be used to decide where your team of coin hunters should focus their efforts.

(a) Your first guess at the density function is to assume that the coins are uniformly distributed between the hiding place and some maximum distance β downstream.

The density is then

$$f(x;\beta) = \begin{cases} \frac{1}{\beta} & \text{if } x \in [0,\beta], \\ 0 & \text{otherwise.} \end{cases}$$

The log-product measure of the fit of this function to your observations is

$$\ell(\beta) = \sum_{i=1}^{10} \log f(x_i;\beta).$$

Find the parameter value $\hat{\beta}$ which maximises this measure. Hint: looking for a stationary point is not the way to solve this problem. (Why not?)

(b) Your second guess at the density function is to assume it takes an exponential form

$$f(x;\beta) = \begin{cases} \beta e^{-\beta x} & \text{if } x \geq 0, \\ 0 & \text{otherwise.} \end{cases}$$

Use the log-product measure to find $\hat{\beta}$ in this case.

(c) Which of these two densities would you prefer to use as a guide for future searches, and why?

(d) Which of the following bits of new information would do most to undermine your confidence in your chosen fitted density function as a guide for your team of coin hunters?

(i) There are unfortunately some disloyal locals living nearby; they may have found and pocketed some of the coins.

(ii) The river currents are unpredictable, changing from day to day.

(iii) Locals visit the stretch of river lying between 0 and 150 feet downstream of the hiding place especially often.

4

Interlude: Probability, Likelihood and Bayes

This chapter provides a gentle introduction to some probabilistic inference ideas. We will use toy problems to illustrate the ideas in the simplest possible ways. Chapters 5 and 6 offer a deeper dive. They will equip you with the mathematical tools you need to apply probabilistic inference ideas to more realistic problems.

We assume that you have studied basic probability before. If you need a refresher, Appendix A provides a brisk review.

4.1 Alien Invasion Detector

You have an alien invasion detector. Each day at noon, the detector completes its daily check. If an alien invasion of Earth is in progress, then with 100% probability the detector's alarm bell rings; there is no chance it fails to notice the invasion. If an alien invasion is not in progress, then with 99% probability the bell stays silent; there is a 1% chance of a false alarm.

It's noon, and the invasion alarm bell rings. Do you think an alien invasion of Earth is in progress?

■ ***Exercise*** 4.1 Before reading on, spend a few seconds thinking about this question.

Let's introduce some formal apparatus. Call the hypothesis than an alien invasion is in progress H_{inv}, and the hypothesis that an alien invasion is not in progress H_{null}. The hypothesis space you're interested in is $\mathcal{S} = \{H_{\text{inv}}, H_{\text{null}}\}$. Your task is to pick one of the hypotheses in $\mathcal{S}$. Your data D is the fact that the invasion alarm bell is ringing.

Both the hypotheses in $\mathcal{S}$ are consistent with D. Aliens could be invading – the bell is ringing, after all – but it might also be a false alarm. That means you are dealing with a risky inference problem. Whichever hypothesis you pick, you might turn out to be wrong.

Although both the hypotheses in $\mathcal{S}$ are consistent with D, there is a sense in which H_{inv} is a better fit to D than H_{null} is. If H_{inv} were true, D would not be at all surprising, because the alarm bell rings with 100% probability when an alien invasion is underway. By contrast, if H_{null} were true, D would be quite surprising, because the probability of the alarm bell ringing when an alien invasion is not underway is only 1%.

4.2 The Principle of Maximum Likelihood

We have just compared what statisticians call the *likelihoods* of the two hypotheses in $\mathcal{S}$. The likelihood of a hypothesis H is, by definition, the **conditional probability** of the data given the hypothesis, $\mathbf{Pr}(D|H)$.[1] Since this is our first 'official' use of the machinery of probability theory (Appendix A), it's worth pausing to make sure everything is in order. Probability measures assign probabilities to *events*. So in writing '$\mathbf{Pr}(D|H)$', we are treating both D and H as events. Does that make sense? Yes: D is the event that the alarm bell rings, and H is the event that a certain hypothesis (also denoted by the symbol H) is true. Events are subsets of a *sample space* of possible outcomes of an experiment, where 'experiment' can be understood very broadly. Here, the experiment is one whose outcome determines both whether the invasion alarm rings and whether an invasion is in fact in progress.

We will often denote the likelihood $\mathbf{Pr}(D|H)$ by the symbol $\mathcal{L}(H; D)$, to show that we are primarily interested in the dependence on H. Why are we more interested in the H dependence than the D dependence? Well, the point of computing likelihoods is to compare how successfully different hypotheses account for the data we actually have, not to play what-if games in which we imagine the data being different. So we will typically look at $\mathcal{L}(H; D)$ for a range of different values of H – every hypothesis in some space of hypotheses – but with D held fixed. When viewed as a function of H alone, $\mathcal{L}(H; D)$ is called the *likelihood function*. For the aliens example, we have

$$\mathcal{L}(H_{\text{inv}}; D) = \mathbf{Pr}(D|H_{\text{inv}}) = 1,$$
$$\mathcal{L}(H_{\text{null}}; D) = \mathbf{Pr}(D|H_{\text{null}}) = 0.01.$$

Thus, the invasion hypothesis H_{inv} is the *maximum likelihood* hypothesis. In other words,

$$\arg\max_{H \in \mathcal{S}} \mathcal{L}(H; D) = H_{\text{inv}}.$$

There is a famous inference principle called the *principle of maximum likelihood*. It recommends that, when choosing among the hypotheses in a hypothesis family, you should pick the one with the highest likelihood. Here, the principle tells you to pick H_{inv} and to conclude that aliens are invading. We will have much more to say about maximum likelihood inference in the remainder of this book. In the meantime, the aliens example should help you get a handle on when maximum likelihood inference is appropriate, and when it isn't.

Suppose that just before hearing the invasion alarm bell, you think an alien invasion today is highly improbable: say, a one-in-a-million chance. Then when you hear the alarm bell you will confidently guess that it is a false alarm. (False alarms are surprising, but they're not as surprising as alien invasions.) You will go right on believing H_{null}, even though its likelihood is a paltry 0.01 while H_{inv}'s is an unimprovable 1. The lesson here is that the maximum likelihood hypothesis is not always the hypothesis you should believe, all things considered. There is nothing paradoxical about this. Remember, the likelihood of a hypothesis is *simply* the conditional probability of the data given the hypothesis. It takes no account of the intrinsic plausibility of the hypothesis, and it certainly is not the same as the conditional probability of the hypothesis given the data. The invasion hypothesis H_{inv} has high likelihood because if

[1] Here we are considering a discrete space of possible datasets: the alarm bell either rings or it doesn't. In many inference problems, though, the space of possible datasets is continuous, and likelihoods must be defined as **conditional probability densities**, not conditional probabilities. We will see examples in Chapter 5.

aliens were invading then the alarm bell would probably (probability 100%!) be ringing. It does not follow that if the alarm bell is ringing then aliens are probably invading.

If you unwisely insist on applying the principle of maximum likelihood to the invasion alarm case, despite your prior belief that an alien invasion is highly improbable, then you fall victim to a kind of overfitting. You contort your beliefs too aggressively to fit a specific dataset (accepting H_{inv} on the strength of D), while giving insufficient weight to relevant prior beliefs. An analogous mistake in the context of Chapter 2's inference task would be picking a very wiggly prediction function simply because it achieves a low RSS, despite your prior belief that the systematic dependence of y on x is piecewise linear.

Let's now look at a couple of cases in which the principle of maximum likelihood gives good advice. Suppose that just before hearing the invasion alarm bell, you think an alien invasion today is a real possibility: say, a one-in-ten chance. After hearing the alarm, your best guess would indeed be that an alien invasion is in progress. (An invasion is slightly surprising, but a false alarm would be *more* surprising.) In this version of the aliens scenario, the maximum likelihood hypothesis really is the hypothesis you should believe. Alternatively, suppose that, as before, you initially think an alien invasion is a one-in-a-million chance, but now you have 50 alien invasion detectors, each with a 1% false positive rate, and each making its occasional mistakes independently of the others. At noon, all 50 of your invasion alarm bells ring. What should you conclude? You should conclude an alien invasion is in progress. That would be astonishing, to be sure, but not as astonishing as 50 independent false positives (probability $0.01^{50} = 10^{-100}$). Here the likelihood of H_{inv} is so much greater than the likelihood of H_{null} that H_{null}'s million-fold advantage in prior plausibility hardly matters. The principle of maximum likelihood tells you to believe H_{inv}, and H_{inv} is indeed the hypothesis you should believe.

When you are trying to make a data-driven choice between hypotheses that were roughly equally plausible before the data arrived, maximum likelihood inference is often a good approach. Even if your hypotheses do vary wildly in prior plausibility, maximum likelihood inference *can* still work well, so long as you have enough data (e.g. 50 independent invasion alarms). In practice, maximum likelihood inference is often useful, and we will be using it in the chapters ahead. In fact, we have used it already.

■ ***Exercise*** 4.2 Look again at the function we used to measure fit in Chapter 3, defined in equation (3.2). Show that it is the logarithm of the relevant likelihood function. (For this exercise, you should assume that hoard score functions are probability density functions.)

The RSS measure we used in Chapter 2 can also be seen as a thinly disguised measure of likelihood, so long as we are willing to make certain assumptions about the dataset-generating process. We will spell out that story in Chapter 5.

■ ***Exercise*** 4.3 All sensible applications of the principle of maximum likelihood are to problems in which a hypothesis is chosen from some specific, limited space of hypotheses, that is, S in the previous examples. Why? What's wrong with the advice 'just pick the highest likelihood hypothesis you can think of'? (Hint: show that if there are no restrictions whatsoever on the candidate hypotheses, then, given any dataset, it will always be possible to think of absurd hypotheses whose likelihoods are 100% – indeed, that it will always be possible to think of many mutually incompatible hypotheses of this sort.)

4.3 A Preview of Bayesian Inference

So far in this chapter, we have framed inference problems as problems of hypothesis selection: given a space of possible hypotheses, pick the 'best' one. But what if there are lots of good hypotheses? Maybe several of them will be tied for 'best'. Or maybe one will be ahead of the others, but only by a narrow margin. Shouldn't we have the option of hedging our bets? Perhaps we should express our attitude towards the hypotheses in our hypothesis space by means of a probability distribution – a probability distribution that represents our *degrees of belief* in the various competing hypotheses.

In fact we have already hinted at something like this. What did it mean to suppose, as we supposed earlier, that you believed an alien invasion was a 'one-in-a-million chance' before you heard the invasion alarm bell? It didn't mean that you believed aliens would decide whether or not to invade today by spinning a giant roulette wheel. Rather, it meant that you were very confident aliens would not invade today. The one-in-a-million 'chance' was really a one-in-a-million degree of belief. Let's spell this out. We can represent your initial degrees of belief in the hypotheses in S by the following probabilities:

$$\mathbf{Pr}(H_{\text{inv}}) = 10^{-6},$$
$$\mathbf{Pr}(H_{\text{null}}) = 1 - 10^{-6}.$$

It is natural to ask how these probabilities ought to be updated once the data D arrives, that is, once you have heard the invasion alarm. The Bayesian answer is that your updated degrees of belief should be

$$\mathbf{Pr}'(H_{\text{inv}}) = \mathbf{Pr}(H_{\text{inv}}|D),$$
$$\mathbf{Pr}'(H_{\text{null}}) = \mathbf{Pr}(H_{\text{null}}|D).$$

Probability measure **Pr** represents your initial belief state, that is, your belief state before the data arrives, while probability measure **Pr**′ represents your updated belief state, that is, your belief state after the data arrives. The intuition behind the Bayesian answer is compelling. Conditional probability $\mathbf{Pr}(H_{\text{inv}}|D)$ is how confident you *would* be in H_{inv} if you suddenly learned that today's data was D. But then, once the alarm rings, you *do* suddenly learn that today's data is D. Your new degree of belief in H_{inv}, then, is $\mathbf{Pr}'(H_{\text{inv}}) = \mathbf{Pr}(H_{\text{inv}}|D)$.

■ ***Exercise*** 4.4 Use the definition of conditional probability (see Appendix A if you need reminding) to prove Bayes' theorem,

$$\mathbf{Pr}(H|D) = \mathbf{Pr}(D|H)\frac{\mathbf{Pr}(H)}{\mathbf{Pr}(D)}.$$

■ ***Exercise* 4.5** Use Bayes' theorem and the Bayesian updating rule to show that

$$\mathbf{Pr}'(H_{\text{inv}}) \approx 10^{-4}.$$

A Bayesian would summarise this as follows: your *prior* probability of H_{inv} is 10^{-6}, but your *posterior* probability of H_{inv} is approximately 10^{-4}. After the invasion alarm goes off, the invasion hypothesis is 100 times more probable than it was before, but still much less probable than H_{null}.

■ ***Exercise* 4.6** Provide a similar Bayesian analysis of the alternative alien scenario in which 50 independent invasion alarms ring. Your prior, as before, is $\mathbf{Pr}(H_{\text{inv}}) = 10^{-6}$.

Bayesian inference provides a principled, quantitative way of combining likelihoods with priors. Of course, there is the question of where the priors come from. In practice, it will sometimes be difficult to say what your priors are or should be. In Chapter 5, we set priors aside and focus on inference methods based on the principle of maximum likelihood. It is worth remembering the lessons of Section 4.2, however. The principle of maximum likelihood can help us choose between competing hypotheses, but it cannot tell us which hypotheses to consider (recall Exercise 4.3). Moreover, its advice is worth heeding only when the competing hypotheses do not differ too greatly in prior plausibility. There is a sense, therefore, in which one implicitly signals one's priors whenever one uses the principle.

Methods based on the principle of maximum likelihood fall under the broader umbrella of *frequentist* inference. Chapter 6 takes up the comparison between frequentist and Bayesian methods in more depth. From that point on, we'll draw on both traditions, though we defer discussion of more advanced Bayesian methods to Chapter 10.

4.4 Chapter Summary

- The **likelihood** of a hypothesis is the conditional probability of the data given the hypothesis.
- One way of using data to choose a hypothesis from a hypothesis space is to pick the hypothesis with the greatest likelihood. This is known as **maximum likelihood inference**.
- It is possible for wildly implausible hypotheses to have high likelihoods. Maximum likelihood inference is unreliable when used to choose between hypotheses that differ greatly in intrinsic plausibility.
- **Bayesian inference** takes likelihoods into account but is also sensitive to the intrinsic plausibility of hypotheses.

5

Probabilistic Modelling

5.1 Overview: From Supervised Learning to Probabilistic Modelling

5.1.1 Probabilistic Hypotheses

Think back to the prediction task from Chapter 2. Imagine that you have picked what you judge is the best prediction function – that is, the best rule for mapping x values to estimated y values. Call this function $\hat{g}$. You communicate $\hat{g}$ to the Queen's inventors, cross your fingers, and wait. You hope your function performs well in the days ahead. Question: have you just endorsed a scientific hypothesis? It *feels* a bit like you have. After all, you've committed yourself to a particular prediction strategy, and that strategy will soon be tested against new data. But if you have endorsed a scientific hypothesis, its content is unclear. Certainly you don't believe every future observation will be a datapoint of the form $(x, \hat{g}(x))$, lying precisely on your prediction curve. Perhaps you think future observations will lie *near* that curve, but 'near' is vague, and scientific hypotheses ought to have well-defined truth conditions. That is, the stories they tell about the world ought to be sufficiently clear-cut that we can understand what it would mean for them to be true or false.

In this chapter, we introduce a conceptual toolkit, *probabilistic modelling*, that lets us reframe inference tasks as hypothesis selection problems. The reframing will often help us think more clearly. For example, it will help us justify our choices of objective functions – Chapter 2's RSS, Chapter 3's 'log-product' measure of fit and others still to come – in intellectually satisfying ways.

To see how probabilities enter the picture, consider what features a scientific hypothesis 'in the spirit of' your proposed prediction function $\hat{g}$ ought to have.

- It should be genuinely predictive, so it can't just be a claim about the data you have now.
- It should allow for uncertainty: you think x fixes y roughly, but not exactly.
- It shouldn't be vague: it should be a claim with well-defined truth conditions.

The second and third requirements look like they might be in tension, but in fact there is a way to satisfy them both. The hypothesis can be a *probabilistic* claim about the *process that generated the dataset.*

The concept of a dataset-generating process is implicit whenever we imagine repeating an experiment. What we actually have is a single dataset, D. But the process that generated it could be run again and again, at least in principle. If you managed to do that, you would obtain a sequence of datasets: $(D_1, D_2, D_3, \ldots)$. Every interesting example of a dataset-generating process involves some randomness, so there is no fact of the matter about what, precisely, the third or the seventeenth dataset in this sequence would look like if you really did repeat the experiments. But there are facts about the long-run statistics of the sequence. As a simple illustration, consider tossing a fair coin. You could toss the coin again and again for decades. You don't, and there is no fact about what the millionth toss would be if you did; but there is a fact about what the long-run fraction of heads would be. It would be 0.5. According to the frequentist interpretation of probability, that is all it means to say that the probability of any given toss yielding heads is 0.5. Notice that probabilistic hypotheses about dataset-generating processes are not vague: they have well-defined truth conditions (which we can spell out by talking about the long-run statistics of sequences of trials). Nevertheless, they allow for uncertainty at the level of individual trials.

5.1.2 Defining a Model

Let's get back to our prediction problem and start to bring some probabilistic tools to bear. We'll assume that each of the 40 datapoints in Chapter 2's dataset was generated independently, so that a single run of the dataset-generating process is equivalent to 40 runs of a datapoint-generating process. Focus on the datapoint-generating process. We can represent it mathematically by a **random vector** (X, Y), whose components X and Y are **random variables**. The 40 datapoints plotted in Fig 2.1 are 40 realisations of this random vector. Suppose that the physics of the situation is such that Y is equal to a function of X plus some noise,

$$Y = \mathfrak{g}(X) + \mathcal{E}^{\text{noise}}. \tag{5.1}$$

Here $\mathcal{E}^{\text{noise}}$ is a random variable whose **conditional expectation** $\mathbb{E}(\mathcal{E}^{\text{noise}}|X = x)$ is zero for every x. In other words, the value Y takes when X takes the value x is $\mathfrak{g}(x)$ plus a random nudge $\mathcal{E}^{\text{noise}}$. The nudge can be positive or negative, but its expectation is 0: over a long sequence of trials that happen to yield $X = x$, the average of the realisations of $\mathcal{E}^{\text{noise}}$ will tend to zero. The expected value of Y given any particular value x of X is therefore just $\mathfrak{g}(x)$. The function $\mathfrak{g}$ represents the true systematic dependence of Y on X, whatever that turns out to be; and the random variable $\mathcal{E}^{\text{noise}}$ represents the true noise, whatever *that* turns out to be.

If you take equation (5.1) seriously, then the framework we presented in Chapter 2 should strike you as incomplete. We want to predict Y from X, so estimating $\mathfrak{g}$ is a good start. But shouldn't we also be trying to learn something about $\mathcal{E}^{\text{noise}}$? At the very least its variance will be important whenever we need to put error bars on our predictions. The way forward here is to fit a *conditional probabilistic model* to our data. To keep things simple, we will focus on models that assume the noise is independent of X. That means models with the structure

$$Y = g(X; \boldsymbol{\theta}) + \mathcal{E}_{\boldsymbol{\theta}}, \tag{5.2}$$

where $\boldsymbol{\theta}$ is a parameter vector chosen from some space of possibilities, and $\mathcal{E}_{\boldsymbol{\theta}}$ is (for any given $\boldsymbol{\theta}$) a zero-mean random variable independent of X. The notation '$g(X; \boldsymbol{\theta})$', with the parameter vector coming after the semicolon, indicates that we are primarily thinking of g as a function of X; *which* function of X is controlled by $\boldsymbol{\theta}$. To define a specific model, we say what the space of possibilities for the parameter vector is, and we explain how different choices of $\boldsymbol{\theta}$ pick out different functions $g(-; \boldsymbol{\theta})$ and different zero-mean random variables $\mathcal{E}_{\boldsymbol{\theta}}$. Making an appropriate choice of $\boldsymbol{\theta}$ is then a further step, known as *fitting* the model. Models with the structure of equation (5.2) are *conditional* probabilistic models because they purport to describe only the conditional distribution of Y given $X = x$ (for any particular value of x). They do not purport fully to describe the distribution of random vector (X, Y).

Equations (5.1) and (5.2) look similar, but conceptually they are very different. Equation (5.1) is the unknown truth. Equation (5.2) is a model, a mathematical contraption we invent in an attempt to describe the truth. Let's consider a concrete example. Suppose we are interested in quadratic approximations to $\mathfrak{g}$ and Gaussian approximations to the true noise $\mathcal{E}^{\text{noise}}$. Then we could take our parameter vector to have the form $\boldsymbol{\theta} = (\beta_0, \beta_1, \beta_2, \sigma) = (\boldsymbol{\beta}, \sigma)$ with $\sigma > 0$, and define our model by writing

$$\begin{aligned} Y &= g(X; \boldsymbol{\beta}) + \mathcal{E}, \\ g(X; \boldsymbol{\beta}) &= \beta_0 + \beta_1 X + \beta_2 X^2, \\ \mathcal{E} &\sim \mathcal{N}(0, \sigma^2). \end{aligned} \tag{5.3}$$

The first line of (5.3) says that Y is the sum of two terms: a term fixed by X through prediction function $g(-; \boldsymbol{\beta})$, and a random variable $\mathcal{E}$; the latter is there to account for noise. The second and third line of (5.3) spell out how the parameter vector controls the prediction function and the noise term, respectively. In this case, the prediction function depends only on the first three components of $\boldsymbol{\theta}$, so we write '$g(X; \boldsymbol{\beta})$' rather than '$g(X; \boldsymbol{\theta})$'. The noise term is a Gaussian random variable – see Box 5.1 – with mean zero and variance σ^2. Noise terms of this particular sort crop up very frequently in statistics and machine learning, and people usually denote them with simple symbols. Thus in (5.3), we have written '$\mathcal{E}$' rather than '$\mathcal{E}_{\boldsymbol{\theta}}$' or '$\mathcal{E}_\sigma$'.

★ Box 5.1 The normal distribution

A random variable Z which is *normally distributed* (or *Gaussian*) with mean μ and variance σ^2 has probability density function

$$f_Z(z; \mu, \sigma) = \frac{1}{\sigma\sqrt{2\pi}} \exp\left(-\frac{(z-\mu)^2}{2\sigma^2}\right).$$

A shorthand way to indicate that Z has this distribution is to write $Z \sim \mathcal{N}(\mu, \sigma^2)$.

Notice that (5.3) does not say anything at all about the distribution of X. Nor, in fact, does it offer a complete story about the distribution of Y. What it does offer (for any particular choice of $(\boldsymbol{\beta}, \sigma)$) is a complete story about the conditional distribution of Y given $X = x$, for any x. Conditional on $X = x$, the random variable Y is just $\mathcal{E} + g(x; \boldsymbol{\beta})$. This is a Gaussian random variable plus a constant: that is, it is another Gaussian random variable, with the same variance as $\mathcal{E}$ but with its mean shifted by $g(x; \boldsymbol{\beta})$. So an alternative way of expressing the content of model (5.3) is

$$\begin{aligned} Y|\{X = x\} &\sim \mathcal{N}(g(x; \boldsymbol{\beta}), \sigma^2), \\ g(x; \boldsymbol{\beta}) &= \beta_0 + \beta_1 x + \beta_2 x^2. \end{aligned}$$

According to the model, then, the conditional probability density of Y given X is

$$f_{Y|X}(y|x; \boldsymbol{\theta}) = \frac{1}{\sigma\sqrt{2\pi}} \exp\left(-\frac{(y - g(x; \boldsymbol{\beta}))^2}{2\sigma^2}\right).$$

(Recall that in our notation $\boldsymbol{\theta} = (\boldsymbol{\beta}, \sigma)$.)

Figuratively, a model of this sort is a contraption with a number of tunable dials: one dial for each component of the parameter vector $\boldsymbol{\theta}$. As we turn the dials, the hypothesis our model makes about the dependence of Y on X changes. Any specific choice of dial settings amounts to a specific probabilistic hypothesis about the datapoint-generating process. If we specify the model but don't specify the dial settings, we have picked out a hypothesis space.

5.1.3 Fitting a Model

Fitting a model to data is traditionally a matter of finding the 'best' dial settings, that is, the best hypothesis in the model's hypothesis space. (We will encounter a more sophisticated conception of model fitting when we discuss Bayesian inference in Chapters 6 and 10.) At the very least, we want to find a $\boldsymbol{\theta}$ such that equation group (5.3) is not too badly wrong. The truth, remember, is given by equation (5.1). But in practice we can't measure how well a particular $\boldsymbol{\theta}$ does by comparing (5.3) to (5.1), because the function $\mathfrak{g}$ and the random variable $\mathcal{E}^{\text{noise}}$ that feature in (5.1) are unknown. Making informed guesses about them is precisely what we are trying to do in fitting a model. So how *do* we measure how well a particular $\boldsymbol{\theta}$ does? We faced a similar question in Chapter 2. There we needed to come up with a measure of how well a prediction function approximated a dataset. We settled on the RSS, but we didn't make a principled case for it over other possible measures. With our new conceptual technology – conditional probabilistic models – we can do better.

Model (5.3), with any particular choice of $\boldsymbol{\theta}$, makes claims about how probable different Y values are given an X value. It therefore makes a claim about how probable the actually observed set of Y values was, given the corresponding set of X values. For some choices of $\boldsymbol{\theta}$ (i.e. for some dial settings on our contraption), the observed Y values will seem very surprising. For other choices, less so. One natural model-fitting strategy is to pick a $\boldsymbol{\theta}$ that makes the data seem as unsurprising as possible. This, of course, is just the *maximum likelihood inference* strategy we saw in Chapter 4. It turns out that using it to fit model (5.3) is mathematically equivalent to minimising the RSS; and the same goes for many similar models. We will spell out the details in Section 5.2, but the basic idea is straightforward: the RSS is a disguised measure of how surprising the data are from the perspective of a certain kind of conditional probabilistic model.

5.1.4 Models and Truth

It makes sense to speak of the best choice of $\boldsymbol{\theta}$, given a specific loss function and a specific dataset.[1] We know what 'best choice' means here: it means the choice that minimises the given loss function on the given dataset. By contrast, it's not clear that it makes sense to speak of the *true* value of $\boldsymbol{\theta}$ – for such talk presumes that there *is* a true value of $\boldsymbol{\theta}$. In other words, it presumes that for some $\boldsymbol{\theta}$, equation group (5.3), our model, is mathematically equivalent to equation (5.1), the true conditional dependence of Y on X. That presumption is almost

[1] We are assuming here that the loss-minimising choice exists and is unique. If several choices are tied for best, we can speak of the best choices (plural), or a best choice. If the loss function has no minimum, there is no such easy fix; but it may be that there is a better way of specifying the inference problem, for example, by removing problematic regions from the parameter space.

certain to be false. Real-world processes are messy, especially when they involve blasting stone balls out of metal tubes. It would be very surprising if $\mathcal{E}^{\text{noise}}$ was exactly Gaussian and perfectly independent of X; it would be very surprising if $\mathfrak{g}$ was exactly quadratic.

Nevertheless, for certain purposes, it might be useful to *pretend* that there is a true value of $\boldsymbol{\theta}$. We will then denote the fictional true value by the symbol $\boldsymbol{\theta}^*$. How could this pretence be useful? Here is one illustration. Imagine that we fit model (5.3) to dataset D, and we find that the loss-minimising parameter vector is $\hat{\boldsymbol{\theta}}$. We thereby pick out a hypothesis about the long-run statistics of the dataset-generating process (the hypothesis is model (5.3) with $\hat{\boldsymbol{\theta}}$ substituted for $\boldsymbol{\theta}$). We can ask: *if this hypothesis were true*, that is, if $\boldsymbol{\theta}^* = \hat{\boldsymbol{\theta}}$, what would the statistical properties of our estimation procedure be? Suppose we ran the dataset-generating process from scratch many times, with the predictors taking the values actually observed in D.[2] That would yield a sequence of datasets $D_1, D_2, D_3, \ldots$. For each dataset, we could compute a separate loss-minimising parameter vector; that would yield a sequence of estimates $\hat{\boldsymbol{\theta}}_1, \hat{\boldsymbol{\theta}}_2, \hat{\boldsymbol{\theta}}_3, \ldots$. Would these estimates be tightly clustered around the (imagined) true value $\boldsymbol{\theta}^* = \hat{\boldsymbol{\theta}}$? Or would they cluster around some other value? Or would they bounce wildly around? We can answer those questions using model (5.3), and the answers we give have implications *outside the thought experiment* for how we think about the uncertainty in our actual estimate, $\hat{\boldsymbol{\theta}}$.[3] For example, if our fitted probabilistic model itself implies that some component of the actual estimate could easily have been 5% higher or 5% lower, then we shouldn't take the third (or even second) significant figure of that component very seriously.

When we speak of 'estimating a parameter vector', the context will sometimes be a thought experiment like the one just envisaged. We imagine that our model is correct, and that the goal of the estimation game is to come up with a parameter vector estimate that is close to the true value. In those contexts, we will sometimes use the '*' notation, and speak (for example) of 'estimating $\boldsymbol{\theta}^*$'. But we will only do that when we are determined to draw attention to the imagined true value of the parameter vector. By default, we will drop the asterisk and speak of 'estimating $\boldsymbol{\theta}$'. This default formulation does not presuppose model correctness. It emphasises our most basic predicament vis-à-vis a model's tunable dials: we need to decide, somehow, what to do with them.

5.2 A Closer Look

5.2.1 Model Fitting

Consider the set of 10 (x, y) predictor–response pairs plotted as blue dots in Figure 5.1a. This is the toy dataset that we'll work with in this section. We write it

$$D = \{(x_k, y_k)\}_{k=1}^{n},$$

2 We imagine fixing the x values because our conditional probabilistic model doesn't tell us how to generate them. It only tells us how to generate realisations of random variable Y given $X = x$.

3 The method we are alluding to here is known as parametric bootstrapping. We introduced the *non*-parametric version of bootstrapping in Section 2.7. We will have more to say about both, and about other ways of quantifying uncertainty, in Chapter 6.

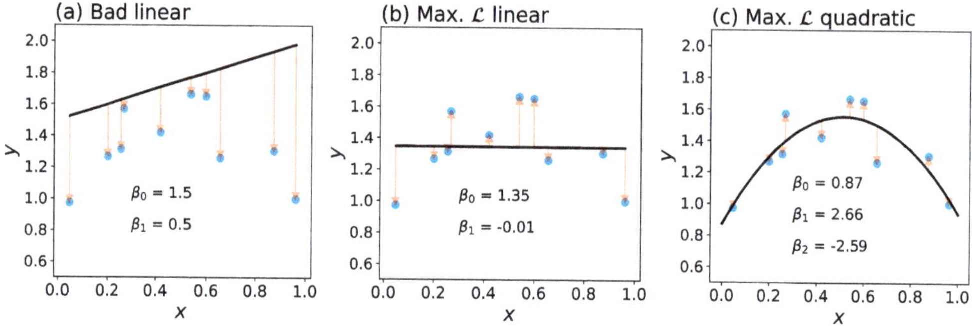

Figure 5.1 Three prediction functions (black lines) based on the same dataset (blue dots), and their residuals (red arrows). (a) A bad linear model. (b) Maximum likelihood linear model. (c) Maximum likelihood quadratic model.

where in this case $n = 10$. We'll tackle the supervised problem of learning to predict y values from x values. The y_k in D were artificially generated using a model of the form

$$Y = \mathfrak{g}(X) + \mathcal{E}^{\text{noise}}, \tag{5.4}$$

where $\mathfrak{g}$ is an unknown-to-us function that we wish to estimate from the data, and $\mathcal{E}^{\text{noise}}$ is a zero-mean normal random variable.[4] To create D, realisations of X and $\mathcal{E}^{\text{noise}}$ were generated first, and then plugged in to (5.4) to create Y. Different realisations of X and $\mathcal{E}^{\text{noise}}$ were generated independently. As a result, D is a set of i.i.d. realisations of the pair (X, Y). Since our task is predicting y values from x values, the nature of the process that generated the x values is unimportant.

Let's begin by trying to fit a *very* simple conditional probabilistic model,

$$\begin{aligned} Y &= \beta_0 + \beta_1 X + \mathcal{E}, \\ \mathcal{E} &\sim \mathcal{N}(0, \sigma^2). \end{aligned}$$

We'll call this model L for linear.[5] The model has three parameters, β_0, β_1 and σ, which we package into a single parameter vector $\boldsymbol{\theta} = (\beta_0, \beta_1, \sigma)$. The black line in Figure 5.1a plots the prediction function $\beta_0 + \beta_1 x$ against x when β_0 and β_1 are set to 1.5 and 0.5, respectively. Those values weren't picked with any care, and the result is a poor fit to the data: notice the large residuals, $r_i = y_i - \hat{y}_i$, all pointing in the same direction.

One way to assess the quality of the fit achieved by a particular $\boldsymbol{\theta}$ is to consider the hypothesis that the observed y values were generated from the observed x values by model L with parameter vector $\boldsymbol{\theta}$. If the observed y values are very improbable on this hypothesis, then $\boldsymbol{\theta}$ is a poor fit to the data. To calculate the relevant probability density, we first note what

[4] In fact, $\mathfrak{g}(x) = 1 + 2x - 2x^2$ and $\mathcal{E} \sim \mathcal{N}(0, 0.1^2)$, but the inference situation we are interested in is one in which we are not given this information.

[5] The term 'linear model' can be used in different ways. Here, we are interested in the dependence of Y on X for given β_0, β_1 and $\mathcal{E}$. Thus, we call our present model $Y = \beta_0 + \beta_1 X + \mathcal{E}$ linear, in contrast to (for example) a quadratic model $Y = \beta_0 + \beta_1 X + \beta_2 X^2 + \mathcal{E}$. But in other contexts, we might be more interested in the dependence of Y on the parameters β_0, β_1 and β_2. Then both the models mentioned in this footnote would count as linear.

the model claims about Y's conditional density given $X = x$. It says this is just a Gaussian with variance σ^2 centred on $\beta_0 + \beta_1 x$. In other words,

$$f_{Y|X}(y|x;\boldsymbol{\theta}) = \frac{1}{\sigma\sqrt{2\pi}} \exp\left(-\frac{(y - \beta_0 - \beta_1 x)^2}{2\sigma^2}\right).$$

The probability density of a single response y_k given a predictor x_k is then $f_{Y|X}(y_k|x_k;\boldsymbol{\theta})$. Now consider a sequence of n predictor–response observations as follows:

$$(x_1, y_1), (x_2, y_2), \ldots, (x_n, y_n).$$

It is useful to write the collected predictor and response values as vectors as follows:

$$\begin{aligned} \underline{x} &= (x_1, x_2, \ldots, x_n) \\ \underline{y} &= (y_1, y_2, \ldots, y_n). \end{aligned}$$

In this book, we will use underlined vectors to represent sets of observations. All other vectors will be written using bold face. The details and motivation for this approach are explained in Box 5.2.

★ Box 5.2 Vector notation

When building statistical models, it is useful to be able to distinguish visually between vectors of observations and vectors of parameters or features. If a vector contains a set of distinct *observations*, then it will be underlined; otherwise it will be in boldface. To denote a set of observations of a vector quantity, we combine both underline and bold face. We also deploy this combined bold–underline notation to denote matrices, which may be viewed as vectors of rows or of columns. In summary:

- $\underline{x}$ is a vector of observations of a scalar;
- $\boldsymbol{x}$ is a vector; and
- $\underline{\boldsymbol{x}}$ is a vector of observations of a vector (also a matrix).

The sequence of predictor values in $\underline{x}$ may be understood as the realisation of a random vector $\underline{X} = (X_1, X_2, \ldots, X_n)$, where the X_i are i.i.d. copies of X. The response values may be viewed as the realisation of a random vector $\underline{Y} = (Y_1, Y_2, \ldots, Y_n)$. Within our model, we assume that noise variables corresponding to different observations are independent. The joint density of all the responses, given the predictors, is therefore a product of the individual conditional densities,

$$\begin{aligned} f_{\underline{Y}|\underline{X}}(\underline{y}|\underline{x};\boldsymbol{\theta}) &= f_{Y|X}(y_1|x_1;\boldsymbol{\theta}) \times f_{Y|X}(y_2|x_2;\boldsymbol{\theta}) \times \ldots \times f_{Y|X}(y_n|x_n;\boldsymbol{\theta}) \\ &= \prod_{k=1}^{n} f_{Y|X}(y_k|x_k;\boldsymbol{\theta}). \end{aligned} \tag{5.5}$$

Viewed as a function of $\boldsymbol{\theta}$, this is the (conditional) *likelihood function*. It is often written as $\mathcal{L}(\boldsymbol{\theta})$. If the set of predictor values are a sample from a multivariate density $f_{\underline{X}}$ according to which they are not i.i.d., then the conditional likelihood is unchanged. This is relevant, for example, when an experimenter decides on a set of X values in advance of collecting the data. In this case, we may think of $f_{\underline{X}}$ as a generative model of the experimenter's decision.

The product in equation (5.5) tends to be a very small or very large number, so we work with its logarithm

$$\log \mathcal{L}(\boldsymbol{\theta}) = \log f_{\underline{Y}|\underline{X}}(\underline{y}|\underline{x};\boldsymbol{\theta}) = \sum_{k=1}^{n} \log f_{Y|X}(y_k|x_k;\boldsymbol{\theta}).$$

This is the *log-likelihood function*, sometimes written as $\ell(\boldsymbol{\theta})$.

If we use log-likelihood as our measure of fit, then it seems sensible to choose our model parameters so as to maximise it. As noted in Section 5.1, this approach to model fitting is known as maximum likelihood inference. We state the guiding principle for a general conditional probabilistic model in Box 5.3.

★ Box 5.3 Principle of maximum likelihood (conditional models)

Given a conditional probabilistic model of response data $\underline{Y}$ given predictors $\underline{X}$, with parameter vector $\boldsymbol{\theta} = (\theta_0, \theta_1, \ldots, \theta_m)$, the conditional probability (density) of the data according to the model, viewed as a function of $\boldsymbol{\theta}$, is the (conditional) **likelihood function**

$$\mathcal{L}(\boldsymbol{\theta}) = f_{\underline{Y}|\underline{X}}(\underline{y}|\underline{x}) = \prod_{k=1}^{n} f_{Y|X}(y_k|x_k;\boldsymbol{\theta}),$$

where the decomposition into a product holds provided the responses are conditionally independent given the predictors. The *log-likelihood* is

$$\ell(\boldsymbol{\theta}) = \log \mathcal{L}(\boldsymbol{\theta}).$$

The principle of maximum likelihood tells us to choose the model parameters so as to maximise the likelihood, that is, to pick parameter vector

$$\hat{\boldsymbol{\theta}} = \arg\max_{\boldsymbol{\theta}} \mathcal{L}(\boldsymbol{\theta}) = \arg\max_{\boldsymbol{\theta}} \ell(\boldsymbol{\theta}).$$

Finding the parameters that maximise likelihood for our linear model yields the prediction function shown in Figure 5.1b. The maximum likelihood estimate of the noise parameter is $\hat{\sigma} = 0.23$. We will consider the practicalities of the maximisation process later. Looking at Figure 5.1b, we might reasonably conclude that our model lacks sufficient flexibility to capture the systematic component of the predictor–response relationship. We can move to a more flexible model by adding a quadratic term to the model,

$$Y = \beta_0 + \beta_1 X + \beta_2 X^2 + \mathcal{E}.$$

For this new model, the parameter vector has the form $(\beta_0, \beta_1, \beta_2, \sigma)$. We can apply the principle of maximum likelihood again to find the quadratic prediction function shown in Figure 5.1c. From the figure, we note that by introducing only one additional parameter, we have substantially reduced the residuals and consequently reduced our estimate of the variance of the noise to $\hat{\sigma} = 0.12$.

5.2.2 Maximum Likelihood and Least Squares

We now look at the process of maximising likelihood in more detail, which will reveal its connection to the least-squares method we developed in Chapter 2. We consider a conditional probabilistic model in which the prediction function is allowed to depend on m parameters, which we write as a vector $\boldsymbol{\beta} = (\beta_0, \ldots, \beta_m)$, so

$$Y = g(X; \boldsymbol{\beta}) + \mathcal{E},$$

where $\mathcal{E} \sim \mathcal{N}(0, \sigma^2)$, independent of X. Here we have separated the parameter argument of the prediction function from the predictor variable by a semicolon, to indicate that they are different types of quantity. The parameters of the model as a whole form the vector

$$\boldsymbol{\theta} = (\beta_0, \beta_1, \ldots, \beta_m, \sigma).$$

Under this model, the probability density of the response given the predictor is

$$f_{Y|X}(y|x; \boldsymbol{\theta}) = \frac{1}{\sigma\sqrt{2\pi}} \exp\left(-\frac{(y - g(x; \boldsymbol{\beta}))^2}{2\sigma^2}\right),$$

and the log-likelihood is

$$\begin{aligned} \ell(\boldsymbol{\theta}) &= \sum_{k=1}^{n} \log f_{Y|X}(y_k|x_k; \boldsymbol{\theta}) \\ &= -\frac{1}{2\sigma^2} \sum_{k=1}^{n} (y_k - g(x_k; \boldsymbol{\beta}))^2 - n \log(\sqrt{2\pi}\sigma), \end{aligned} \tag{5.6}$$

where n is the number of datapoints in the dataset.

■ ***Exercise*** 5.1 Let $a_1, a_2, \ldots, a_n$ be a sequence of positive numbers.

(a) Use summation notation to write the 'log product' $\log\left(\prod_{i=1}^{n} a_i\right)$ in terms of the logarithm of the individual terms of the sequence $\log(a_1), \log(a_2), \ldots, \log(a_n)$.

(b) Write the log product

$$\log\left(\prod_{i=1}^{n} \frac{e^{\frac{a_i}{b}}}{c}\right)$$

in terms of b, c and n and the sum $S = \sum_{i=1}^{n} a_i$.

■ ***Exercise*** 5.2 Check that equation (5.6) is correct. Don't read on until you have satisfied yourself that it is!

To maximise the log-likelihood we look for a stationary point, where the partial derivatives of ℓ with respect to the model parameters are all zero. Let's consider the noise parameter first. We have

$$\frac{\partial \ell(\boldsymbol{\theta})}{\partial \sigma} = \frac{1}{\sigma^3} \sum_{k=1}^{n} (y_k - g(x_k; \boldsymbol{\beta}))^2 - \frac{n}{\sigma}.$$

Setting this to zero and solving for σ, we find

$$\hat{\sigma}^2 = \frac{1}{n}\sum_{k=1}^{n}(y_k - g(x_k;\hat{\boldsymbol{\beta}}))^2,$$

where $\hat{\boldsymbol{\beta}}$ is the maximum likelihood estimate of the parameters of the prediction function, to be determined. For given σ, maximising the likelihood with respect to $\boldsymbol{\beta}$ requires us to maximise

$$-\sum_{k=1}^{n}(y_k - g(x_k;\boldsymbol{\beta}))^2.$$

We can always turn maximisation into minimisation by changing the sign of the quantity we are trying to maximise so, according to maximum likelihood, the prediction function is found by minimising

$$\sum_{k=1}^{n}(y_k - g(x_k;\boldsymbol{\beta}))^2 = \mathrm{RSS}(\boldsymbol{\beta}).$$

Therefore, the maximum likelihood estimates coincide with the minimum RSS estimates used in least squares. After we have found $\hat{\boldsymbol{\beta}}$, our prediction function is

$$\hat{g}(x) = g(x;\hat{\boldsymbol{\beta}}).$$

Notice that this result does not depend on $\hat{\sigma}$, so we do not need to find it in order to make predictions.[6] However, its value is still revealing, as we will see shortly.

Maximum likelihood produces the same estimates as least squares in this case only because we made certain assumptions about the noise. Specifically, we assumed that it was normal, additive, with variance independent of the predictor ('homoscedastic') and with independent realisations for each datapoint. Had we made different assumptions, we would have obtained a different likelihood function and hence different maximum likelihood estimates. For example, in some situations our observational data might suggest that the noise is *heteroscedastic*, meaning that its variance depends on the value of the predictors. Following the maximum likelihood recipe with a heteroscedastic noise model will *not* lead us to minimise the RSS. Instead, we will end up minimising a more complicated function that penalises residuals differently at different predictor values.

■ ***Exercise*** 5.3 Consider the following dataset, collected by observing the pair of random variables (X, Y) four times.

X	1	2	3	4
Y	0.5	4.8	8.3	17.5

(5.7)

Suppose Y and X are related by

$$Y = \beta X^2 + \mathcal{E},$$

[6] Model parameters whose values are not needed for the inference tasks we happen to care about are sometimes called *nuisance parameters*.

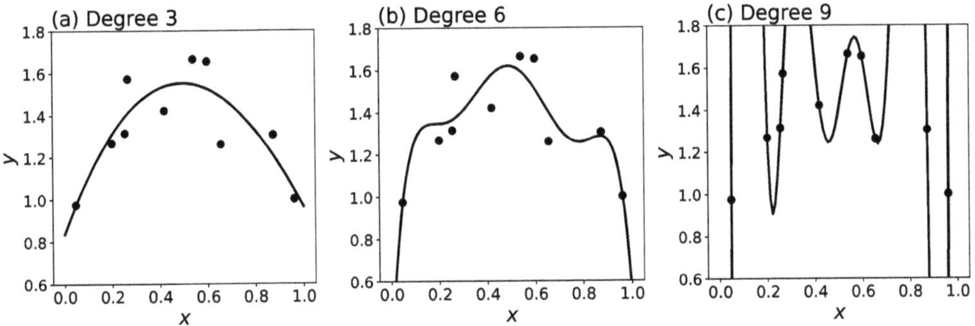

Figure 5.2 Maximum likelihood prediction functions of polynomial degree (a) $p = 3$, (b) $p = 6$ and (c) $p = 9$.

where $\mathcal{E} \sim \mathcal{N}(0, \sigma^2)$ and $\mathcal{E}$ and X are independent.

(a) What is the parameter vector $\boldsymbol{\theta}$ of this model?
(b) Find the log-likelihood $\ell(\boldsymbol{\theta})$ given arbitrary dataset $D = \{(x_i, y_i)\}_{i=1}^n$.
(c) Obtain an expression for the maximum likelihood estimate $\hat{\beta}$ of β in terms of D.
(d) Calculate $\hat{\beta}$ using the dataset given in the table above. Plot both the data and the curve $y = \hat{\beta}x^2$ on the same pair of axes.

5.3 Maximum Likelihood Can Lead to Overfitting

In Section 5.2.1, we used maximum likelihood to fit two different models to our dataset D. The quadratic model produced a fit with smaller residuals. Should we go further and try an even more flexible model? To address this question, we consider a sequence of increasingly flexible models, with the pth model in the sequence taking the form

$$Y = \sum_{i=0}^{p} \beta_i X^i + \mathcal{E},$$

where $\mathcal{E} \sim \mathcal{N}(0, \sigma^2)$, independent of X, and p is the polynomial degree of the model. Figure 5.2 shows the maximum likelihood prediction functions for three models. As flexibility increases, these functions contort in increasingly complex ways in order to get closer to the data. When $p = 9$, our prediction function has 10 parameters, allowing it to fit the data perfectly, but with wiggles that stray far out of the range of the observed responses. This should ring loud alarm bells.

Since we used a probabilistic method to fit our models, perhaps probability can also help us choose between them. Given a maximum likelihood parameter estimate $\hat{\boldsymbol{\theta}}$, we can evaluate the log-likelihood $\ell(\hat{\boldsymbol{\theta}})$, which gives the logarithm of the probability density of the data according to the fitted model. Figure 5.3a shows these log densities plotted against p. As p increases, so does the probability density of the data according to the model. Does this mean that the more flexible models are more plausible descriptions of the process that generated D? **No!** It is always possible to find a probabilistic model and a parameter vector $\boldsymbol{\theta}$ such

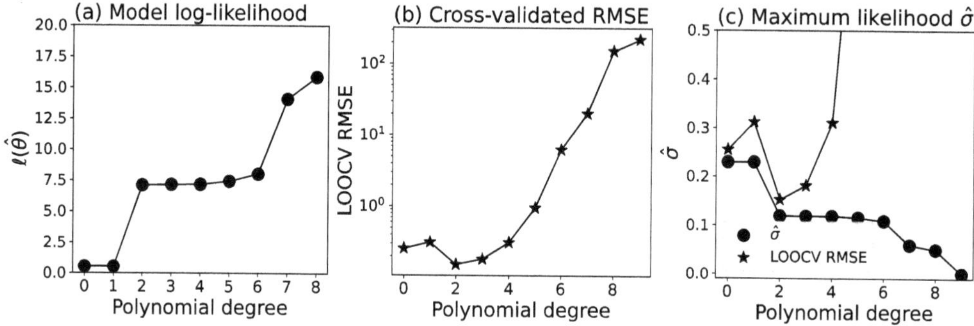

Figure 5.3 (a) Log-likelihood of fitted models, (b) cross-validated RMSE and (c) maximum likelihood estimates of σ, all plotted against polynomial degree.

that the model's conditional probability density function $f_{Y|X}(y|x;\boldsymbol{\theta})$ achieves very large values when evaluated on individual datapoints in D. The problem is that when a new point (x_0, y_0) is observed, the model will likely assign it a very low probability density. If we were to attempt to predict y_0 from x_0, such a model would do a poor job because the conditional density function $f_{Y|X}(-|x_0;\boldsymbol{\theta})$ would be sharply peaked in the wrong place.

The extent to which the probability density is concentrated around the prediction function is illustrated by Figure 5.3c, which shows the relationship between p and the maximum likelihood estimate of σ. As p increases, $\hat{\sigma}$ shrinks, eventually reaching zero when $p = 9$. As we increase flexibility, the systematic component of the model takes over as the explanatory mechanism for the patterns in the data. Where we *should* be modelling the scatter in the data using the noise component $\mathcal{E}$ we are *actually* using the systematic component $\sum_{i=0}^{p} \beta_i X^i$. We are *fitting to the noise* or, in other words, overfitting. This story should sound familiar. We found in Chapter 2 that smaller and smaller RSS values could be achieved by considering more and more flexible prediction function families. Given that the RSS and negative log-likelihood are linearly related for given σ (see equation (5.6)), it should come as no surprise that the peak likelihood fares no better as a means to select model flexibility.

As we saw in Chapter 2, cross-validation can help us detect overfitting. Letting D_{-i} be our dataset with point (x_i, y_i) removed, and letting $\hat{g}_{-i}(x)$ be the prediction function fitted to D_{-i}, the leave-one-out mean squared cross-validation error is

$$\text{MSE}_{\text{CV}} = \frac{1}{n}\sum_{i=1}^{n}(y_i - \hat{g}_{-i}(x_i))^2.$$

Figure 5.3b shows the square root of this error – the RMSE – for different model degrees. The minimum error is obtained when $p = 2$, and then increases very rapidly with complexity. The degree-2 model therefore represents our best guess at the optimal trade-off between simplicity and complexity. It is interesting to compare the cross-validated RMSE with our estimate $\hat{\sigma}$, which is equal to the RMSE calculated using the same data that was used to fit ('train') the model. Both errors are shown in Figure 5.3c. When the model has low complexity the two errors are of similar size because there is little scope for overfitting. However, as model flexibility grows the training error shrinks to zero, while the error on unseen data becomes large. This divergence of performance metrics based on seen and unseen data is a classic sign of overfitting.

5.4 The Unsupervised Case

We remarked in Section 5.1 that model (5.3) says nothing about the distribution of X. However, we twist its dials, the model makes claims only about the conditional distribution of Y given $X = x$. For Chapter 2's supervised task – 'predict Y from X!' – this is appropriate. But it won't do for Chapter 3's unsupervised task. Recall that in Chapter 3 we were trying to craft a 'hoard score function' whose value at a point (x, y) was proportional to the estimated chance of finding a new hoard when digging at that point. To guide us, we had a dataset of 90 hoard locations (assumed to be a random sample from the full population of locations). Let's once again assume that the datapoints were generated independently, and represent the datapoint-generating process by a random vector (X, Y).[7] So far so familiar. Now, though, we are interested in the **joint distribution** of X and Y – or in other words, the distribution of random vector (X, Y). That is because we want to learn where in two-dimensional space hoards are likely to be found. Learning the conditional distribution of Y given X would not be sufficient.

What exactly do we mean by 'the datapoint-generating process' in the hoard discovery case? For talk of long-run statistics to make any sense, we have to imagine a repeatable process – one that can be run again and again without limit. So if we think of the process as *picking a hoard at random from the population of buried hoards*, we'd better add *and then putting it back afterwards*, otherwise the datapoint-generating process will eventually run out of hoards. This is a perfectly well-defined process – a statistician would describe it as sampling with replacement – and we might be tempted to say that the task of Chapter 3 was to describe it. If we go that route, we'll have to say that the objectively correct hoard score function is a probability distribution that puts a dense blob of probability on each of the initially buried hoards, and spreads no probability anywhere else. (There is no chance of finding a hoard at a location that does not contain a hoard.) Sadly we can't hope to learn this distribution from the data, because that would require us to infer the exact locations of all the remaining buried hoards from the locations of the hoards unearthed so far – clearly an impossible task. But there is a less disheartening way of defining the datapoint-generating process. We can think back to the people who buried the hoards in the first place. They picked locations 'at random', that is, according to some probability distribution or other. Our dataset comprises 90 locations sampled from that distribution.[8] If the distribution was something fairly simple – say, a weighted sum of a few Gaussians – we really might be able to learn it from the data. For the hoard discovery problem, then, it makes sense to take the datapoint-generating process to be the process that generated the locations at which the hoards were buried. Chapter 3's hoard score functions then become attempts to describe the

[7] Of course, this is not the same random vector we were discussing earlier, in relation to model (5.3). We are reusing the earlier notation to make the structural difference between supervised and unsupervised tasks stand out as clearly as possible.

[8] Are we ignoring the role played by the bookworms here? No, because the bookworms destroyed record slips uniformly at random.

long-run statistics of this process; and the objectively correct hoard score function is the one that gets those long-run statistics right.

In Chapter 3 we considered 'k-blob' hoard score function families. Using our new language, we can say that fixing k fixes a *probabilistic model* of the datapoint-generating process. According to that model, the probability density function of the random vector (X, Y) is a weighted sum of k Gaussians. The model's parameter vector $\boldsymbol{\theta}_k$ controls the means, variances and weights of the individual Gaussians. Again, we have a contraption with a number of tunable dials: one for each component of the parameter vector. But this time, the dials control the model's story about the *joint* distribution of X and Y, that is, the distribution of random vector $\boldsymbol{X} = (X, Y)$. That story is encapsulated in a probability density function $f_{\boldsymbol{X}}^{(k)}(-; \boldsymbol{\theta}_k)$ which we can regard equation (3.3) as defining – though the details don't matter for the present discussion.

How should we go about fitting *this* sort of model to our data? Well, one option is to once again try to maximise likelihood, that is, to look for a $\boldsymbol{\theta}_k$ that makes the observed data seem as unsurprising as possible from the perspective of the model. The probability density the model assigns to the ith datapoint is $f_{\boldsymbol{X}}^{(k)}(\boldsymbol{x}_i; \boldsymbol{\theta}_k)$. Thus, the probability density it assigns to the complete sequence of 90 independently drawn datapoints is

$$\prod_{i=1}^{90} f_{\boldsymbol{X}}^{(k)}(\boldsymbol{x}_i; \boldsymbol{\theta}_k).$$

In practice, it will be more convenient to work with the logarithm of this measure. Thus, we recover Chapter 3's 'log-product' measure of fit (with slightly different notation this time, since the probabilistic ideas are now explicit):

$$L_k(\boldsymbol{\theta}_k) = \sum_{\boldsymbol{x} \in D} \log f_{\boldsymbol{X}}^{(k)}(\boldsymbol{x}; \boldsymbol{\theta}_k). \tag{5.8}$$

The higher the value of $L_k(\boldsymbol{\theta}_k)$, the better the fit achieved by parameter vector $\boldsymbol{\theta}_k$. We hit upon this measure in Chapter 3 via a rather unsatisfactory process of trial and error. Now we have motivated it in a principled way. (If you were puzzled by Exercises 3.3 and 4.2, you should be able to answer them confidently now.) The log probability measure in equation (5.8) is an example of a log-likelihood function for an unconditional probabilistic model. Box 5.4 summarises the maximum likelihood principle for unconditional probabilistic models in general.

★ Box 5.4 Principle of maximum likelihood (unconditional models)

Let $\boldsymbol{X}$ be a random vector. Consider a dataset of observations of this vector viewed as a single realisation of the 'vector of random vectors' $\underline{\boldsymbol{X}} = (\boldsymbol{X}_1, \ldots, \boldsymbol{X}_n)$, where the $\boldsymbol{X}_i$ are i.i.d. copies of $\boldsymbol{X}$. Given a probabilistic model of $\underline{\boldsymbol{X}}$ with parameter vector $\boldsymbol{\theta}$, we refer to the probability density (or mass) function of data according to the model, viewed as a function of $\boldsymbol{\theta}$, as the *likelihood function*

$$\mathcal{L}(\boldsymbol{\theta}) = f_{\underline{\boldsymbol{X}}}(\underline{\boldsymbol{x}}) = \prod_{i=1}^{n} f_{\boldsymbol{X}}(\boldsymbol{x}_i; \boldsymbol{\theta}).$$

The *log-likelihood* is $\ell(\boldsymbol{\theta}) = \log \mathcal{L}(\boldsymbol{\theta})$. The principle of maximum likelihood tells us to choose the model parameters so as to maximise the likelihood, that is

$$\hat{\boldsymbol{\theta}} = \arg\max_{\boldsymbol{\theta}} \mathcal{L}(\boldsymbol{\theta}) = \arg\max_{\boldsymbol{\theta}} \ell(\boldsymbol{\theta}).$$

5.5 Chapter Summary

In this chapter, we introduced the idea of a *probabilistic model* of the mechanism that generated our data. Probabilistic models allow us to formulate scientific hypotheses with well-defined truth conditions about mechanisms that are inherently stochastic.

- A **conditional probabilistic model** describes how the conditional probability density of a response variable given a predictor variable depends on the value of the predictor.
- The form of the dependence is controlled by a parameter vector $\boldsymbol{\theta}$. The space of possible values of $\boldsymbol{\theta}$ is the model's hypothesis space.
- To fit a model to data, one must make a data-driven choice of a specific hypothesis in the model's hypothesis space. In other words, one must pick a specific value of $\boldsymbol{\theta}$.
- The **principle of maximum likelihood** offers one way of doing this.
- For a wide class of conditional probabilistic models, maximising likelihood is equivalent to minimising RSS.
- The likelihood is a function of the parameters, not the data. It can be used to measure the plausibility of different parameter values *within a single model family*, but it should not be used – at least in its raw form – for comparing the plausibilities of different families.
- The likelihood principle can also be applied to **unconditional probabilistic models**, where the likelihood function is the probability of the data according to model, viewed as a function of its parameters.

5.6 Further Exercises

■ ***Exercise*** 5.4 A metal sphere sinking through a fluid will eventually reach its terminal velocity when the force due to gravity is equal to the drag force from the fluid plus the buoyancy force. Consider a sphere of radius r and density ρ_s moving with velocity v through a fluid with density ρ_l. The drag force, buoyancy force and gravitational force (weight) on the sphere are

$$F_d = \frac{\pi r^2 \rho_l c_d v^2}{2}, \quad F_b = \frac{4}{3}\pi r^3 \rho_l g, \quad F_g = \frac{4}{3}\pi r^3 \rho_s g,$$

where c_d is the drag coefficient of a sphere and $g = 9.81\ \mathrm{m\,s^{-2}}$ is the acceleration due to gravity. Consider the case where the fluid is water and the sphere is made of steel. Then $\rho_l = 10^3\ \mathrm{kg\,m^{-3}}$ and $\rho_s = 8 \times 10^3\ \mathrm{kg\,m^{-3}}$.

(a) Show that at terminal velocity, where $F_g = F_d + F_b$,

$$v^2 = \frac{8gr}{3c_d}\left(\frac{\rho_s}{\rho_l} - 1\right).$$

(b) Suppose we want to measure the drag coefficient c_d. We can do this by measuring the terminal velocities of spheres of various radii sinking in water. Suppose we model the relationship between the terminal velocity and radius measurements using

$$V = \beta\sqrt{R} + \mathcal{E}, \tag{5.9}$$

where $\mathcal{E} \sim \mathcal{N}(0, \sigma^2)$, independently of R. Given a dataset $D = \{(r_k, v_k)\}_{k=1}^{n}$, find the log-likelihood function $\ell(\boldsymbol{\theta})$, where $\boldsymbol{\theta} = (\beta, \sigma)$, using model (5.9).

(c) Derive an expression for the maximum likelihood estimate $\hat{\beta}$ of β, in terms of D.

(d) The table below gives some experimental results. Use these to calculate $\hat{\beta}$.

k	1	2	3	4
r_k (m)	0.01	0.05	0.1	0.2
v_k (m s^{-1})	1.9	4.4	6.3	8.9

(5.10)

(e) Estimate the drag coefficient of a sphere.

■ ***Exercise*** 5.5 You want to estimate the probability density function $f_X(x)$ of a random variable X using a dataset $D = \{x_i\}_{i=1}^{n}$ which comprises independent observations of X. You assume that $X \sim \mathcal{N}(\mu, \sigma^2)$ (see Box 5.1), where μ and σ are unknown. The parameter vector of your assumed model, then, is $\boldsymbol{\theta} = (\mu, \sigma)$.

(a) Show that the likelihood function is

$$\mathcal{L}(\boldsymbol{\theta}) = \frac{1}{\sigma^n (2\pi)^{\frac{n}{2}}} \exp\left(-\frac{1}{2\sigma^2}\sum_{i=1}^{n}(x_i - \mu)^2\right).$$

(b) What is the corresponding log-likelihood function, ℓ?

(c) Find expressions for the maximum likelihood estimates $\hat{\mu}$ and $\hat{\sigma}$ in terms of D.

6

Frequentist and Bayesian Uncertainty

So far we have focused on computing 'best-guess' parameter vectors. (Statisticians refer to these best guesses as *point estimates*.) We touched briefly on methods for quantifying uncertainty in Sections 2.7 and 5.1.4. It is now time to take a closer look.

The present chapter begins with a brisk review of learning tasks, models, densities and likelihoods. We note that Bayesians can use probability densities over a model's parameter space to represent uncertainty in their estimates, while frequentists cannot. We then tackle a very simple estimation problem – estimating a sex ratio – first by frequentist, then by Bayesian methods. In frequentist mode, we introduce sampling distributions, standard errors and confidence intervals. In Bayesian mode, we introduce conditionalisation, posterior densities and credible intervals. We work slowly and carefully through our simple example because we want to convey the *conceptual* contrasts between the frequentist and Bayesian approaches as clearly as possible. Of course, real-world inference problems tend to be more complicated. Frequentists will often be unable to derive the sampling distribution of an estimator, even when they assume their fitted model is correct; Bayesians will often be unable to compute exact posteriors. We conclude the chapter by surveying some standard tools for overcoming these obstacles. The Bayesian tools are developed in detail in Chapter 10, so we keep the Bayesian part of our survey brief.

6.1 Tasks and Datasets

We have emphasised the distinction between *supervised* and *unsupervised* learning tasks. At first, in Chapters 2 and 3, we tackled these tasks without the explicit use of probabilistic tools. Then, in Chapters 4 and 5, we introduced the idea of a probabilistic model of the process that generated our observations. There we saw that to tackle a *supervised* prediction problem – predicting a response Y given a predictor variable X or vector $\boldsymbol{X}$ – we need a model of the *conditional* probability distribution of Y given X or $\boldsymbol{X}$. In contrast, for an *unsupervised* task we need a model of the *unconditional* probability distribution of a random variable X or vector $\boldsymbol{X}$. Constructing unconditional models is often referred to as *density estimation*.

In both the conditional and unconditional cases, it is useful to think of our *observed* dataset D as the realisation of a collection of random variables or random vectors, written $\mathcal{D}$. We will refer to $\mathcal{D}$ as a *random dataset* to distinguish it from an observed dataset D. For example, a set of predictor–response observations $D = \{(x_i, y_i)\}_{i=1}^n$ is viewed as a single realisation of the random dataset $\mathcal{D} = \{(X_i, Y_i)\}_{i=1}^n$. We will often find it useful to view this random dataset as a pair of random vectors $\mathcal{D} = (\underline{X}, \underline{Y})$, where $\underline{X} = (X_1, \dots, X_n)$ and $\underline{Y} = (Y_1, \dots, Y_n)$.

We denote these vectors using the underline notation to indicate that they represent a set of individual observations from the same dataset-generating process (see Box 5.2). We reserve boldface vector notation for the situation where each individual observation is a vector. For example, to build a probability model of the random vector $\boldsymbol{X}$ – for example, the location of a treasure hoard – we would need a dataset $D = \{\boldsymbol{x}_i\}_{i=1}^n$, viewed as the realisation of the random dataset $\mathcal{D} = \{\boldsymbol{X}_i\}_{i=1}^n$. By combining underline and boldface vector notation, we can write this dataset as a single vector-of-vectors $\mathcal{D} = \underline{\boldsymbol{X}}$. In this chapter, starting in Section 6.3, we will consider a dataset of observations of a single random variable X. That is, $\mathcal{D} = \{X_i\}_{i=1}^n = \underline{X}$.

■ ***Exercise*** 6.1 Suppose you want to build a conditional probabilistic model of a random variable Y based on a predictor vector $\boldsymbol{X}$. What form would $\mathcal{D}$ take in this case?

6.2 Models, Densities, Likelihoods – and Uncertainty

Generically, an unconditional probabilistic model of a dataset-generating process may be expressed as a two-slot function $f_{\mathcal{D}}(-;-)$. The first slot takes possible realisations of the random dataset $\mathcal{D}$; the second slot takes a parameter vector $\boldsymbol{\theta}$. The real number $f_{\mathcal{D}}(D;\boldsymbol{\theta})$ is a probability if $\mathcal{D}$ is collection of discrete random variables or vectors, and a probability density if those variables or vectors are continuous. Specifically, it is the probability (or probability density) that $\mathcal{D}$ takes the value D according to the model when its parameter vector is set to $\boldsymbol{\theta}$.

There are two natural ways of obtaining a one-slot function from the two-slot function $f_{\mathcal{D}}(-;-)$: we can fix the parameter vector, or we can fix the dataset. If we fix the parameter vector, then we get a one-slot function $f_{\mathcal{D}}(-;\boldsymbol{\theta})$. When $\mathcal{D}$ is continuous, this function is a probability density function over possible datasets.[1] If we fix the dataset instead, we get a different one-slot function, $f_{\mathcal{D}}(D;-)$. This is a likelihood function: it represents how the probability density assigned by the model to a specific possible dataset D depends on the model's parameter vector $\boldsymbol{\theta}$. Likelihood functions are *not* probability density functions over possible values of the parameter vector (or indeed over anything else). Exercise 6.2 demonstrates this point.

■ ***Exercise*** 6.2 Consider the data-generating process that consists of a single toss of a particular coin. The result will be either HEADS or TAILS. Define the indicator random variable

$$X = \begin{cases} 1 & \text{if HEADS,} \\ 0 & \text{if TAILS.} \end{cases}$$

Here is a (silly) probabilistic model of that process. The model has a single scalar parameter θ, which can take any real value in the range $(-\infty, +\infty)$. The probability of HEADS according to the model is

$$f_{\mathcal{D}}(\mathcal{D} = \{1\}; \theta) = 1.$$

[1] It's a probability mass function when $\mathcal{D}$ is discrete.

In other words, the model always predicts with 100% confidence that the coin will land HEADS, and it pays no attention to the value of its single parameter. Although the model is silly, it is mathematically well defined.

(a) Suppose the actual dataset is $D = \{1\}$. What is the associated likelihood function?
(b) Suppose the actual dataset is $D = \{0\}$. What is the associated likelihood function?
(c) Explain why neither of the likelihood functions you just wrote down can be interpreted as a probability density function over possible values of parameter θ.

The story is similar for conditional probabilistic models. When the predictor and the response are both scalars, the random dataset is $\mathcal{D} = (\underline{X}, \underline{Y})$.[2] A conditional probabilistic model of the dataset-generating process defines a *three*-slot function $f_{\underline{Y}|\underline{X}}(-|-;-)$. The first slot takes possible realisations of random vector $\underline{Y}$, the second slot takes possible realisations of random vector $\underline{X}$, and the third slot takes a parameter vector $\boldsymbol{\theta}$. If Y is continuous, then the real number $f_{\underline{Y}|\underline{X}}(\underline{y}|\underline{x};\boldsymbol{\theta})$ is the conditional probability density that $\underline{Y} = \underline{y}$ given $\underline{X} = \underline{x}$, according to the model when its parameter vector is set to $\boldsymbol{\theta}$.

Once again, we have two natural ways of obtaining a one-slot function. We can fix the parameter vector and the vector of predictors, or we can fix the dataset (i.e. the predictors and the responses). If we fix the parameter vector and the predictors, we get the function $f_{\underline{Y}|\underline{X}}(-|\underline{x};\boldsymbol{\theta})$. When Y is continuous, this is a probability density function over the set of responses given the observed predictors. If we fix the dataset instead we get a (conditional) likelihood function, $f_{\underline{Y}|\underline{X}}(\underline{y}|\underline{x};-)$. Often we refer to this simply as the likelihood function. In the conditional probabilistic modelling context, a likelihood function represents how the *conditional* probability density assigned by the model to a specific set of responses given a specific set of predictors depends on the model's parameter vector. Once again, this is not a probability density function over possible values of $\boldsymbol{\theta}$.

In Chapter 5, we saw some model-fitting strategies based on the principle of maximum likelihood. Different possible values of a model's parameter vector amount to different hypotheses about the dataset-generating process. The principle of maximum likelihood tells us to set our model's parameter vector to whichever value maximises the likelihood function associated with the actual dataset – that is, the value that would make the data least surprising according to the model. The intuition behind the principle is that we ought to regard high likelihood hypotheses as more plausible than low likelihood ones. As noted in Chapter 4, though, the intuition is not entirely satisfactory, because likelihood is not in general a measure of plausibility. Recall Exercise 4.3: utterly implausible hypotheses can have very high likelihoods! It is therefore tempting to say something like the following.

> What we really want is a way of assigning *probabilities* or probability densities to different hypotheses (in a way that is responsive to the available data). For example, if we assume for the sake of argument that our model is structurally correct, we would like a probability density function defined on the model's parameter space. The point in parameter space that maximised the density would be the most plausible point estimate of the parameter vector. Moreover, the probability density function would be an explicit representation of our *uncertainty* about the true value of the parameter vector.

Bayesians yield to this temptation; frequentists do not – and cannot. From a frequentist perspective, the idea of a probability density function on a model's parameter space does

[2] We'll consider the case of vector predictors in Chapter 7.

not make sense, because the parameter vector is not a random variable. If we imagine that our model is correct, then we must imagine that there is a true value of the model's parameter vector. The true value is an unknown but fixed property of the data-generating process; it would not change from trial to trial if the process were rerun, and so there are no interesting relative frequencies to associate with it. By contrast, from a Bayesian perspective the idea of a probability density function on a model's parameter space *does* make sense. Bayesians do not insist that all probabilities are relative frequencies. They are willing to use probability distributions to represent states of belief – for example, our state of belief about the parameters of a particular probability model. This of course introduces a subjective element. In particular, to use Bayesian techniques, we have to articulate our *priors*. That is, we have to make plain our state of belief *before* we start the process of learning from data. Some people are uncomfortable with this. They feel that scientifically respectable statistical procedures should not begin with demands for psychological introspection. But if we are willing to play along, the payoff of the Bayesian approach is a coherent system for doing inference that harnesses the full power of the probability calculus.

Since frequentists cannot appeal to probability densities on parameter space, they need other strategies for quantifying uncertainty. The core idea behind all frequentist strategies is the same. Consider the process that generated the actual dataset, and imagine an infinite sequence of independent *reruns* of that process, generating hypothetical datasets $D_1, D_2, \ldots$.[3] Each hypothetical dataset is a realisation of $\mathcal{D}$, so questions about the long-run statistics of the sequence are equivalent to questions about the distribution of $\mathcal{D}$. Moreover, questions about the long-run statistics of anything we compute *from* the dataset – whether an estimate of a parameter vector or anything else – are equivalent to questions about the distribution of some *function* of $\mathcal{D}$. When it is clear which function we have in mind, we will refer to its distribution as the *sampling distribution*. Frequentist statements about uncertainty can always be interpreted as statements about some function's sampling distribution. There is a problem, however: the true sampling distribution is unknown, because the true distribution of $\mathcal{D}$ is unknown. (We just have the one actual dataset to go on, after all.) Frequentists therefore need to make an informed guess about the true distribution of $\mathcal{D}$. They will do that by fitting a probabilistic model of some sort to the actual data. Now, at this point you may worry that we are going round in circles. Frequentist uncertainty is a consequence of the fact that the data could have been different from the actual data. Can we really appeal to a model fitted to the actual data to assess the uncertainty in the parameter vector of a model fitted to the actual data? The answer is that we can. Indeed these will often (though not always) be the *same* model.

We have trailed some important ideas in this opening section, but the details are best explained through concrete examples. So let's turn to one. In Section 6.3, we will derive a point-estimation procedure for the parameter of a very simple probabilistic model. Then, in Sections 6.4 and 6.5, we will present frequentist and Bayesian approaches to quantifying uncertainty about the value of this parameter.

[3] When the context is a conditional probabilistic modelling task, we will often (though not always) define the dataset-generating process in a way that fixes the predictors. Then the hypothetical datasets differ only in respect of their response variables.

6.3 Lizard Sex Ratios

The ratio of males to females in most animal species is approximately 1:1. An explanation for this fact, known as *Fisher's principle*,[4] is that deviations from an equal ratio confer a reproductive advantage on the minority sex. Individuals of this sex are in short supply and therefore have a greater choice of mates, and a greater chance of reproducing. Genes that cause individuals to produce more offspring of the minority sex will therefore be reproduced more effectively. Over time the increase in frequency of these genes within the population will restore an equal sex ratio, and their comparative advantage will be lost. Deviations from an equal sex ratio are of interest to evolutionary biologists, because they indicate the presence of evolutionary forces beyond Fisher's principle.

We will address the problem of estimating the birth sex ratio in a population of viviparous lizards, *Niveoscincus ocellatus*, living in Tasmania.[5] Over a 16-year period, researchers observed the sexes of newborn lizards from a lowland population. Consider the sex of the kth newborn lizard observed by the researchers. We define the following indicator random variable:

$$X_i = \begin{cases} 1 & \text{if the } i\text{th lizard is male,} \\ 0 & \text{if it is female.} \end{cases}$$

The researchers observed the sequence $X_1, X_2, X_3, \ldots$ obtaining $x_1, x_2, x_3, \ldots$ Our aim is to *infer* the sex ratio based on these observations. To do this, we need a probability model of the sequence. To keep things simple, we will assume an idealised model according to which the X_is are *independent and identically distributed* (i.i.d.) with probability mass function

$$\mathbf{Pr}(X_i = x; p) = \begin{cases} p & \text{if } x = 1, \\ 1 - p & \text{if } x = 0, \end{cases}$$

where $p \in [0, 1]$ is a parameter of the model. We imagine that there is a true value p^* of the parameter – the true probability that any particular lizard is male. Notice that this true probability is an (imagined) property of the world, not of the model. The model just has a p 'dial' that can be set to any value in the interval $[0, 1]$. By playing around with the dial, we hope to find a value $\hat{p}$ that fits the data well, and hence can serve as an estimate of p^*. We refer to this process as 'estimating p^*' when we want to draw attention to the imagined true value serving as the target. Otherwise, and more commonly, we drop the asterisk and just speak of 'estimating p'. (See Section 5.1.4.)

According to the model, each X_i is an example of a *Bernoulli random variable* with parameter p, written $X_i \sim \text{Bernoulli}(p)$. We can rewrite the Bernoulli probability mass function as

$$f_X(x; p) = p^x(1 - p)^{1-x},$$

where $x \in \{0, 1\}$. Here we have included p as an additional argument of the mass function, but written after a semicolon to indicate that it is a parameter. In other words, we primarily think of f_X as a function of x; the parameter controls the form of that function.

[4] After Ronald Fisher, who played a central role in laying the foundations of statistical science.
[5] The data used in this example is taken from the study in Cunningham et al. (2017).

■ ***Exercise*** 6.3 Confirm that our compact definition of $f_X(x; p)$ is equivalent to the probability mass function $\mathbf{Pr}(X_i = x)$ as it was first defined. Hint: check the two cases $x = 1$ and $x = 0$.

■ ***Exercise*** 6.4 Lizards are born in litters that typically contain between two and six offspring. The researchers observed one litter at a time, with a litter of size n corresponding to a sub-sequence $X_i, X_{i+1}, \ldots, X_{i+n-1}$. Which of our modelling assumptions would be violated if it turned out that each litter was either all-male or all-female?

Our aim now is to apply the principle of maximum likelihood to estimate parameter p from observations of our random variables. In our previous application of the principle, in Chapter 5, our data consisted of pairs of values. The current situation is somewhat simpler, but the principle remains the same. We have a dataset $\underline{x} = (x_1, x_2, \ldots, x_n)$ where x_i represents an observation of the random variable X_i. Because the X_is are assumed to be i.i.d. we can write the joint probability mass function of the whole sequence $\underline{X} = (X_1, X_2, \ldots, X_n)$ as a product of the mass functions of the individual variables:

$$f_{\underline{X}}(\underline{x}; p) = \prod_{i=1}^{n} f_X(x_i; p).$$

With the actual dataset $\underline{x}$ fixed, the function mapping p to the probability of $\underline{x}$ is the *likelihood function*. It is defined by

$$\mathcal{L}(p) = f_{\underline{X}}(\underline{x}; p) = \prod_{i=1}^{n} f_X(x_i; p).$$

The maximum likelihood estimator of p is

$$\hat{p} = \arg\max_{p} \mathcal{L}(p).$$

This is the value of p that maximises the probability of the data, according to the model. The same estimate is obtained by maximising the log-likelihood

$$\ell(p) = \log \mathcal{L}(p),$$

because log is a monotonically increasing function.

Let us now put the principle of maximum likelihood to work. Suppose that the first baby lizard observed is male ($x_1 = 1$). The likelihood function is

$$\mathcal{L}(p) = p.$$

The maximum likelihood estimate (MLE) of p after the first visit is therefore

$$\hat{p}_1 = 1.$$

Here we have added a subscript to $\hat{p}$ to indicate that this parameter estimate was obtained

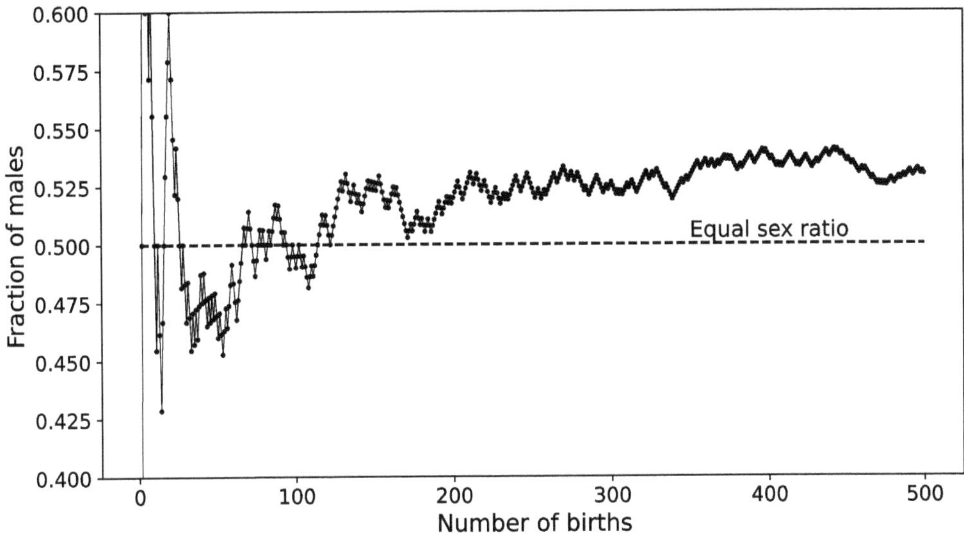

Figure 6.1 The cumulative fraction of males, $n^{-1}\sum_{i=1}^{n} x_i$, calculated from a sequence of observations of offspring of the viviparous lizard *Niveoscincus ocellatus*.

using one datapoint. Clearly this is insufficient to estimate the sex ratio, so let's repeat the calculation using more observations. The likelihood function after n observations is

$$\begin{aligned}\mathcal{L}(p) &= \prod_{i=1}^{n} f_X(x_i; p)\\ &= \prod_{i=1}^{n} p^{x_i}(1-p)^{1-x_i}\\ &= p^{\sum_{i=1}^{n} x_i}(1-p)^{n-\sum_{i=1}^{n} x_i}\\ &= p^{s_n}(1-p)^{n-s_n},\end{aligned}$$

where $s_n = x_1 + x_2 + \cdots + x_n$ is the number of males observed so far. The log-likelihood is

$$\ell(p) = s_n \log p + (n - s_n)\log(1-p).$$

The maximum likelihood occurs where $\ell'(p) = 0$, that is, where

$$\frac{s_n}{p} - \frac{n - s_n}{1-p} = 0.$$

Solving for p we obtain the estimate

$$\hat{p}_n = \frac{s_n}{n}.$$

This seems very reasonable: our estimate for the probability of producing a male lizard is just the fraction male lizards observed so far.

Figure 6.1 shows a sequence of MLEs calculated from the real lizard data. Here we see that for small sample sizes the estimate is very noisy: it jumps around a good deal as further observations are made. As the number of observations increases, the estimate settles down. Intuitively we think of the estimate as becoming more accurate with increasing n. An important question is then: how should we measure this accuracy?

■ ***Exercise*** 6.5 Define the random variable $S_n = X_1 + X_2 + \cdots + X_n$. What is the probability distribution of S_n (assuming that our i.i.d. Bernoulli model for the X_is is correct)? What are the mean and variance of this distribution?

6.4 Frequentist Uncertainty: The Basics

6.4.1 The Standard Error

Our confidence in $\hat{p}_n$ as an estimate of p grows as we observe more lizards. But how should we quantify this confidence? The classical frequentist approach is as follows. We have a single sequence of observations $\underline{x} = (x_1, \ldots, x_n)$, and we have an algorithm for estimating a parameter based on these observations. (In the present example, the algorithm is maximum likelihood.) It is useful to think of the algorithm as a function that maps datasets to estimates. We'll call the function t. In our lizards example, it is defined by

$$t(\underline{X}) = \frac{1}{n} \sum_{i=1}^{n} X_i. \tag{6.1}$$

The function t is an *estimator*. Given the dataset $\underline{x}$ the estimate it produces is

$$\hat{p} = t(\underline{x}).$$

Now imagine we had a large number of hypothetical datasets, $\underline{x}_1, \underline{x}_2, \ldots$, generated by the same process that generated the actual dataset. The hypothetical datasets are i.i.d. samples from some unknown probability distribution $F_{\underline{X}}$ representing the data-generating process.[6] If we plug the hypothetical datasets into our estimator, we get a sequence of estimates for $\hat{p}$,

$$t(\underline{x}_1), t(\underline{x}_2), \ldots$$

The probability distribution of these estimates is known as the *sampling distribution*. It is the probability distribution of the random variable $t(\underline{X})$, where $\underline{X} \sim F_{\underline{X}}$. The estimate of p we have made using our dataset represents a *single realisation* of the random variable $t(\underline{X})$. If we re-ran the dataset-generation process from scratch (and re-computed our estimate), we'd get another realisation. From the frequentist perspective, the probability distribution from which these realisations are drawn provides an objective basis for statements about the uncertainty in our actual estimate. For example, the *standard error* of the estimator is defined to be $\sqrt{\mathrm{Var}(t(\underline{X}))}$. Since the true sampling distribution is unfortunately unknown, we must estimate it somehow. Box 6.1 summarises this perspective, and Box 6.2 presents one way of estimating the sampling distribution.

[6] Formally, the symbol $F_{\underline{X}}$ denotes the cumulative distribution – or simply the *distribution function* – of $\underline{X}$. We can also interpret $F_{\underline{X}}$ as a symbol representing the distribution of $\underline{X}$ in an abstract sense.

★ Box 6.1 Frequentist uncertainty and the standard error

Let $\underline{x}$ be the realisation of a dataset random vector $\underline{X}$ with unknown distribution $F_{\underline{X}}$.

Suppose we have an estimator function t for a parameter θ. The random variable

$$\hat{\Theta} = t(\underline{X})$$

is called the **estimator**, and its distribution is the estimator's **sampling distribution**. The word 'estimator' is correctly applied both to the function t and to the random variable $\hat{\Theta}$. A realisation of $\hat{\Theta}$ – for example, $t(\underline{x})$ – is called an **estimate**.

Frequentist uncertainty in the estimate $\hat{\theta} = t(\underline{x})$ would ideally be represented by the estimator's sampling distribution. For example, the estimator's **standard error** is

$$\text{se} = \text{se}(t) = \sqrt{\text{Var}(\hat{\Theta})} = \sqrt{\text{Var}(t(\underline{X}))}.$$

However, because $F_{\underline{X}}$ is unknown, we do not know the sampling distribution, and hence we do not know the standard error. We can try to estimate it by computing what the standard error *would be* if $\underline{X}$ had some particular distribution $\hat{F}_{\underline{X}}$ that we *are* able to pin down. Our estimate of the standard error is then

$$\hat{\text{se}} = \sqrt{\text{Var}_{\underline{X} \sim \hat{F}_{\underline{X}}}(t(\underline{X}))}.$$

Obviously this procedure makes sense only if $\hat{F}_{\underline{X}}$ is a plausible proxy for $F_{\underline{X}}$.

★ Box 6.2 The plug-in estimate of the standard error

A special case of the predicament described in Box 6.1 arises when θ is the sole parameter of a model family $\{F_{\underline{X},\theta}\}$ that we are hoping to use to approximate $F_{\underline{X}}$. Then there is a natural choice of proxy for $F_{\underline{X}}$: we can use $F_{\underline{X},\hat{\theta}}$. If the model family is simple enough, it may be possible to derive a formula for $\sqrt{\text{Var}_{\underline{X} \sim F_{\underline{X},\theta}}(t(\underline{X}))}$ as a function of θ. Plugging $\hat{\theta}$ (our estimate of θ) into the formula gives the **plug-in estimate** of the standard error.

Before we return to the lizards, it is worth stressing the generality of the framework we have just sketched. The central idea is that of *hypothetical reruns of the process that generated the dataset*. Sometimes – as with the lizards – it will make sense to think of the dataset-generating process as n independent 'copies' of a single datapoint-generating process. Then the dataset random vector will be a bundle of n i.i.d. datapoint random variables/vectors. This structure is common, but there are plenty of other possibilities. In a regression context, the dataset random vector might represent measurements of a response variable for n fixed, non-random values of a predictor variable. In a computer-vision context, it might represent an n-pixel image. In those cases, it would *not* make sense to think of the dataset random vector as a bundle of n i.i.d. datapoint random vectors; but we remain free to imagine sequences of hypothetical datasets as i.i.d. samples from a single probability distribution (the probability distribution of the data*set* random vector).

Let's now return to the lizard sex ratio example and give a concrete demonstration of the plug-in approach. We are supposing, recall, that the individual sex indicator random variables X_i are i.i.d. Bernoulli with unknown parameter p^*. That means the dataset random vector $\underline{X}$ is a bundle of n independent Bernoulli(p^*) random variables. Our task is to estimate the unknown parameter. We have an estimator, $t(\underline{X}) = \frac{1}{n}\sum_{i=1}^n X_i$. Its true variance is

$$\text{Var}\left(\frac{1}{n}\sum_{i=1}^{n} X_i\right) = \frac{1}{n^2}\sum_{i=1}^{n}\text{Var}(X_i) = \frac{1}{n}p^*(1-p^*).$$

Therefore,

$$\text{se} = \sqrt{\frac{p^*(1-p^*)}{n}}.$$

We don't know p^* so we can't compute the standard error. But we can estimate it by plugging our estimate $\hat{p} = t(\underline{x})$ into the above equation in p^*'s place. Hence,

$$\widehat{\text{se}} = \sqrt{\frac{\hat{p}(1-\hat{p})}{n}}.$$

For example, using the first $n = 500$ births we obtain $\hat{p} = 0.531$ (see Figure 6.1), giving an approximate standard error estimate of $\widehat{\text{se}} = 0.022$. It is conventional to use the '±' symbol to present standard error estimates alongside estimates of the corresponding parameters. After observing the sexes of 500 baby lizards, we have $\hat{p} = 0.531 \pm 0.022$.

■ ***Exercise* 6.6** A coin has an unknown probability p of turning up heads. You flip the coin twenty times and obtain 14 heads. Compute an MLE of p and a plug-in estimate of your estimator's standard error.

6.4.2 Bias and Consistency

In Box 6.2, we considered the (common) situation in which we want to estimate the parameter θ of a probabilistic model of the dataset-generating process. According to the model, the distribution of the dataset random vector $\underline{X}$ is $F_{\underline{X},\theta}$, for some unknown value of θ. To motivate the plug-in approach to estimating standard errors, we noted that it might be possible to calculate the variance our estimator $t(\underline{X})$ would have if the model were correct – as a function of θ, of course. It is natural to consider the analogous construction for expectation, that is, $\mathbb{E}_{\underline{X}\sim F_{\underline{X},\theta}}(t(\underline{X}))$. This construction lets us define the **bias** of estimator t as follows:

$$b(\theta) = \mathbb{E}_{\underline{X}\sim F_{\underline{X},\theta}}(t(\underline{X})) - \theta.$$

Notice that the bias is a function of θ. If $b(\theta) = 0$ for all possible values θ, then the estimator is said to be *unbiased*; otherwise it is *biased*. Unbiased estimators of model parameters have an attractive property: if the model was correct, then whatever the true value of θ turned out to be, the *expected* value of the estimator would match it exactly.

■ ***Exercise* 6.7** Is the lizard sex ratio estimator defined in equation (6.1) unbiased?

Many useful estimators are biased. In practice, it is often more important to ask whether an estimator would approach the true value of the parameter as the sample size is increased, if the model were correct. Estimators with this property are said to be *consistent*.

6.4.3 Confidence Intervals

The standard error is a measure of the variation we would expect to see in our parameter estimates if we repeatedly ran our estimation procedure on multiple different datasets drawn from the same source. A closely related measure is the *confidence interval*. Given a data-generating mechanism with an unknown parameter, we can devise a method for constructing intervals based on our observed data which have a given chance, specified by us, of containing the true value the parameter. Statisticians sometimes speak of intervals *trapping* the true value: a confidence interval is like a carefully placed box within which we hope our prey (the true value of the parameter) is hiding. The concept is defined precisely in Box 6.3.

★ Box 6.3 Definition of a confidence interval

Let $\underline{X}$ be a vector of i.i.d. random variables. Let θ be a parameter that we want to estimate. Suppose we define a procedure for constructing intervals of the form

$$C(\underline{X}) = [a(\underline{X}), b(\underline{X})]$$

such that for some number $\alpha \in [0, 1]$,

$$\mathbf{Pr}(\theta \in C(\underline{X})) = 1 - \alpha. \tag{6.2}$$

If $\underline{x}$ is a single realisation of $\underline{X}$ then the interval $C(\underline{x}) = [a(\underline{x}), b(\underline{x})]$ is called a $1 - \alpha$ **confidence interval**. The number $1 - \alpha$ is called the *coverage* of the interval. Often we set $\alpha = 0.05$ giving a 95% confidence interval.

The probability statement (6.2) is open to misinterpretation. For a given dataset $\underline{x}$, it does *not* mean that $\theta \in C(\underline{x})$ with probability $1 - \alpha$. In fact, from a frequentist perspective, the statement '$\theta \in C(\underline{x})$' does not refer to any random quantities: we already observed the data, and the parameter θ is a fixed but unknown number. Our confidence in the truth or otherwise of the statement is therefore not something that we can express a view on using probability. What, then, do we mean by *confidence* when we talk about confidence intervals?

Remember that for frequentists to attach a probability to an event A its occurrence must be decided by the outcome of an infinitely repeatable experiment: $\mathbf{Pr}(A)$ is the fraction of times A occurs in the limit of large numbers of experimental repeats. So, probability statement (6.2) means that if we ran the dataset-generating process many times, and computed a series of intervals

$$C(\underline{x}_1), C(\underline{x}_2), \ldots,$$

then a fraction $1 - \alpha$ of them would contain the true parameter.[7] The interval we compute from the data is the random quantity here, not the parameter. We will see later that Bayesian inference allows us to make statements *superficially* similar to (6.2) in which the parameter (not the interval) is the random quantity.

Let's turn now to practicalities. Often we will want symmetric confidence intervals. The goal then will be to establish a procedure for finding intervals of the form $C(\underline{x}) = [a(\underline{x}), b(\underline{x})]$

[7] More precisely: with probability 1 (i.e. almost surely), $\lim_{n\to\infty} \frac{1}{n} \sum_{i=1}^{n} \mathbf{1}\{\theta \in C(\underline{x}_1)\} = 1 - \alpha$. Here $\mathbf{1}$ is the indicator function that maps events to $\{0, 1\}$. If event A occurs, then $\mathbf{1}(A) = 1$; otherwise $\mathbf{1}(A) = 0$.

such that

$$\mathbf{Pr}(\theta < a(\underline{X})) = \mathbf{Pr}(\theta > b(\underline{X})) = \frac{\alpha}{2}. \tag{6.3}$$

Notice that these conditions involve the random vector $\underline{X}$ and not the observed dataset $\underline{x}$.

■ ***Exercise* 6.8** Demonstrate that if conditions (6.3) hold then

$$\mathbf{Pr}(\theta \in [a(\underline{X}), b(\underline{X})]) = 1 - \alpha.$$

Are conditions (6.3) the only conditions consistent with a $1 - \alpha$ confidence interval?

In order to establish such a procedure, we first need to approximate the sampling distribution. To simplify our presentation, we'll make a small abuse of notation by using the same symbol, $\hat{\theta}$, for both the estimator function and the estimate. That is, we'll write

$$\hat{\theta} = \hat{\theta}(\underline{x}),$$
$$\hat{\Theta} = \hat{\theta}(\underline{X}).$$

Now we make use of an important fact, whose derivation we postpone to Section 6.6.3: maximum likelihood estimators are consistent, and their sampling distributions are (given large enough datasets) approximately normal. This implies that, when data is plentiful, $\hat{\theta}(\underline{X}) \sim \mathcal{N}(\theta, \text{se}^2)$, or equivalently that

$$\hat{\theta}(\underline{X}) \overset{d}{\approx} \theta + \text{se} \times Z, \tag{6.4}$$

where $Z \sim \mathcal{N}(0, 1)$ and se is the estimator's standard error. Now suppose that we set the upper limit of our interval to be

$$b(\underline{X}) = \hat{\theta}(\underline{X}) + c \times \widehat{\text{se}}(\underline{X}),$$

where c is a constant to be determined and $\widehat{\text{se}}(\underline{X})$ is our estimate of the standard error based on dataset $\underline{X}$. We then have

$$\mathbf{Pr}(\theta > b(\underline{X})) = \mathbf{Pr}\left(\frac{\hat{\theta}(\underline{X}) - \theta}{\text{se}(\hat{\underline{X}})} < -c\right).$$

Using equation (6.4) and assuming that $\widehat{\text{se}}(\underline{X}) \approx \text{se}$, we obtain

$$\mathbf{Pr}(\theta > b(\underline{X})) \approx \mathbf{Pr}(Z < -c).$$

We determine c by imposing the condition

$$\mathbf{Pr}(Z < -c) = \frac{\alpha}{2}.$$

For a 95% interval, this yields $c = 1.96$. A similar argument can be used to show that the appropriate lower limit is $a(\underline{X}) = \hat{\theta}(\underline{X}) - c \times \widehat{\text{se}}(\underline{X})$. Hence, having observed a dataset $\underline{x}$, the interval constructed according to the rule

$$C(\underline{x}) = [\hat{\theta}(\underline{x}) - 1.96 \times \widehat{\text{se}}(\underline{x}), \hat{\theta}(\underline{x}) + 1.96 \times \widehat{\text{se}}(\underline{x})]$$

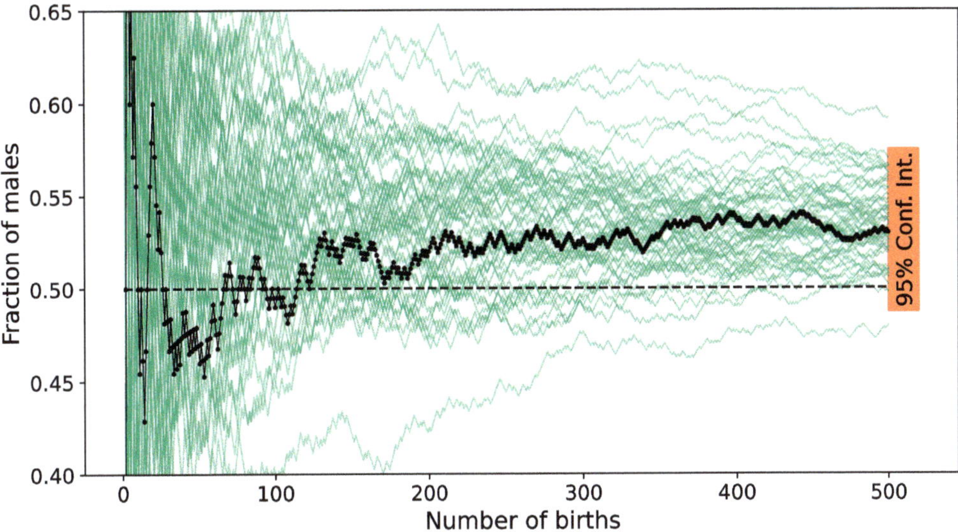

Figure 6.2 Green lines show cumulative fractions of males calculated from 50 synthetic datasets generated using the i.i.d. Bernoulli model with $\hat{p} = 0.531$. Black curve shows real data, and orange bar shows the 95% confidence interval calculated from the real data.

is an approximate 95% confidence interval. A convenient abbreviated way of writing a 95% interval, using the fact that $c = 1.96 \approx 2$, is

$$\hat{\theta}(\underline{x}) \pm 2 \times \hat{\text{se}}(\underline{x}).$$

Returning to lizard sex ratios, after observing $n = 500$ births we obtain the approximate 95% confidence interval for p of 0.531 ± 0.044, or equivalently $[0.487, 0.575]$. To visualise the sampling distribution and the confidence interval, in Figure 6.2 we have overlaid the cumulative fraction of males obtained from the real data with fractions obtained from 50 'synthetic' datasets. These samples were generated using our model of the data-generating process with p set equal to its estimated value from the real data (a plug-in approximation to the true distribution). The orange bar in Figure 6.2 shows the 95% confidence interval after $n = 500$ births. We see that 48 of the 50 synthetic samples (96%) lie within the interval. The green lines in Figure 6.2 give us a useful sense of the range of datasets which we might have observed if p were equal to the value we predicted.

■ ***Exercise*** 6.9 Of the first $n = 1000$ baby lizards observed, 549 were male. Compute a 95% confidence interval for p using this new data.

6.4.4 Hypothesis Testing and p-Values

Sometimes we wish to declare whether or not we accept a hypothesis, even if its truth is uncertain. For example, in a criminal trial the jury are not allowed to declare the defendant 'probably guilty' or 'most likely innocent'. They must make a firm decision one way or the other. In this situation, *not guilty* is the default position – the *null hypothesis* – and the evidence must be very strong ('beyond reasonable doubt') for it to be rejected. The special status of the 'not guilty' hypothesis is enshrined in the principle that a defendant is 'innocent until *proven* guilty'.

Similar situations arise in scientific and industrial settings. Suppose a new drug is invented which is intended to reduce blood pressure. New drugs are expensive and carry risks, so the default conclusion – the null hypothesis – is that they are useless or make things worse. Very strong evidence of their is efficacy required in order to reject the null hypothesis. Because humans are complex biological machines, the drop in blood pressure after administering the drug to a patient will be a random variable, which we'll call X. Of course, if the blood pressure increases, X will be negative. Let us introduce the parameter $\mu = \mathbb{E}(X)$ equal to the expected drop in blood pressure. We can formulate the null hypothesis, H_0, as a statement about the value of μ

$$H_0 : \mu \leq 0.$$

The complement of H_0 is the *alternative hypothesis* $H_1 : \mu > 0$. Formulating the null hypothesis in this way allows us to use probabilistic reasoning to decide whether it should be rejected. Let $\underline{X} = (X_1, \ldots, X_n)$ be a random vector of i.i.d. copies of X. A dataset of blood pressure drops would constitute a single observation of $\underline{X}$. We now define a function of $\underline{X}$ called the *test statistic*, which is designed to test H_0. In the current context, a suitable choice is the sample mean

$$T = t(\underline{X}) = \frac{1}{n}\sum_{i=1}^{n} X_i.$$

We then use this statistic to define a set R, known as the *rejection region*, which our data would be unlikely to belong to if the null hypothesis were true. Often R is of the form

$$R = \left\{\underline{x} \middle| t(\underline{x}) > c\right\},$$

where the constant c is known as the *critical value* of the statistic. Having observed a dataset $\underline{x}$, if we find that it lies in R then we *reject* H_0 and support the alternative H_1. In the current context, the larger the critical value, the bigger the average drop in blood pressure required to endorse the drug. In probabilistic terms, the larger the critical value, the more improbable our observations need to be according to the null hypothesis in order for us to reject it. We give the technical details of hypothesis testing for a single parameter in Box 6.4, before working through some examples.

★ Box 6.4 Hypothesis testing for a single parameter

Let X be a random variable whose distribution belongs to a family with a single unknown parameter, θ. The family may depend on other parameters, but we assume they are known. A null hypothesis that θ takes a specific value θ_0 is known as *simple* and the test $H_0 : \theta = \theta_0$ versus $H_1 : \theta \neq \theta_0$ is a *two-sided test*. A null hypothesis that θ satisfies an inequality is known as *composite* and the tests $H_0 : \theta \leq \theta_0$ versus

$H_1 : \theta > \theta_0$ or $H_0 : \theta \geq \theta_0$ versus $H_1 : \theta < \theta_0$ are known as *one-sided*. We write Ω_0 for the set of parameter values consistent with H_0, and Ω_1 for the set consistent with H_1.

Let $\underline{x}$ be a dataset of i.i.d. observations of X. We test H_0 by finding a set R of possible datasets which would be unlikely to be observed if H_0 were true. If $\underline{x} \in R$ we reject H_0, otherwise we retain it. Rejecting H_0 when it is true is a **type I error**, and retaining H_0 when it is false is a **type II error**. The **power** function of a test is defined

$$\pi(\theta) = \mathbf{Pr}_\theta(\underline{X} \in R) = \begin{cases} \mathbf{Pr}_\theta(\text{type I error}) & \text{when } \theta \in \Omega_0, \\ 1 - \mathbf{Pr}_\theta(\text{type II error}) & \text{when } \theta \in \Omega_1, \end{cases}$$

where $\mathbf{Pr}_\theta$ denotes probability under the model with parameter value θ. When H_1 is true, the power gives the probability of *correctly* rejecting H_0. When H_0 is true, it gives the probability of *incorrectly* rejecting H_0. The **size** of a test is the highest probability of a type I error out of all parameter values consistent with H_0

$$\alpha = \sup_{\theta \in \Omega_0} \pi(\theta),$$

where sup stands for 'supremum'.[a] For a simple null hypothesis, $\Omega_0 = \{\theta_0\}$ so the size is the probability of a type I error. Due to the special status of H_0, we usually want α to be small. That is, we want the probability of incorrectly rejecting the null hypothesis to be small. A test is said to have **significance level** α if its size is less than or equal to α. In most circumstances, size and significance level are used interchangeably. For a given significance level, we would like a test to be as powerful as possible under H_1. Searching for the most powerful test is challenging, and beyond the scope of this book.

[a] The supremum of a real-valued function f on a set A is the least upper bound of the set $\{f(a) | a \in A\}$.

Suppose we have a large set of observations $\underline{x} = (x_1, \ldots, x_n)$ of a random variable X drawn from a *known* distribution family with a single *unknown* parameter θ. Let us derive a (two-sided) test for the simple hypothesis $H_0 : \theta = \theta_0$ versus $H_1 : \theta \neq \theta_0$. First, recall the crucial distributional fact about the maximum likelihood estimator $\hat{\theta}(\underline{X})$ of θ:

$$\hat{\theta}(\underline{X}) \overset{d}{\approx} \theta + \text{se} \times Z,$$

where $Z \sim \mathcal{N}(0, 1)$ (see equation (6.4) and Section 6.6.3). An appropriate test statistic is the deviation of this estimate from θ_0,

$$T = t(\underline{X}) = \left|\hat{\theta}(\underline{X}) - \theta_0\right|,$$

giving the rejection region

$$R = \left\{\underline{x} \,\middle|\, \left|\hat{\theta}(\underline{x}) - \theta_0\right| > c\right\},$$

where c is a critical value determined by our significance level. The power of this test is

$$\pi(\theta) = \mathbf{Pr}_\theta(\underline{X} \in R) \approx \mathbf{Pr}_\theta(|\theta - \theta_0 + \text{se} \times Z| > c),$$

and its size is

$$\alpha = \pi(\theta_0) = \mathbf{Pr}\left(|Z| > \frac{c}{\text{se}}\right).$$

At this point it is useful to define the *Z-critical value*. It is the value z_α such that

$$\mathbf{Pr}(Z > z_\alpha) = \alpha.$$

Defining the standard normal cumulative distribution function $\Phi(z) = \mathbf{Pr}(Z \leq z)$, we have $z_\alpha = \Phi^{-1}(1-\alpha)$, where Φ^{-1} is the inverse function of Φ.

■ ***Exercise*** **6.10** Show that if $Z \sim \mathcal{N}(0,1)$ and $\mathbf{Pr}(Z > z_\alpha) = \alpha$, then $z_\alpha = \Phi^{-1}(1-\alpha)$.

From the definition of the Z-critical value, we see that a test of size α will have critical value $c = z_{\alpha/2}\text{se}$. Therefore, at significance level α,

$$\left|\frac{\hat{\theta}(\underline{x}) - \theta_0}{\hat{\text{se}}}\right| > z_{\alpha/2} \implies \text{reject } H_0,$$

$$\left|\frac{\hat{\theta}(\underline{x}) - \theta_0}{\hat{\text{se}}}\right| \leq z_{\alpha/2} \implies \text{retain } H_0,$$

where $\hat{\text{se}}$ is an approximation of the true standard error. That is, if the probability of seeing a deviation $|\hat{\theta} - \theta_0|$ at least big as ours is less than α, then we reject H_0.

As a concrete example, let us test the null hypothesis that a strict form of Fisher's principle holds for our lizards. That is, we test

$$H_0 : p = \frac{1}{2} \text{ versus } H_1 : p \neq \frac{1}{2}.$$

Using the first $n = 500$ births, we have $\hat{p} = 0.531$ and $\hat{\text{se}} = 0.022$, so the (standardised) deviation of $\hat{p}$ from p_0 is

$$\left|\frac{\hat{p} - \frac{1}{2}}{\hat{\text{se}}}\right| = 1.41.$$

Suppose we test at the $\alpha = 5\%$ significance level, for which the corresponding Z-critical value is $z_{\alpha/2} = z_{0.025} = 1.96$. Since our standardised deviation is less than this critical value, then we accept H_0. That is, we conclude that there is insufficient evidence to reject Fisher's principle at the 5% level. We emphasise that we performed this test to illustrate the methodology, not to draw any scientific conclusions. In fact our null hypothesis seems impossibly unlikely, given the complex nature of biological organisms and the many factors which affect their reproduction.

A measure closely related to hypothesis testing is the ***p*-value**. The p-value is the probability of observing a test statistic *at least as extreme* as the observed statistic, assuming the null hypothesis is true. A small p-value implies that our observations would be surprising under H_0, casting doubt on its truth. The p-value is defined formally in Box 6.5.

★ Box 6.5 *p*-value

Suppose we have a test statistic $t(\underline{x})$ for a hypothesis H_0 and a test at significance level α with rejection region

$$R(\alpha) = \left\{\underline{x} \middle| t(\underline{x}) > c(\alpha)\right\}.$$

Here the critical value $c(\alpha)$ will be a decreasing function of α. That is, the smaller the significance level (the chance of falsely rejecting H_0), the more extreme the statistic needs to be for us to reject H_0. Given a dataset $\underline{x}$, the p-value of the corresponding test statistic is the smallest significance level at which we would reject H_0.

Let us calculate the p-value of our lizard data under the null hypothesis $H_0 : p = 1/2$. We want to choose α just large enough so that H_0 is rejected. That is,

$$\left| \frac{\hat{p} - \frac{1}{2}}{\hat{\text{se}}} \right| = 1.41 = z_{\alpha/2}.$$

From the definition of z_α, we have $\alpha = 2\Phi(-z_{\alpha/2})$, so

$$p\text{-value} = 2\Phi(-1.41) = 0.159.$$

The p-value may be used as a measure of the evidence against H_0, The smaller the p-value, the stronger the evidence. Typically, a p-value over 0.1 is considered to provide very little evidence against H_0.

■ ***Exercise*** 6.11 Of the first $n = 1000$ baby lizards observed, 549 were male. Test the null hypothesis that there are more female lizards born than male at the 5% significance level. Calculate the p-value corresponding to this null hypothesis.

We end this section with a word of warning. Hypothesis testing and p-values are widely used for deciding when there is enough statistical evidence to support a scientific claim or hypothesis. However, the approach has been heavily criticised. When many hypotheses are tested and only significant results are published, the proportion of false positives in the literature may be high.[8]

6.5 Bayesian Uncertainty: The Basics

6.5.1 The Bayesian Way

The frequentist definition of probability using the concept of an infinitely repeatable trial is attractive because it is tangible and transparent. However, it is also restrictive, meaning that there are important kinds of uncertainty it cannot describe. For example, in Section 6.4.3, we found that our level of confidence in the truth of propositions such as

$$\theta \in [a, b],$$

where θ is an unknown parameter and a and b are fixed values, cannot be characterised using frequentist probability. The problem is that θ either lies in the interval or it does not, so there is no way to associate a non-trivial relative frequency with the proposition. There are plenty of other examples of situations in which we might want to attach probabilities to different hypotheses, but where these probabilities can't be thought of as relative frequencies. Such hypotheses include possible answers to questions such as: Where in the ocean did a particular ship sink? Did the accused man steal the jewels? Do gravitons exist?

[8] See McShane et al. (2017) for an in-depth critique of significance testing.

If we drop the frequentist interpretation of probability, we are left with a subjective notion of the probabilities of propositions or events as measures of our *degree of belief* in their truth or their occurrence. This is the Bayesian interpretation of probability. In some cases, it makes little practical difference whether we regard a probability as a relative frequency or a degree of belief. For example, the probability that a fair coin will land heads when next tossed is 0.5. That number is *both* the relative frequency of heads you would observe if you tossed the coin a very large number of times *and* the only sensible degree of belief you could have now in the proposition 'the coin will land heads when next tossed'. The advantage of the Bayesian approach is that it allows us to handle uncertainties in hypotheses and parameter values using the same mathematical apparatus – probability theory – that we use to handle uncertainties about future observations.

To get Bayesian reasoning started, we must declare what our *prior* degrees of belief regarding model parameter values or hypotheses are (or were) before we make (or made) any observations. This is where the subjectivity comes in. We will cover the choice of prior in Chapter 10, including methods for minimising the subjective element.

6.5.2 Bayesian Inference for Lizard Sex Ratios: One Observation

Let us bring a Bayesian perspective to the lizard sex ratio estimation problem. We start by promoting the parameter p, the probability that a baby lizard will be male, to a random variable P. Although we cannot observe P – it is a *latent variable* – we imagine that it still takes a specific value, determined by the complex evolutionary process which has created the current population of lizards. It doesn't matter whether we regard this process as repeatable-in-principle or not, because the distribution of a Bayesian random variable is not a summary of the long-run statistics of a hypothetical sequence of repeated experiments. It is instead a representation of *our uncertainty* about the variable's true value.

We declare our initial state of knowledge about the value of P using a probability density function $f_P(p)$, which we call the *prior*. To be precise, the quantity

$$\mathbf{Pr}(P \in [a, b]) = \int_a^b f_P(p)dp$$

is our degree of belief *before we have observed any baby lizards* that the realised value of P lies in the interval $[a, b]$.

Now suppose we observe the random variable X_1 indicating the sex of the first baby lizard. Since both X_1 and P are random variables, they have a joint probability mass-density function $f_{X,P}(x, p)$, which can be written in terms of the prior using the definition of conditional probability

$$f_{X,P}(x, p) = f_{X|P}(x|p)f_P(p),$$

where $f_{X|P}(x|p)$ is the mass function of X_1 given that $P = p$. This is just our original (frequentist) probability model for X_1,

$$\underbrace{f_{X|P}(x|p)}_{\text{Bayesian}} = p^x(1-p)^{1-x} = \underbrace{f_X(x; p)}_{\text{Frequentist}}.$$

Notice the contrast between the Bayesian notation and the frequentist notation. For the frequentist, p is a parameter – a fixed constant. For the Bayesian, it is the value of a random variable. Conditioning on the event $P = p$ would make no sense for the frequentist; it makes perfectly good sense for the Bayesian. How does our degree of belief regarding the different possible values of P change after observing X_1? The probability density of P given that $X_1 = x$ may be obtained using Bayes' rule,

$$\underbrace{f_{P|X}(p|x)}_{\text{Posterior}} = \frac{f_{X|P}(x|p)f_P(p)}{f_X(x)}. \tag{6.5}$$

The conditional density function on the left is called the *posterior* meaning 'coming after'. It is an updated version of our beliefs about P, having made an observation of X_1. Calculating the posterior is the central goal of Bayesian inference.

In applying Bayes' rule, we introduced a new mass function, $f_X(x)$, the marginal distribution of X_1. It is defined by

$$f_X(x) = \int_0^1 f_{X,P}(x, p)dp,$$

and it gives the distribution of X_1 without reference to the value of P. Since $f_{X,P}(x, p) = f_{X|P}(x|p)f_P(p)$, we can also express the marginal as the average of all possible conditional probability models for X_1 weighted by our prior beliefs,

$$f_X(x) = \int_0^1 f_{X|P}(x|p)f_P(p)dp.$$

Like the prior, the marginal has no counterpart in the frequentist world. It provides the normalising constant for the posterior distribution of P in equation (6.5). Noting that $f_{X|P}(x|p)$ is the likelihood of p (based on one observation), we can summarise the relationship between prior, posterior and likelihood as

$$\text{Posterior} \propto \text{likelihood} \times \text{prior}.$$

We will see below (when we calculate the posterior based on many observations) and in Chapters 9 and 10 that this relationship holds in a much more general sense.

■ ***Exercise*** 6.12 Suppose that $f_{X,P}(x, p) = 2p^{1+x}(1-p)^{1-x}$. The marginal probability mass function of X is given by

$$f_X(x) = \int_0^1 f_{X,P}(x, p)dp.$$

Show that $f_X(0) = \frac{1}{3}$ and $f_X(1) = \frac{2}{3}$.

Let us put the above theory into practice. Our prior knowledge of sex ratios in lizards and other species might lead us to believe that the fraction of males in the population we are about to observe will be close to 50%. This belief would correspond to a prior that was quite sharply peaked around $p = \frac{1}{2}$. Instead let us take an open-minded approach and assume that

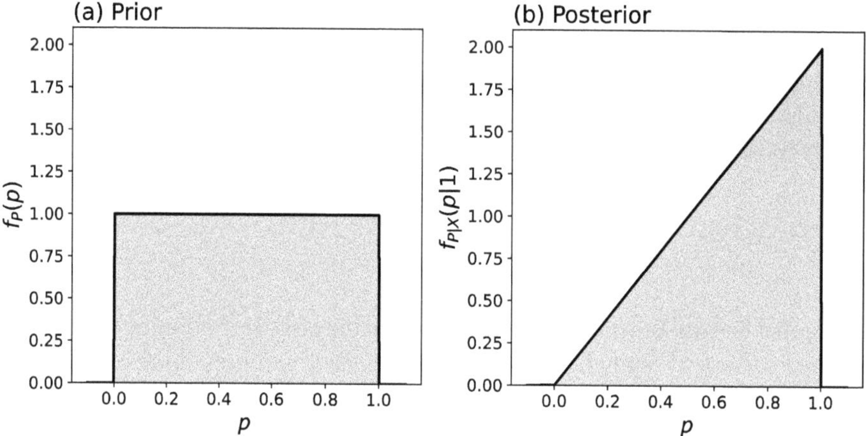

Figure 6.3 (a) Uniform prior distribution of the probability that a lizard will be born male. (b) Posterior distribution after observing a single male baby lizard.

all values of P are equally plausible, corresponding to a *uniform* prior

$$f_P(p) = \begin{cases} 1 & \text{if } p \in [0, 1], \\ 0 & \text{otherwise.} \end{cases}$$

To simplify our notation, we will assume from here on that expressions for prior and posterior densities of P hold for $p \in [0, 1]$ and are zero outside this interval. The first baby lizard observed is male ($x_1 = 1$). Expression (6.5) for the posterior then becomes

$$\begin{aligned} f_{P|X}(p|x_1) &= \frac{f_{X|P}(1|p) f_P(p)}{f_X(1)} \\ &= \frac{p \times 1}{f_X(1)} \\ &= 2p. \end{aligned} \tag{6.6}$$

■ ***Exercise*** 6.13 Show that the marginal probability $f_X(1)$ in the denominator of equation (6.6) is equal to $\frac{1}{2}$.

Figure 6.3 shows the uniform prior, and the posterior we have just calculated. Observing a male has shifted our beliefs towards higher values of p. If we wanted to provide a *point estimate* of p, then there are two options. We could report the *most plausible* value, known as the maximum a posteriori (MAP) estimate,

$$\hat{p}^{\text{MAP}} = \arg\max_p f_{P|X}(p|x_1).$$

Alternatively, we could report the expected value of the random variable P according to the posterior distribution, known as the *posterior mean* (PM),

$$\hat{p}^{\mathrm{PM}} = \int_0^1 p f_{P|X}(p|x_1)dp.$$

■ ***Exercise*** 6.14 Show that $\hat{p}^{\mathrm{MAP}} = 1$ and $\hat{p}^{\mathrm{PM}} = \frac{2}{3}$.

6.5.3 A Sequence of Predictions

We are about to observe another baby lizard. At this stage, our state of knowledge about P is based on our original prior *and* the sex of the first lizard we observed. Combining these two gave us our first posterior, which we now think of as a new 'second' prior, $f_P^{(2)}(p)$. Now we observe the next lizard. Again it is a male, so $x_2 = 1$. Applying Bayes' rule, we obtain our second posterior,

$$\begin{aligned} f_{P|X}^{(2)}(p|x_2) &= \frac{f_{X|P}(1|p)f_P^{(2)}(p)}{f_X^{(2)}(1)} \\ &= \frac{2p^2}{f_X^{(2)}(1)}. \end{aligned}$$

Here, the marginal probability in the denominator has changed from its previous form because the joint distribution of X_1 and P is based on a new prior. One way to calculate $f_X^{(2)}(1)$ is by integration, most straightforwardly by viewing it as a factor that ensures that the posterior is normalised, so

$$f_X^{(2)}(1) = \int_0^1 2p^2 dp = \frac{2}{3}.$$

Thus,

$$f_{P|X}^{(2)}(p|x_2) = 3p^2.$$

We could continue with this iterative process, progressively incorporating more observations and updating our posterior. However, this would be rather time consuming. It would be useful to be able to learn from many observations at once. Suppose we want to learn from the first two observations in one step. Let $f_{P|X_2,X_1}(p|x_2,x_1)$ be the posterior distribution of P given X_1 and X_2. According to Bayes' rule we have

$$f_{P|X_2,X_1}(p|x_2,x_1) = \frac{f_{X_2,X_1|P}(x_2,x_1|p)f_P(p)}{f_{X_2,X_1}(x_2,x_1)}.$$

Now, since we assumed that the observations are independent and identically distributed, we have

$$f_{X_2,X_1|P}(x_2,x_1|p) = f_{X|P}(x_2|p)f_{X|P}(x_1|p);$$

so our posterior obeys

$$f_{P|X_2,X_1}(p|x_2,x_1) \propto \underbrace{f_{X|P}(x_2|p)\ \underbrace{f_{X|P}(x_1|p)\ \underbrace{f_P(p)}_{\text{first prior}}}_{\text{first posterior = second prior}}}_{\text{second posterior= third prior}} .$$

Here our labelling of terms ignores normalising factors. We can repeat this process as many times as we like to include more observations. After n observations, written as the vector $\underline{X} = (X_1, X_2, \ldots, X_n)$, we have

$$\begin{aligned} f_{P|\underline{X}}(p|\underline{x}) &\propto f_{X|P}(x_n|p) \times f_{X|P}(x_{n-1}|p) \times \cdots \times f_{X|P}(x_1|p) \times f_P(p) \\ &= \left(\prod_{i=1}^{n} f_{X|P}(x_i|p)\right) f_P(p) \\ &= \mathcal{L}(p) f_P(p), \end{aligned}$$

where $\mathcal{L}(p)$ is the likelihood function given n observations. Therefore, we again have

$$\text{Posterior} \propto \text{likelihood} \times \text{prior}.$$

We could actually have jumped directly to this final result without needing to consider the observations in sequence. Since they are independent and identically distributed, their joint probability mass function is

$$f_{\underline{X}|P}(\underline{x}|p) = \prod_{i=1}^{n} f_{X|P}(x_i|p),$$

which is just the likelihood function $\mathcal{L}(p)$. A single application of Bayes' rule gives us the posterior

$$f_{P|\underline{X}}(p|\underline{x}) = \frac{f_{\underline{X}|P}(\underline{x}|p) f_P(p)}{f_{\underline{X}}(\underline{x})} \propto \mathcal{L}(p) f_P(p).$$

If we think of the marginal in the denominator purely as a normalising constant, we can write

$$f_{P|\underline{X}}(p|\underline{x}) = \frac{\mathcal{L}(p) f_P(p)}{c},$$

where c is chosen so that $\int_0^1 f_{P|\underline{X}}(p|\underline{x}) dp = 1$. This is often the most practically useful way to write the relationship between prior, posterior and likelihood.

Now let us apply this result to a set of observations of n baby lizards. When applying the maximum likelihood principle in Section 6.3, we found that $\mathcal{L}(p) = p^{s_n}(1-p)^{n-s_n}$, where $s_n = \sum_{i=1}^n x_i$. With a uniform prior, the posterior is just a normalised version of this likelihood,

$$f_{P|\underline{X}}(p|\underline{x}) = \frac{p^{s_n}(1-p)^{n-s_n}}{c}.$$

For some purposes, for example finding $\hat{p}^{\text{MAP}}$, we won't need to evaluate c at all. If we do need to evaluate it, we can do so by integration: $c = \int_0^1 p^{s_n}(1-p)^{n-s_n} dp$. Alternatively, and

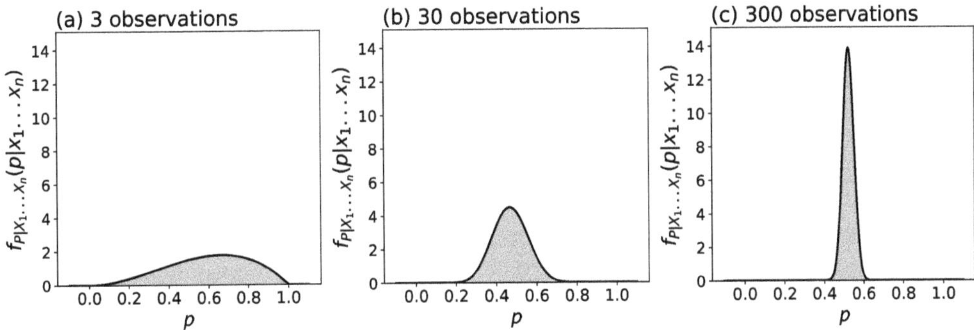

Figure 6.4 Posterior distribution of the male probability P with increasing number of observations: (a) 3 observations, (b) 30 observations and (c) 300 observations.

more simply, we can recognise that the posterior is a member of a well-known family: the beta distributions. The properties of beta distributions are summarised in Box 6.6.

★ Box 6.6 The beta distribution family

Distributions in the beta family describe random variables taking values in the interval $[0, 1]$. If P is beta distributed, we write $P \sim \text{Beta}(\alpha, \beta)$, where $\alpha > 0$ and $\beta > 0$ are the parameters of the distribution. The probability density function of P is

$$f_P(p) = \frac{p^{\alpha-1}(1-p)^{\beta-1}}{B(\alpha,\beta)},$$

where the normalising constant $B(\alpha, \beta) = \int_0^1 p^{\alpha-1}(1-p)^{\beta-1}dp$ defines the beta function. The beta function may also be expressed in terms of the gamma function,

$$\Gamma(z) = \int_0^\infty t^{z-1}e^{-t}dt,$$

as $B(\alpha,\beta) = \Gamma(\alpha)\Gamma(\beta)/\Gamma(\alpha+\beta)$. For integer arguments, $\Gamma(n) = (n-1)!$ and $B(m, n) = (m-1)!(n-1)!/(m+n-1)!$. The mean and mode of the beta distribution are $\mathbb{E}(P) = \alpha/(\alpha+\beta)$ and $\text{mode}(P) = (\alpha-1)/(\alpha+\beta-2)$, respectively.

By comparison to the beta probability density, we see that our posterior is a beta distribution with $\alpha = s_n + 1$ and $\beta = n - s_n + 1$, or equivalently,

$$P \sim \text{Beta}(s_n + 1, n - s_n + 1). \tag{6.7}$$

Figure 6.4 shows three examples of these posteriors. As the number of observations increases, the posterior becomes more concentrated, meaning that we are able to be more precise in our estimates of p.

Once we have the posterior, we can obtain parameter estimates. The posterior mean is the mean of the beta distribution in (6.7),

$$\hat{p}^{\text{PM}} = \frac{\sum_{i=1}^n x_i + 1}{n+2},$$

while the MAP estimate is the mode,

$$\hat{p}^{\text{MAP}} = \frac{1}{n}\sum_{i=1}^{n} x_i = \hat{p}^{\text{ML}},$$

where $\hat{p}^{\text{ML}}$ is the MLE. The fact that the MAP and MLE coincide in this case should come as no surprise: we selected a uniform prior, so the posterior density is proportional to the likelihood. In order for the MAP estimate to depart from the MLE, we would need to specify a non-uniform prior. Even then, in most situations all three estimates – MAP, maximum likelihood and posterior mean – will converge as the number of observations increases.

6.5.4 Bayesian Credible Intervals

In Section 6.4.3, we showed how to define a procedure for constructing frequentist confidence intervals which would succeed in trapping the true parameter value a certain fraction of the time. One cannot attach a frequentist probability to the event that a *particular* confidence interval contains the true parameter value, only to the event that the interval generating procedure will trap that value. In Bayesian mode, however, it *is* possible to define a particular interval which has a specified probability of containing the parameter, according to the posterior distribution. To do this, we first specify the desired probability $1 - \alpha$. We then find a and b such that

$$\int_{-\infty}^{a} f_{\Theta|\underline{X}}(\theta, \underline{x})d\theta = \int_{b}^{\infty} f_{\Theta|\underline{X}}(\theta, \underline{x})d\theta = \frac{\alpha}{2}.$$

Let $C = [a, b]$, then

$$\mathbf{Pr}(\Theta \in C|\underline{x}) = \int_{a}^{b} f_{\Theta|\underline{X}}(\theta, \underline{x})d\theta = 1 - \alpha.$$

In other words, the probability (our degree of belief) that Θ lies in C is $1 - \alpha$. An interval containing $1 - \alpha$ of the posterior probability weight is called a $1 - \alpha$ *credible interval*. In this example, we chose a symmetric interval where the probability density outside the interval was shared equally on either side of it. An alternative credible interval is the 'highest density interval' (HDI), for which every point inside must have higher probability density than every point outside.

Figure 6.5 shows the posterior distribution of P in our lizard sex ratio example, based on $n = 500$ observations and a uniform prior. The symmetric 95% credible interval is also shown. Whereas confidence intervals are used to express probability statements about the estimator, credible intervals express probability statements about the parameter. Also shown in Figure 6.5 is the plug-in normal approximation to the frequentist sampling distribution. In this case, the sampling and posterior distributions are almost identical. We explain why in Section 6.8. One consequence is that the Bayesian 95% credible interval computed here is almost identical to the frequentist 95% confidence interval we obtained in Section 6.4.3.

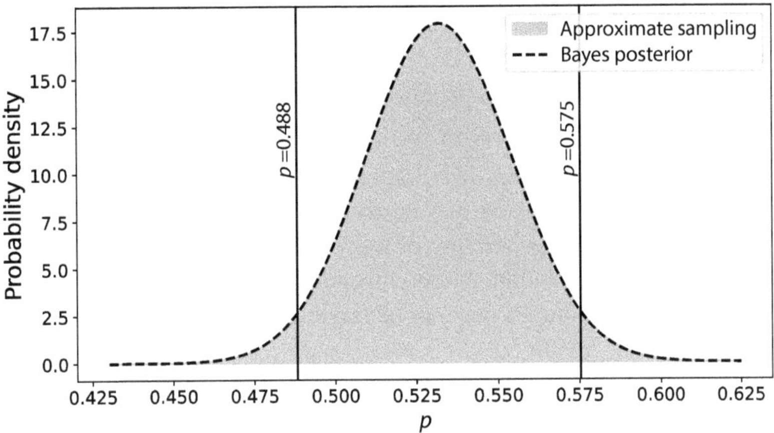

Figure 6.5 Dotted line is posterior of P (uniform prior) from $n = 500$ observations. Vertical lines show 95% Bayesian credible interval. Grey distribution is normal (plug-in) approximation to the frequentist sampling distribution of $\hat{p}^{\mathrm{ML}}$.

6.6 Frequentist Tools

We have already discussed one frequentist strategy for estimating uncertainty in an estimate: the plug-in approach (Section 6.4.1). Quite strict conditions must be met for the plug-in approach to be viable. First, the estimate in question must correspond to a (fitted) probabilistic model of the dataset-generating process. For example, our lizard sex ratio estimate $\hat{p}$ corresponded to a model in which the indicator variables are i.i.d. Bernoulli with parameter $\hat{p}$. Second, it must be possible – without too much difficulty – to calculate what the variance of the estimator would be if the probabilistic model picked out by the estimate were correct. That too was straightforward in the lizard sex ratio case, because sums of independent Bernoulli random variables are easy to handle. What if these two conditions are not met?

6.6.1 Parametric Bootstrap

Let's start by imagining a situation in which the first condition is met but the second is not. We have a dataset D and an estimate $\hat{\theta} = t(D)$. The estimate picks out a fitted probabilistic model of the dataset-generating process. According to the fitted model, D is a realisation of a random dataset $\mathcal{D}$ with distribution $F_{\mathcal{D},\hat{\theta}}$. However, we don't know how to calculate what the variance of our estimator $t(\mathcal{D})$ would be if the model were correct. (Perhaps the model is quite complicated.) We might nevertheless be able to *simulate* the dataset-generating process on the assumption that the model is correct. In other words, we might still be able to sample from $F_{\mathcal{D},\hat{\theta}}$. If we can do that, then we can generate a set of simulated datasets $D_1, D_2, \ldots, D_B$. These are called **parametric bootstrap replica datasets**, and the corresponding simulated estimates $t(D_1), t(D_2), \ldots, t(D_B)$ are called **parametric bootstrap replica estimates**. Being able to generate these replicas is almost as good as being able to carry out the calculations required for the plug-in approach. The empirical standard deviation of the replica estimates can serve as an estimate of $\sqrt{\text{Var}_{\mathcal{D}\sim F_{\mathcal{D},\hat{\theta}}}(t(\mathcal{D}))}$; and if B is large, it will be a reliable estimate.

Conceptually, a parametric bootstrap uncertainty estimate is very similar to a plug-in uncertainty estimate. The only difference is that instead of calculating $\sqrt{\text{Var}_{\mathcal{D}\sim F_{\mathcal{D},\hat{\theta}}}(t(\mathcal{D}))}$ exactly, we estimate it by sampling. It is important to note, however, that getting reliable estimates this way will sometimes be very costly. To obtain each replica estimate $t(D_i)$, we must first simulate the dataset-generating process to obtain a replica dataset D_i, and then compute $t(D_i)$ from it. Either or both of these operations may be computationally expensive. Repeating them thousands of times may not be feasible.

6.6.2 Non-parametric Bootstrap

Now we'll imagine a situation in which the first of the two conditions mentioned at the start of this section is not met. Suppose you wish to estimate median personal income from a large random sample of incomes $\underline{x} = (x_1, ..., x_n)$ drawn from the population of interest. The sample median is the obvious estimator of the population median. It is easy to compute, but it does not correspond to any particular probabilistic model of the dataset-generating process. How, then, can you estimate the standard error of your estimator?

First, let's restate your predicament. You have a random sample of incomes. This implies that $\underline{x}$ is a realisation of a random vector $\underline{X} = (X_1, ..., X_n)$ where the X_i are i.i.d. with some *unknown* common distribution F_X. To estimate the standard error in your median estimator, you will need to *estimate* F_X. One option would be to fit a parametric probabilistic model to the dataset, purely as a tool for estimating the standard error in your median estimate. For example, if the income distribution looked roughly log-normal, you could try fitting a log-normal distribution and using that as your estimate of F_X. You could then sample from it to generate replica datasets.[9] But what if, as you eyeball the data, no particular parametric family of distributions suggests itself? There is an alternative: you can treat the **empirical distribution** as your estimate of F_X.

The empirical distribution is a probability mass function that assigns non-zero probability to all and only the values that occur in the actual dataset. A value that occurs m times is assigned probability $\frac{m}{n}$. We'll denote the empirical distribution by $\hat{F}_{X,n}$; the subscript n is a reminder that the distribution is based on n datapoints, and that it parcels out probability in units of $\frac{1}{n}$. Simulating the dataset-generating process on the assumption that the X_i are i.i.d. with distribution $\hat{F}_{X,n}$ is easy: you just sample uniformly at random *with replacement* from the actually observed datapoints. Once you have sampled n datapoints, you have a (non-parametric) **bootstrap replica dataset**, from which you can compute a (non-parametric) **bootstrap replica estimate** of the median. (We have written 'non-parametric' in parentheses because 'bootstrap' refers to the non-parametric method by default.) Let $\hat{F}_{\underline{X},n}$ be the distribution of the vector $\underline{X}$ of n i.i.d. copies of $X \sim \hat{F}_{X,n}$. The empirical standard deviation of these replica estimates, computed from B independently resampled replica datasets, serves

[9] This would be a variant of the parametric bootstrap method. It differs from the 'classic' parametric bootstrap described in Section 6.6.1 in only one respect: the parameters of the model used to simulate the estimator's sampling distribution are not supplied by the estimate itself in this case.

as an estimate of $\sqrt{\text{Var}_{\underline{X} \sim \hat{F}_{\underline{X},n}}(\text{median}(\underline{X}))}$. We previewed this elegant strategy for estimating uncertainty back in Section 2.7. There we gave an intuitive description, with no explicit talk of random variables or distributions. Now we can tell a more satisfying story.

The empirical distribution $\hat{F}_{X,n}$ turns out to be the maximum likelihood estimate of the true distribution F_X, where the maximum is taken over all discrete distributions.[10] To see this, notice first that if you are crafting a distribution to maximise the probability of observing the values $(x_1, ..., x_n)$, you do not want to waste any of your probability budget on *other* values. The only remaining question is how to share out the probability among the actually observed values. It is plausible that the empirical distribution does this in the optimal way; Exercise 6.15 proves it using tools from Appendix C.

■ ***Exercise*** **6.15** Denote the number of *distinct* values in dataset $\underline{x} = (x_1, ..., x_n)$ by $\tilde{n}$, the distinct values themselves by $(\xi_1, ..., \xi_{\tilde{n}})$, and the number of times value ξ_i occurs by m_i. So $\sum_{i=1}^{\tilde{n}} m_i = n$. A probability mass function over the $\tilde{n}$ distinct values can be represented by a probability vector $\boldsymbol{p} = (p_1, ..., p_{\tilde{n}})$. The empirical distribution is represented by $\boldsymbol{p}^* = (m_1/n, ..., m_{\tilde{n}}/n)$.

Let $H_{\boldsymbol{p}}$ be the following hypothesis about the dataset-generating process: the X_i are i.i.d., and X_i takes value ξ_j with probability p_j.

(a) Show that the log-likelihood of $H_{\boldsymbol{p}}$ is

$$\begin{aligned} \ell(\boldsymbol{p}) &= \sum_{i}^{\tilde{n}} m_i \log p_i \\ &= \sum_{i=1}^{\tilde{n}} m_i \log m_i - n \log n - n\mathbb{KL}(\boldsymbol{p}^* \| \boldsymbol{p}). \end{aligned}$$

(b) Use the result in (a) to show that setting $\boldsymbol{p} = \boldsymbol{p}^*$ maximises the likelihood of $H_{\boldsymbol{p}}$.

Among all hypotheses that represent the X_i as i.i.d., the hypothesis that they are i.i.d. with common distribution $\hat{F}_{X,n}$ has the highest likelihood. Of course, it is not plausible that this hypothesis is *true*. (If you obtained a new sample of personal incomes from the population, you would expect to see some values that differed, at least slightly, from the values that occurred in your original dataset.) Nevertheless, it is useful for certain purposes, and one of those purposes is estimating the standard error in your median estimator.

The median estimation example involved datapoints that were just real numbers, but the non-parametric bootstrap method works in exactly the same way when datapoints are multidimensional, as they were in Section 2.7. Generating replica datasets by resampling datapoints from the actual dataset is a conceptually simple, computationally cheap and widely applicable technique. Even when a fitted probabilistic model of the dataset-generating process is available, the non-parametric bootstrap can be useful, because it lets you estimate

[10] We restrict our attention to discrete distributions here because likelihoods are defined differently for discrete and continuous distributions; strictly speaking, they cannot be compared. However, notice that any hypothesis according to which $\underline{X}$ has a continuous distribution assigns zero probability *mass* to the event $\underline{X} = \underline{x}$. There is therefore a sense in which it implies that the dataset is very surprising.

sampling distributions in a way that is *independent of that model*. That's especially valuable if you suspect the model might be a bad one.

■ ***Exercise*** 6.16 What will happen if you try to use the parametric bootstrap to measure the uncertainty in the estimated parameters of an *overfitted* probabilistic model? (Hint: you will get false reassurance.) Why might the non-parametric bootstrap be more useful in such cases?

We'll conclude by flagging some limitations of the non-parametric bootstrap. One is practical. Although *generating* replica datasets by resampling from the actual dataset is computationally cheap, obtaining replica estimates from them may not be. (The fact that model-fitting sometimes involves expensive optimisations is a problem for both versions of the bootstrap.) Another limitation is conceptual. The non-parametric bootstrap is appropriate only if the datapoints in the dataset can reasonably be treated as realisations of i.i.d. random variables (or random vectors). This will not always be the case. For example, suppose you are fitting a conditional probabilistic model to a dataset in which the predictor variable is controlled by the experimenters. The dataset is of the form $((x_1, y_1), (x_2, y_2), \ldots, (x_n, y_n))$, but the x_i values were *chosen*, not observed. If the scientists reran the experiment, they would reuse precisely the same set of predictor values. The datapoints in this dataset are not realisations of i.i.d. random vectors, and you cannot hope to simulate reruns of the experiment using the non-parametric bootstrap. The *parametric* bootstrap might work well, however. If your fitted conditional probabilistic model says that $Y_i = \hat{\beta}_0 + \hat{\beta}_1 X_i + \epsilon_i$, where the ϵ_i are i.i.d. normal random variables with variance $\hat{\sigma}^2$, then you can obtain a non-parametric bootstrap replica dataset by (i) copying the x_i values that occurred in the actual dataset and (ii) sampling Y_i from $\mathcal{N}(\hat{\beta}_0 + \hat{\beta}_1 x_i; \hat{\sigma}^2)$ independently for $i = 1, \ldots, n$.

■ ***Exercise*** 6.17 Show that in the special case of the lizard sex ratio estimation problem, the parametric and non-parametric bootstraps are exactly equivalent. To estimate the standard error of the sex ratio estimator, would you favour the bootstrap method or the plug-in method? Explain your answer.

6.6.3 Fisher Information: Approximate Sampling Distribution of Maximum Likelihood Estimators

The parametric bootstrap is a useful all-purpose tool for estimating the sampling distribution, but its computational costs will sometimes be prohibitive. In this section, we present an *analytical* strategy for approximating sampling distributions. It is useful in practice; we will find that maximum likelihood estimators have sampling distributions that are approximately normal (a result we assumed in Sections 6.4.3 and 6.4.4); and in Section 6.8, it will help us see connections between frequentist and Bayesian approaches to uncertainty.

Suppose we have a set of n observations of a random variable, X, which may be either continuous or discrete. In the simplest case, we have a one-parameter family of candidate density (or mass) functions for X

$$\{f_X(-;\theta) | \theta \in S\},$$

where S is the set of possible values that the parameter θ can take within the family. We imagine that there is one particular value of this parameter, θ^*, which gives the true distribution of X. Our aim is to estimate θ^*.

Our dataset may be viewed as a single observation of $\underline{X} = (X_1, \ldots, X_n)$, a vector of i.i.d. variables with common density (or mass) function $f_X(-;\theta^*)$. The log-likelihood function $\ell(-;\underline{x})$ is defined by

$$\ell(\theta;\underline{x}) = \sum_{i=1}^{n} \log f_X(x_i;\theta),$$

and the MLE of θ^* is

$$\hat{\theta} = t(\underline{x}) = \arg\max_{\theta} \ell(\theta;\underline{x}).$$

We have included the dataset as an explicit argument of the log-likelihood function to emphasise that each observed dataset determines a particular function of θ, and therefore a particular MLE. Our goal is to construct an approximation to the sampling distribution of the estimator – that is, an approximation to the probability distribution of the random variable $t(\underline{X})$.

Our strategy is to approximate the probability distribution of the deviation of our estimate from the true parameter value, θ^*. To simplify our notation, we will use the following shorthand for derivatives of ℓ:

$$\ell'(a;\underline{X}) := \left.\frac{\partial \ell(\theta;\underline{X})}{\partial \theta}\right|_{\theta=a},$$
$$\ell''(a;\underline{X}) := \left.\frac{\partial^2 \ell(\theta;\underline{X})}{\partial \theta^2}\right|_{\theta=a}.$$

That is, $\ell'(a;\underline{X})$ and $\ell''(a;\underline{X})$ are the first and second derivatives of $\ell(\theta;\underline{X})$ with respect to θ evaluated at $\theta = a$. Approximating the derivative of ℓ at $t(\underline{X})$ as a Taylor series about θ^*, we obtain

$$\ell'(t(\underline{X});\underline{X}) \approx \ell'(\theta^*;\underline{X}) + (t(\underline{X}) - \theta^*)\ell''(\theta^*;\underline{X}).$$

Since $t(\underline{X})$ is a maximum likelihood estimator, by definition $\ell'(t(\underline{X});\underline{X}) = 0$. We can therefore rearrange the above equation to obtain

$$t(\underline{X}) - \theta^* \approx -\frac{\ell'(\theta^*;\underline{X})}{\ell''(\theta^*;\underline{X})}.$$

The left-hand side of this expression gives the deviation of our estimate from the true parameter value. The right-hand side is a random variable whose distribution we can approximate. We first define the *score function*, s, and the *Fisher information*, $\mathcal{I}$:

$$s(X;\theta) := \frac{\partial}{\partial \theta} \log f_X(X;\theta),$$
$$\mathcal{I}(\theta) := \mathrm{Var}(s(X;\theta)),$$

where the variance is taken with respect to the density function $f_X(-;\theta)$. (Notice that for any given θ, $s(X;\theta)$ is a random variable, while $\mathcal{I}(\theta)$ is just a real number.) We will see that the

Fisher information can be regarded as a measure of the information that a single observation of X contains about the value of θ. Two important properties of the score function are

$$\mathbb{E}(s(X;\theta)) = 0, \tag{6.8}$$

$$\mathrm{Var}(s(X;\theta)) = -\mathbb{E}\left(\frac{\partial s(X;\theta)}{\partial \theta}\right), \tag{6.9}$$

where both the expectation and the variance are taken with respect to $f_X(-;\theta)$. Exercise 6.18 invites you to derive these equations.

■ ***Exercise*** 6.18 Suppose X is a continuous random variable with density $f_X(x;\theta)$.

(a) Show that

$$\int \frac{\partial \log f_X(x;\theta)}{\partial \theta} f_X(x;\theta)dx = \frac{\partial}{\partial \theta}\int f_X(x;\theta)dx = 0.$$

(b) Show that

$$\int \frac{\partial^2 \log f_X(x;\theta)}{\partial \theta^2} f_X(x;\theta)dx = -\int \left(\frac{\partial \log f_X(x;\theta)}{\partial \theta}\right)^2 f_X(x;\theta)dx.$$

(c) Use the two identities established earlier to establish the relationships (6.8) and (6.9).

We can write the derivatives of the log-likelihood at θ^* in terms of the score as follows:

$$\ell'(\theta^*;\underline{X}) = \sum_{i=1}^{n} s(X_i;\theta^*), \tag{6.10}$$

$$\ell''(\theta^*;\underline{X}) = \sum_{i=1}^{n} \left.\frac{\partial s(X_i;\theta)}{\partial \theta}\right|_{\theta=\theta^*}. \tag{6.11}$$

Both of these expressions are sums of i.i.d. random variables. We will approximate their distributions via the central limit theorem (CLT) given in Box 6.7.

★ Box 6.7 Central limit theorem

Let $X_1, \ldots, X_n$ be i.i.d. copies of the random variable X with mean μ and variance σ^2. Define a standardised sum of these variables,

$$Z_n := \frac{\sum_{i=1}^n X_i - n\mu}{\sigma\sqrt{n}}.$$

By definition, $\mathbb{E}(Z_n) = 0$ and $\mathrm{Var}(Z_n) = 1$. The CLT states that

$$Z_n \rightsquigarrow Z, \quad \text{as } n \to \infty,$$

where $Z \sim \mathcal{N}(0,1)$ and $\rightsquigarrow$ denotes 'convergence in distribution'.[a] Informally we will write

$$\frac{1}{\sqrt{n}}\sum_{i=1}^{n} X_i \overset{d}{\approx} \mu\sqrt{n} + \sigma Z,$$

where $\overset{d}{\approx}$ means 'approximately equal in distribution'. This approximation becomes increasingly accurate for increasing n.[b]

[a] Let $Y_1, Y_2, \ldots$ be a sequence of random variables and Y be another random variable. We say that Y_n 'converges in distribution' to Y, written $Y_n \rightsquigarrow Y$, if for every real number a we have $\mathbf{Pr}(Y_n < a) \to \mathbf{Pr}(Y < a)$ as $n \to \infty$.

[b] Strictly, the CLT tells us *only* about the behaviour of sums of random variables in the limit of large n. The *rate* at which the distribution function of Z_n converges to the standard normal distribution function is quantified by the Berry–Esseen theorem: $\mathbf{Pr}(Z_n < a) = \mathbf{Pr}(Z < a) + O\left(\frac{\mathbb{E}(|X|^3)}{\sqrt{n}}\right)$.

Defining the random variable $Y_i = s(X_i; \theta^*)$, we have $\mathbb{E}(Y_i) = 0$ and $\text{Var}(Y_i) = \mathcal{I}(\theta^*)$. Hence, from Box 6.7, we have

$$\frac{1}{\sqrt{n}}\ell'(\theta^*; \boldsymbol{X}) = \frac{1}{\sqrt{n}}\sum_{i=1}^{n} Y_i \overset{d}{\approx} \sqrt{\mathcal{I}(\theta^*)}Z,$$

where $Z \sim \mathcal{N}(0, 1)$. Similarly,

$$\frac{1}{\sqrt{n}}\ell''(\theta^*; \underline{X}) \overset{d}{\approx} -\mathcal{I}(\theta^*)\sqrt{n} + O(1).$$

Hence,

$$t(\boldsymbol{X}) - \theta^* \approx -\frac{\ell'(\theta^*; \underline{X})}{\ell''(\theta^*, \underline{X})} \overset{d}{\approx} \frac{Z}{\sqrt{n\mathcal{I}(\theta^*)}} + O(n^{-1}).$$

From this we see that the maximum likelihood estimator is approximately normal with mean θ^* and standard error

$$\text{se} = \frac{1}{\sqrt{n\mathcal{I}(\theta^*)}}.$$

Since θ^* is unknown, we use the plug-in principle to estimate the standard error,

$$\widehat{\text{se}} = \frac{1}{\sqrt{n\mathcal{I}(\hat{\theta})}}.$$

Asymptotically, as $n \to \infty$ we have

$$t(\underline{X}) \rightsquigarrow \theta^* + \frac{Z}{\sqrt{n\mathcal{I}(\theta^*)}},$$

where $Z \sim \mathcal{N}(0, 1)$. From this we see that the larger the Fisher information, the narrower the sampling distribution, meaning that each observation provides more 'information' about the unknown parameter.

Let us apply this method to an explicit example. Suppose that we have n observations, $\underline{x} = (x_1, \ldots, x_n)$, of the random variable X with density

$$f_X(x; \tau) = \frac{\exp\left(-\frac{x}{\tau}\right)}{\tau}.$$

The variable X is said to be *exponentially distributed* with scale parameter τ. The log-likelihood function is

$$\ell(\tau) = -\frac{1}{\tau}\sum_{i=1}^{n} x_i - n\log\tau,$$

yielding the MLE

$$\hat{\tau} = \frac{1}{n}\sum_{i=1}^{n} x_i.$$

The score function and Fisher information are

$$s(X;\tau) = -\frac{1}{\tau} + \frac{X}{\tau^2},$$
$$\mathcal{I}(\tau) = \frac{1}{\tau^2},$$

yielding the plug-in estimate of the standard error

$$\widehat{\text{se}} = \frac{1}{\sqrt{n\mathcal{I}(\hat{\tau})}} = \frac{\hat{\tau}}{\sqrt{n}}.$$

■ ***Exercise*** 6.19 The maximum likelihood estimator of the scale parameter of the exponential distribution is

$$t(\underline{X}) = \frac{1}{n}\sum_{i=1}^{n} X_i.$$

Calculate the standard error of this estimator exactly in terms of τ, and compare it to the normal approximation to the standard error.

■ ***Exercise*** 6.20 Let $X \sim \text{Bernoulli}(p)$. The probability mass function of X may be written $f_X(x;p) = p^x(1-p)^{1-x}$, where $x \in \{0,1\}$ and $p \in [0,1]$.

(i) Calculate the score function and Fisher information for the Bernoulli distribution.
(ii) Calculate the normal approximation to the standard error for the maximum likelihood estimator of p. How does your result compare with the exact expression for the standard error calculated in Section 6.4.1?

6.7 Bayesian Tools

The primary goal of Bayesian inference is to compute the posterior distribution. In some cases, this can be done analytically, but more often we must resort to computational methods. In this section, we set out an important class of models for which analytical calculation is possible, before previewing the computational approach.

6.7.1 Conjugate Priors

Suppose we want to construct a statistical model of a random variable X depending on a single parameter θ. The Bayesian approach to this problem is to view θ as the realised value of a latent random variable Θ, and to define a probabilistic model of X given the value of Θ, specified as a conditional probability density (or mass function) $f_{X|\Theta}(x|\theta)$. Given a prior, $f_\Theta(\theta)$, on Θ, and a dataset, $\underline{x} = (x_1, \ldots, x_n)$ of i.i.d. observations of X, the posterior may be found using the relationship

$$\underbrace{f_{\Theta|\underline{X}}(\theta|\underline{x})}_{\text{posterior}} \propto \underbrace{\left(\prod_{i=1}^{n} f_{X|\Theta}(x_i|\theta)\right)}_{\text{likelihood}} \underbrace{f_\Theta(\theta)}_{\text{prior}} .$$

Determining the posterior is straightforward if the product of the likelihood and the prior can be recognised as a member of a known distribution family (up to a normalising constant). This turns out to be possible when $f_{X|\Theta}$ belongs to the (one parameter) *exponential family*. Probability distributions in this family take the form

$$f_{X|\Theta}(x|\theta) = h(x) \exp\left(\eta(\theta)T(x) - A(\theta)\right), \tag{6.12}$$

where the functions $h(x) > 0$, $\eta(\theta)$ and $T(x)$ are referred to as the *base measure*, *natural parameter* and *canonical statistic*. The function $A(\theta)$ (the *log-normaliser*) ensures that the distribution is normalised. Many well-known distributions belong to this family. For example, consider the *Poisson distribution* with parameter θ, which has probability mass function

$$f_{X|\Theta}(x|\theta) = \frac{\theta^x e^{-\theta}}{x!} = \frac{1}{x!} \exp\left(x \log\theta - \theta\right),$$

where $x \in \{0, 1, 2, \ldots\}$. By comparison with equation (6.12), we see that this is a member of the exponential family with $h(x) = 1/x!$, $\eta(\theta) = \log\theta$, $T(x) = x$ and $A(\theta) = \theta$.

■ ***Exercise* 6.21** Show that the Bernoulli mass function may be written $f_{X|\Theta}(x|\theta) = \exp\left(x \log \frac{\theta}{1-\theta} + \log(1-\theta)\right)$. Hence show that the Bernoulli distribution is a member of the exponential family and identify the functions h, η, T and A.

■ ***Exercise* 6.22** Show that the exponential distribution is in the exponential family.

■ ***Exercise* 6.23** Consider the exponential family distribution with $h(x) = \binom{N}{x}$, $\eta(\theta) = \log(\theta/(1-\theta))$, $T(x) = x$ and $A(\theta) = -N\log(1-\theta)$, where N is a known positive integer. Show that this is the binomial distribution $f_{X|\Theta}(x|\theta) = \binom{N}{x}\theta^x(1-\theta)^{N-x}$.

For a distribution in the exponential family, the likelihood is

$$\mathcal{L}(\theta) = \left(\prod_{i=1}^{n} h(x_i)\right) \exp\left(\eta(\theta) \sum_{i=1}^{n} T(x_i) - nA(\theta)\right) \propto \exp\left(\eta(\theta) \sum_{i=1}^{n} T(x_i) - nA(\theta)\right).$$

Here the final proportional relationship arises because the product of the base measures depends only on the data and not on θ. Note that the value of $\sum_{i=1}^{n} T(x_i)$ is all we need to know about the data in order to compute the likelihood. For that reason, this sum is called the *sufficient statistic*. (The canonical statistic is the sufficient statistic in the special case of a dataset consisting of just one datapoint.) Now suppose that our prior belongs to the family of distributions with general form

$$f_\Theta(\theta) \propto \exp\left(\nu\eta(\theta) - \lambda A(\theta)\right),$$

where ν and λ are the parameters of the distribution. Since this is the density of Θ rather than X, the functions η and A are no longer the natural parameter and log-normaliser. With a prior in this form, the posterior satisfies

$$f_{\Theta|\underline{X}}(\theta|\underline{x}) \propto \exp\left(\left(\nu + \sum_{i=1}^{n} T(x_i)\right)\eta(\theta) - (\lambda + n)A(\theta)\right). \tag{6.13}$$

Thus, the posterior belongs to the same family as the prior, but with different parameters. When the prior and the posterior are in the same family, the prior is said to be *conjugate* to the likelihood.

Let us work through a concrete example. Suppose you are an ecologist counting orchids in each of a large number of 1m^2 quadrats in a forest. Let X be the number of orchids in a randomly selected quadrat. Assume that $X|\{\Theta = \theta\} \sim \text{Poisson}(\theta)$, so $\eta(\theta) = \log\theta$, $T(x) = x$ and $A(\theta) = \theta$. The conjugate prior is then

$$f_\Theta(\theta) = \frac{1}{c}\exp(\nu\log\theta - \lambda\theta) = \frac{1}{c}\theta^{\nu}e^{-\lambda\theta},$$

where c is a normalising constant. This density belongs to the *gamma distribution* family, whose density function has standard form

$$f_\Theta(\theta; a, b) = \frac{b^a\theta^{a-1}e^{-b\theta}}{\Gamma(a)},$$

where $\Gamma(a)$ is the gamma function (see Box 9.3 for further details). We indicate that Θ follows a gamma distribution with parameters a and b by writing $\Theta \sim \text{Gamma}(a, b)$. The gamma density matches the form of our conjugate prior when $\nu = a - 1$ and $\lambda = b$. Suppose we observe n quadrats, yielding a dataset $\underline{x} = (x_1, x_2, \ldots, x_n)$. If we use a Gamma$(a, b)$ prior, we know that our posterior will also be a gamma distribution. Making use of the general form (6.13) for the posterior, we have

$$\Theta|\{\underline{X} = \underline{x}\} \sim \text{Gamma}\left(a + \sum_{i=1}^{n} x_i, b + n\right).$$

The fact that the posterior is a distribution with known analytical form, obtainable with only minimal calculation, reveals the power of using conjugate priors. In this example, we deduced the form of the conjugate prior from first principles, but to save time we often make use of tables of well-known conjugate prior-likelihood pairs.

■ *Exercise* 6.24 Suppose the numbers of orchids found in 10 different quadrats were $\underline{x} = (0, 1, 2, 0, 1, 1, 0, 3, 1, 2)$. Assuming that $X|\{\Theta = \theta\} \sim \text{Poisson}(\theta)$ and a Gamma(1, 1) prior, find the posterior. The mean of the Gamma(a, b) distribution is a/b. Find the posterior mean estimate of θ.

■ *Exercise* 6.25 Find the conjugate prior to the binomial likelihood.

In this section, we considered models with one unknown parameter, but the definition of the exponential family extends naturally to multiple parameters. Rather than developing the theory of the multi-parameter exponential family and its conjugate priors, we will focus on methods that can be used to explore the posterior for arbitrary likelihoods and priors.

6.7.2 Markov Chain Monte Carlo: A Preview

Conjugate priors are a powerful tool for quickly finding posteriors in analytical form. But in many cases, using one is impractical, inappropriate or impossible. It is therefore useful to have a general method for exploring the posterior that applies to any model. Suppose our model has a parameter vector $\boldsymbol{\theta} = (\theta_1, \ldots, \theta_p)$. The posterior takes the form

$$f_{\Theta|\underline{X}}(\boldsymbol{\theta}|\underline{x}) = \frac{\mathcal{L}(\boldsymbol{\theta}) f_{\Theta}(\boldsymbol{\theta})}{c},$$

where c is an unknown normalising constant. In principle, the statistical properties of this distribution – including c – can be computed by numerical integration (quadrature), but when we have more than a few parameters this becomes prohibitively slow. An alternative is to generate random samples from the posterior and to use those samples to estimate and explore the model's properties. In Chapter 10, we will introduce a powerful, generally applicable method for doing this that requires only the ability to evaluate the likelihood and the prior. The idea is to perform a 'random walk' through parameter space such that the locations visited collectively form a sample from the posterior. The technique is known as Markov chain Monte Carlo.

6.8 Connections between the Sampling and Posterior Distributions

We have seen that frequentists and Bayesians differ in some important respects. They understand probability differently, and they use different tools. However, they do both use probability distributions to represent uncertainty, and there are situations in which they will end up using more or less the *same* distribution.

Consider a random variable X with probability density depending on one parameter, θ. As usual we suppose there is a dataset, $\underline{x}$, which represents a single realisation of the random vector $\underline{X} = (X_1, \ldots, X_n)$, where the X_i are i.i.d. 'copies' of X. Let us summarise the distinct goals of frequentist and Bayesian inference in this scenario.

- In frequentist mode, we write the density of X as $f_X(x; \theta)$, where θ is fixed but unknown. Having made some observations of X, we want to find an estimator for this parameter and the sampling distribution of the estimator.

- In Bayesian mode, we imagine that X has conditional density $f_{X|\Theta}(x|\theta)$, where Θ, a latent variable, has been sampled (once, before the dataset was created) from some unknown distribution. Having made some observations of X, we want to summarise our beliefs – in the form of a posterior – about the value of the parameter that was sampled, and possibly provide a point estimate.

Given the dataset $\underline{x}$, the MLE of θ is

$$\hat{\theta} = t(\underline{x}) = \arg\max_{\theta} \ell(\theta; \underline{x}),$$

where t is the estimator function and ℓ is the log-likelihood. Frequentists characterise uncertainty in the estimate by way of the sampling distribution, that is, the distribution of

$$\hat{\Theta} = t(\underline{X}).$$

In Section 6.6.3, we derived the plug-in normal approximation

$$\hat{\Theta} \sim \mathcal{N}\left(\hat{\theta}, \frac{1}{n\mathcal{I}\,(\hat{\theta})}\right), \tag{6.14}$$

where $\mathcal{I}$ is the Fisher information. The more data we collect (the larger the value of n), the more accurate this approximation becomes.

Now consider the Bayesian approach. Assuming a flat prior, the posterior is a normalised version of the likelihood,

$$f_{\Theta|\underline{X}}(\theta|\underline{x}) = \frac{1}{c}\exp\left(\ell(\theta; \underline{x})\right),$$

and the MAP and MLE estimates of θ coincide. Expanding ℓ as a Taylor series about $\hat{\theta}$ we obtain

$$\ell(\theta; \underline{x}) \approx \ell(\hat{\theta}; \underline{x}) + \frac{(\theta - \hat{\theta})^2}{2}\ell''(\hat{\theta}; \underline{x}).$$

There is no linear term in the series because $\hat{\theta}$ is a stationary point. For large n, we expect the ℓ to be concentrated around $\hat{\theta}$ meaning that higher-order terms can be neglected. Now suppose that the value of θ used to generate the data was θ^*. We showed in Section 6.6.3 that

$$\ell''(\theta^*; \underline{X}) \approx -n\mathcal{I}(\theta^*).$$

Assuming that $\hat{\theta} \approx \theta^*$ then $\mathcal{I}(\theta^*) \approx \mathcal{I}(\hat{\theta})$ and the posterior may be written

$$f_{\Theta|\underline{X}}(\theta|\underline{x}) = \frac{1}{c}\exp\left(\ell(\hat{\theta}; \underline{x})\right)\exp\left(-\frac{(\theta - \hat{\theta})^2}{2(n\mathcal{I}\,(\hat{\theta}))^{-1}}\right),$$

which we recognise as a normal density density function with mean $\hat{\theta}$ and variance $\left(n\mathcal{I}\,(\hat{\theta})\right)^{-1}$. According to this approximation,

$$\Theta|\{\underline{X} = \underline{x}\} \sim \mathcal{N}\left(\hat{\theta}, \frac{1}{n\mathcal{I}\,(\hat{\theta})}\right).$$

Comparing to equation (6.14), we find $\Theta \overset{d}{=} \hat{\Theta}$. So, when our prior is uniform, the normal approximations to the sampling and posterior distributions coincide, explaining the close match between the two distributions in Figure 6.5. A non-uniform prior will change the form of the posterior, but the influence of most reasonable priors declines as n becomes large.

In the above calculations, we assumed that the likelihood function was concentrated in a single location. This need not necessarily be the case for more complex families of models which may contain distinct but similarly plausible explanations of the observed data. In such cases, we would see posterior probability (or likelihood) concentrated in more than one location in parameter space. The Bayesian approach to inference provides a natural framework for analysing such models due to its focus on the structure of the posterior as a whole, rather than the highest peak of the likelihood function.

6.9 Chapter Summary

In this chapter, we investigated ways to quantify our uncertainty about the values of model parameters. There are two distinctive approaches, depending on whether we interpret probabilities in a Bayesian or frequentist way.

- We can model a data-generating mechanism by defining a parametric family of models, with different parameter values corresponding to different models of the mechanism.
- Assuming our family contains models capable of describing the mechanism, we use data to *infer* the most plausible parameters and to quantify uncertainties in our inferences.
- There are two conceptually distinct inference methods: **frequentist** and **Bayesian**.
- In frequentist inference, we construct parameter **estimators** and estimate their **sampling distribution** over multiple repeats of the data collection and estimation process.
- In frequentist inference, only the parameter estimators can be described probabilistically. The parameters are viewed as fixed unknown constants.
- In Bayesian inference, parameters are viewed as *latent* (unobservable) random variables. This allows us to express our uncertainty about their values using probability.
- In Bayesian inference, we assume a **prior** probability distribution of the parameters, reflecting our state of knowledge *before* making observations. Bayes' rule combines prior knowledge with data to form a **posterior** parameter distribution.
- The posterior distribution and the sampling distribution play analogous roles in Bayesian and frequentist inference and, in some situations, are very similar.
- **Bootstrapping** and **Markov chain Monte Carlo** are computational tools for estimating, respectively, the sampling and posterior distributions.

6.10 Further Exercises

■ ***Exercise*** 6.26 The daily changes in price (in \$) of a financial asset may be modelled as i.i.d. realisations of a zero-mean normal random variable $X \sim \mathcal{N}(0, \sigma^2)$. We will treat σ^2 (rather than σ) as the model's parameter.

(a) Given a dataset of observations $D = \{x_i\}_{i=1}^n$, find expressions for the likelihood $\mathcal{L}(\sigma^2)$ and log-likelihood $\ell(\sigma^2)$ in terms of the sum $s_n = \sum_{i=1}^n x_i^2$.

(b) Find an expression for $\widehat{\sigma^2}$, the MLE of σ^2.
(c) Suppose that $D = \{1.1, -0.5, 3.4, -2.7, -0.9\}$. Find $\widehat{\sigma^2}$.
(d) Find an expression for the standard error of the sampling distribution of your estimator, assuming that D was generated according to the $\mathcal{N}(0, \sigma^2)$ model.
(e) Use the plug-in method to estimate the standard error and calculate an approximate 95% confidence interval, assuming that the sampling distribution is approximately normal. Is there anything odd about this confidence interval? What has gone wrong?

■ ***Exercise*** 6.27 A biased coin has probability p of coming up heads. Let X be the indicator variable for heads, so $X(H) = 1$ and $X(T) = 0$. Suppose you flip the coin 10 times, producing a series of i.i.d. indicators $X_1, \ldots, X_{10}$, where $X_i \overset{d}{=} X$. Let x_i be the observed value of X_i. Suppose that

$$\underline{x} = (x_1, \ldots, x_{10}) = (1, 0, 1, 0, 0, 1, 1, 1, 0, 1).$$

(a) Write down the probability mass function of X.
(b) Determine $\mathrm{Var}(X)$.
(c) Determine the likelihood function $\mathcal{L}(p)$ for an arbitrary dataset $\underline{x}$.
(d) Find a formula for the MLE of p and its standard error.
(e) Evaluate $\hat{p}$ and $\widehat{\mathrm{se}}$ (via the plug-in method) based on the observations.
(f) Determine an approximate 95% confidence interval for p.

■ ***Exercise*** 6.28 Let X be a random variable representing the waiting time between successive major storm events. Suppose that X has probability density function

$$f_X(x; \tau) = \frac{1}{\tau} e^{-x/\tau}.$$

Suppose you observe 10 such times, producing a series of i.i.d. variables $X_1, \ldots, X_{10}$ where $X_i \overset{d}{=} X$. Let x_i be the observed value of X_i. Suppose that

$$\underline{x} = (x_1, \ldots, x_{10}) = (0.1, 1.1, 1, 0.6, 2.1, 2, 3, 5, 10, 9).$$

(a) Determine $\mathbb{E}(X)$ in terms of τ.
(b) Determine $\mathrm{Var}(X)$ in terms of τ.
(c) Determine the likelihood function $\mathcal{L}(\tau)$ for an arbitrary dataset $\underline{x}$.
(d) Find a formula for the MLE of τ and its standard error.
(e) Evaluate $\hat{\tau}$ and $\widehat{\mathrm{se}}$ (via the plug-in method) based on the observations.
(f) Determine an approximate 95% confidence interval for τ.

■ ***Exercise*** 6.29 We now return to the biased coin scenario from Exercise 6.27 (same model, same dataset) but take a Bayesian approach. We will assume the prior

$$f_P(p) = \begin{cases} 2p & \text{if } p \in [0, 1], \\ 0 & \text{otherwise.} \end{cases}$$

(a) Find the posterior after one observation, up to a normalising constant.
(b) Find the posterior after two observations, up to a normalising constant.
(c) Find the MAP estimate of p after the second observation.
(d) Find the posterior after 10 observations, including the normalising constant. Compute both the MAP estimate of p and the PM.

7

Frequentist Linear Regression

In Chapter 2, we tackled the problem of building a predictive model of a response variable Y based on a single predictor variable X. In Chapter 5, we showed how to address this task using a *conditional* probabilistic model of Y given X. In the current chapter, we will develop conditional probabilistic models of a response variable Y based on an arbitrary number of different predictors $X_1, X_2, \ldots, X_p$. In these models, we'll assume that the response variable depends linearly on the predictors. This technique, known as *linear regression*, is a workhorse of predictive modelling.

In Chapter 6, we learned how to quantify uncertainty in the parameters of a probability model using both frequentist and Bayesian methods. In the current chapter, we take a frequentist perspective, treating the model parameters as fixed but unknown constants and measuring uncertainty using bootstrap sampling. We will explore important concepts of model performance and selection, overfitting, model interpretation, explanatory power and the connections between variable correlations and model parameters. We will expand our treatment of linear regression to the Bayesian setting in Chapter 9, where we will also introduce a systematic way to control model complexity known as *regularisation*. The topics of the current chapter are important precursors to understanding both Bayesian linear regression and regularisation.

7.1 Predicting Body Fat

The question of how much of a person's body is made up of fat is of interest to athletes, doctors and dieters. Accurately measuring body fat percentage (BFP) requires specialist equipment, but might it be possible to estimate it using other, simpler measurements? We will explore that question with the help of an illustrative dataset. The dataset contains measurements of BFP, height, body mass index (BMI), age and various body part circumferences for a group of 248 men.[1] The complete list of variables can be seen in Figure 7.2.

We view our dataset as a set of predictor–response pairs,

$$D = \{(\boldsymbol{x}_i, y_i)\}_{i=1}^{n},$$

where $n = 248$. Response y_i is the BFP of the ith individual in the dataset. Predictor $\boldsymbol{x}_i$ is a

[1] The original data in Johnson (1996) contained a suspicious participant with BMI = 165.6, which was excluded. Three further observations with BMI values greater than three standard deviations from the mean were also excluded.

vector representing all the other measurements for the ith individual, that is,

$$\boldsymbol{x}_i = (x_{i1}, x_{i2}, \ldots, x_{ip}),$$

where x_{ij} is the value of the jth measurement of the ith individual. We use the letter p to denote the number of predictors (13 in our case). We now introduce the set of random variables $Y, X_1, X_2, \ldots, X_p$, representing the 14 measurements obtained from a man selected uniformly at random from the population. For convenience, we also define the random vector $\boldsymbol{X} = (X_1, X_2, \ldots, X_p)$. Since we are dealing with a *supervised* learning problem (a regression), we don't need to estimate the full joint density of all the measurements. All we need is the conditional probability density of Y given $\boldsymbol{X}$. To obtain this, we hypothesise some relationship between Y and $\boldsymbol{X}$. One simple possibility is a *multiple linear model*,

$$\begin{aligned} Y &= \beta_0 + \beta_1 X_1 + \cdots + \beta_p X_p + \mathcal{E}, \\ \mathcal{E} &\sim \mathcal{N}(0, \sigma^2). \end{aligned}$$

The parameters of the model are the components of the parameter vector $\boldsymbol{\theta} = (\beta_0, \beta_1, \ldots, \beta_p, \sigma)$. Under this model, the probability density of the response given the predictors is as follows:

$$f_{Y|\boldsymbol{X}}(y|\boldsymbol{x}; \boldsymbol{\theta}) = \frac{1}{\sigma\sqrt{2\pi}} \exp\left(-\frac{1}{2\sigma^2}\left(y - \beta_0 - \sum_{j=1}^{p} \beta_j x_j\right)^2\right),$$

where $\boldsymbol{x} = (x_1, \ldots, x_p)$. Assuming that our dataset consists of independent realisations of $(\boldsymbol{X}, Y)$ (in fact the predictor vectors need not be independent), then the likelihood function is

$$\mathcal{L}(\boldsymbol{\theta}) = \frac{1}{\sigma^n (2\pi)^{\frac{n}{2}}} \exp\left(-\frac{1}{2\sigma^2} \sum_{i=1}^{n} \left(y_i - \beta_0 - \sum_{j=1}^{p} \beta_j x_{ij}\right)^2\right),$$

and maximum likelihood parameter estimates are given by

$$\hat{\boldsymbol{\theta}} = \arg\max_{\boldsymbol{\theta}} \mathcal{L}(\boldsymbol{\theta}).$$

There are efficient computational methods for obtaining these estimates, but we will not go into details here. There are many excellent and easy-to-use computer packages that implement them. Having computed the estimates, we typically make predictions of the value of Y using the *regression function* $r(\boldsymbol{x}) = \mathbb{E}(Y|\boldsymbol{X} = \boldsymbol{x})$. That is, if $\boldsymbol{x}$ is a new observation of the predictor vector, then our prediction of the response variable is

$$\hat{y} = \hat{\beta}_0 + \hat{\beta}_1 x_1 + \cdots + \hat{\beta}_p x_p.$$

Note that for models of the form $Y = g(\boldsymbol{X}) + \mathcal{E}$, the regression function, r, and the function g which specifies the $\boldsymbol{x}$ dependence of the model, often called the 'prediction function', are the same. In other classes of model, in particular if the distribution of Y given X were not symmetric, then the *most probable* value of Y given X might be a more appropriate way to make predictions.

Modelling the response variable as a linear function of multiple predictors is known as *multiple linear regression*. The model we have just defined uses all predictor variables, but we will explore models that use various subsets of the available predictors. Apart from using

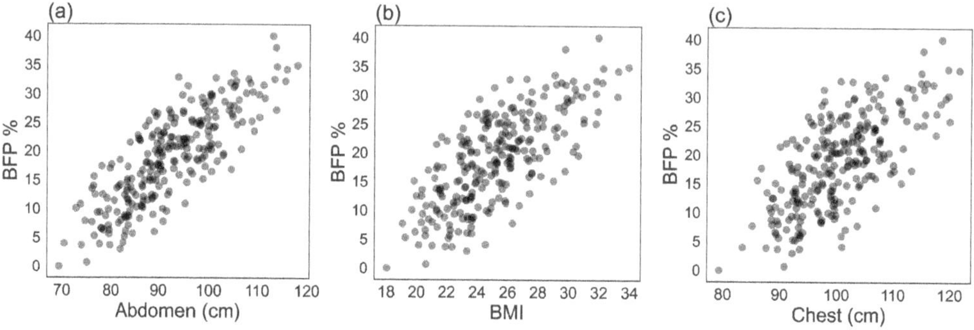

Figure 7.1 Scatter plots of body fat percentage against (a) abdomen circumference (cm), (b) body mass index and (c) chest circumference (cm).

lower-dimensional predictor vectors, these models can be formulated in the same way as above.

7.2 Data Exploration

Before attempting to fit a model, we should get a rough sense of the main patterns in the data. If our predictor vector were one-dimensional, we would plot all the datapoints and stare at the resulting figure. Since the predictor vector is 13-dimensional, that simple strategy is not available. However, we can use scatter plots to visualise the relationships between individual components of the predictor vector and the response variable. Figure 7.1 does this for three predictor variables that we might expect to be strongly associated with body fat: abdomen circumference, BMI and chest circumference. In these examples, the relationship between predictor and response is approximately *linear*. That is, we can write it in the form:

$$Y = a + bX + \text{`noise'},$$

where X is a single predictor variable and a and b are constants. 'Noise' is in scare quotes because we expect $Y - a - bX$ to be correlated with (and so to some extent predictable from) the other predictor variables.

The scatter plots in Figure 7.1 summarise relationships between single predictors and the response variable. We will see below that relationships between pairs (or triples or quadruples) of predictors are also important when building a model. Considering all 14 body measurements, we have 91 pairs. To summarise all pairwise relationships in a concise way, it is useful to compute correlations. The correlation between random variables X_i and X_j is defined as

$$\rho(X_i, X_j) = \frac{\text{Cov}(X_i, X_j)}{\sqrt{\text{Var}(X_i)\text{Var}(X_j)}}. \tag{7.1}$$

The Pearson correlation coefficient between measurement i and measurement j is a sample approximation to the correlation, given by

$$\hat{\rho}(X_i, X_j) = \frac{\sum_{k=1}^{n}(x_{ki} - \bar{x}_i)(x_{kj} - \bar{x}_j)}{\sqrt{\sum_{k=1}^{n}(x_{ki} - \bar{x}_i)^2}\sqrt{\sum_{k=1}^{n}(x_{kj} - \bar{x}_j)^2}} \in [-1, 1],$$

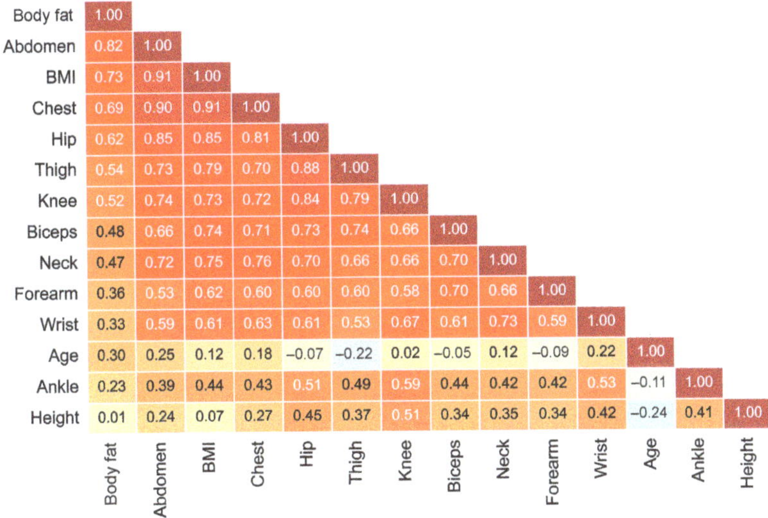

Figure 7.2 Pearson sample correlation coefficients between all measurements in our dataset.

where

$$\bar{x}_j = \frac{1}{n} \sum_{k=1}^{n} x_{kj}.$$

Figure 7.2 shows the correlation coefficients for all pairs of measurements in our dataset. The Pearson coefficient is a measure of the strength of a *linear* relationship between two variables with $|\hat{\rho}(X_i, X_j)| = 1$ indicating a relationship of the form $X_i = a + bX_j$, with $\text{sign}(\rho) = \text{sign}(b)$. If we were to weaken this relationship by adding noise $X_i = a + bX_j + \mathcal{E}$, then $|\hat{\rho}(X_i, X_j)|$ would shrink. In general, the Pearson coefficient is not a reliable way to measure nonlinear relationships,[2] but for approximately linear relationships, it is a useful measure. From Figure 7.2, we see that all circumference measurements are, to varying degrees, positively correlated with body fat, with the abdomen circumference having the strongest association. This makes sense biologically – most parts of the body can store fat – but suggests that the abdomen measurement is likely to be a particularly important predictor variable. We also notice that age is negatively correlated with limb circumference, which again makes sense because muscle mass tends to reduce as we age. The lack of correlation between height and body fat is also not terribly surprising. Finally, we note that some of the strongest correlations are *between* predictors. Once we begin fitting models, we will see that this has important consequences.

7.3 Model Interpretation

Statistical models are essential tools for *describing* and *explaining* patterns in data. Some models describe but do not explain. For example, a model of the relationship between raincoat

[2] Suppose $X \sim \text{Uniform}(-1, 1)$ and $Y = X^2$. Then $\rho(X, Y) = 0$, even though Y is a simple function of X.

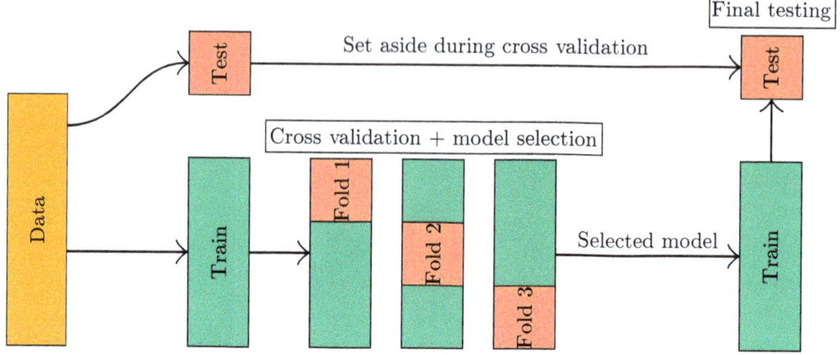

Figure 7.3 The cross-validation, model selection and final testing pipeline. The pipeline consists of (i) performing a train-test split, (ii) selecting a model using cross-validation, (iii) training the selected model on the full training set and (iv) testing the model on the held-out test set.

sales and umbrella sales would (if it tracked no other variables) merely describe an empirical association. Other models do both. For example, a model that predicts the range of cannon based on volume of gunpowder used, firing angle and wind velocity is explanatory. In this case, we already know *roughly* how the variables in question hook up causally; when we fit the model to data, we nail down the quantitative details. The body fat models we build in this chapter are best seen as descriptive. They are not inspired by hunches about what is causing what, let alone by detailed knowledge of physiological mechanisms. Still, the patterns they reveal might be useful evidence for scientists researching such mechanisms.

7.4 Measuring Performance and Comparing Models

The correlations in Figure 7.2 suggest that abdomen circumference and BMI are both individually likely to be useful predictors of BFP, but almost all variables seem to have some kind of association with it. Should we use them all? Will our predictions always improve if we use more predictors? To begin addressing these questions, we will start by fitting single predictor linear models to provide baseline performance measures. We will then compare these with the performance of models that use multiple predictors.

When comparing models, we need methods for measuring performance which are as objective and unbiased as possible. For regression models, the metric of choice is usually the mean squared error (MSE) or its square root (RMSE). In Chapters 2, 3 and 5, we estimated model performance on unseen data using cross-validation. That is, we repeatedly removed ('held out') a fraction of the training data, fit the model to what was left, and then tested it on the held-out set. We also saw in Chapter 2 that errors estimated by cross-validation can be misleading. If we try enough models, then the one with the lowest cross-validation error may not be the one that performs best on new data: winning at cross-validation is a combination of true predictive performance and luck. Our error estimate for the winning model will therefore be biased downward compared to its true expected value on unseen data. While there are ways to estimate this bias,[3] one way to avoid it altogether is to set aside a 'test set' of data, D_{test}, *before* fitting any models. The remaining data – the 'training set', D_{train} – is then used both to compare the performance of different models via cross-validation and to estimate the parameters of the winning model. A final objective measure of the performance of the model we have selected is obtained by applying it to make predictions on the test set. Figure 7.3 summarises this process of cross-validation, model selection and final testing.

The pipeline illustrated in Figure 7.3 uses three-fold cross-validation. That is, we split the data into three *folds*, which each take their turn as the held-out set. Using k folds is referred to as 'k-fold' cross-validation. Leave-one-out cross-validation (LOOCV) is $|D_{\text{train}}|$-fold cross-validation. One disadvantage of using a small number of folds is that we have less data to train each version of the model. Consequently, the cross-validation error tends to be an overestimate of the error we would obtain using the whole dataset. Using more folds is therefore desirable if we want to reduce this bias. However, bias is not the only factor we need to consider. Suppose we re-ran the dataset-generating process many times and applied the same cross-validation procedure on each dataset. Each time we regenerated the data, we would get a different MSE on the ith fold. We can view these MSEs as realisations of a random variable M_i. Our cross-validation estimate of the MSE of a model trained on a single dataset can therefore be viewed as a random variable

$$t(\boldsymbol{M}) = \frac{1}{k}\sum_{i=1}^{k} M_i,$$

where $\boldsymbol{M} = (M_1, M_2, \ldots, M_k)$. The function t is an *estimator* of the MSE of our model on a very large unseen test set. For a particular dataset, we obtain fold MSEs $\boldsymbol{m} = (m_1, m_2, \ldots, m_k)$ yielding a cross-validation estimate of the true MSE, $\widehat{\text{MSE}} = t(\boldsymbol{m})$. There are two ways this estimator may be inaccurate. First, as we already know, it can be biased. Second, the sampling distribution of the estimator can have a high *variance*. Increasing the number of folds means that the datasets used to train each model have more points in common. This in turn means that their errors (the M_i) have a higher positive correlation. It is a consequence of Bienaymé's identity, explained in Box 7.1, that the variance of a sum of correlated random variables is an increasing function of their correlation. Thus, increasing k increases the variance of the estimator. Selecting k is therefore a trade-off between low bias (large k) and low variance (small k). Using $k = 5$ or $k = 10$ folds is recommended as a good compromise. In what follows we will use 10-fold cross-validation and a 60:40 train:test split.[4]

★ Box 7.1 Bienaymé's identity

Suppose we have a set of random variables $X_1, X_2, \ldots, X_n$. The variance of their sum is given by Bienaymé's identity:

$$\text{Var}\left(\sum_{i=1}^{n} X_i\right) = \sum_{i=1}^{n}\sum_{j=1}^{n} \text{Cov}(X_i, X_j) = \sum_{i=1}^{n} \text{Var}(X_i) + 2\sum_{i=1}^{n-1}\sum_{j=i+1}^{n} \text{Cov}(X_i, X_j).$$

[3] See Tibshirani and Tibshirani (2009), for example.

[4] We are setting aside slightly more test data than is conventional, for pedagogical reasons: a smaller training set lets us illustrate overfitting and model selection more clearly. Common splits are 80:20 and 70:30.

For example, suppose all the variables have the same variance σ^2 and the average correlation between pairs of variables is ρ. In that case, the variance of the mean of all the variables is

$$\mathrm{Var}\left(\frac{1}{n}\sum_{i=1}^{n} X_i\right) = \frac{\sigma^2}{n}(1 + (n-1)\rho).$$

From this we see that for large n, the variance is $\approx \sigma^2\rho$.

■ ***Exercise*** 7.1 Prove Bienaymé's identity. (If you need a refresher on the definitions of variance and covariance, see Appendix A.)

An important principle when selecting models is that we don't run through the pipeline in Figure 7.3 more than once. Doing so would amount, at best, to using the test set for deciding between models, and at worst to gaming the train-test split to produce the results we want. We should be mentally prepared for the possibility that the final test error will not match our cross-validation error. Indeed, the test error of the selected model may be higher than the cross-validation errors of some of the other models that it defeated. This does *not* imply that we made the wrong model choice. All the reported errors are noisy estimates of expected errors on unseen data, and test errors tend to be noisier than most (because test sets tend to be smaller than training sets). The key point is that we use the test set *purely* to provide a final objective performance measure, not to decide between models.[5]

7.5 Fitting Single-Predictor Models

Suppose we select a single predictor and fit a model of the form:

$$\begin{aligned} Y &= \beta_0 + \beta_1 X + \mathcal{E}, \\ \mathcal{E} &\sim \mathcal{N}(0, \sigma^2). \end{aligned}$$

Here X represents one of the available predictors $X_1, X_2, \ldots, X_p$. We will write our observations of this variable $x_1, x_2, \ldots, x_n$, so

$$D_{\text{train}} = \{(x_i, y_i)\}_{i=1}^{n}.$$

Maximising the likelihood computed using D_{train}, we obtain the following closed form coefficient estimates:

$$\hat{\beta}_1 = \frac{\sum_{i=1}^{n}(x_i - \bar{x})(y_i - \bar{y})}{\sum_{i=1}^{n}(x_i - \bar{x})^2}, \tag{7.2}$$

$$\hat{\beta}_0 = \bar{y} - \hat{\beta}_1\bar{x}. \tag{7.3}$$

[5] For many 'benchmark' machine learning datasets (e.g. ImageNet, Deng et al. (2009)), an *official* test set is specified, and researchers are expected to report errors on this standard set. This prevents 'gaming' the train-test split, but we still need to be careful about ranking models published by different researchers based on these test errors; we should consider both the error *and* our uncertainty in its magnitude.

Table 7.1 *Coefficients, bootstrap standard errors (500 replicas), cross-validated RMSE and R^2 for simple linear models (one predictor). Starred parameters, β_1^*, are based on standardised predictors (see Section 7.5.3).*

Predictor	$\hat{\beta}_1$	$\hat{\text{se}}(\hat{\beta}_1)$	$\hat{\beta}_1^*$	$\hat{\text{se}}(\hat{\beta}_1^*)$	RMSE	R^2
Abdomen	0.636	0.034	6.160	0.341	4.792	0.634
BMI	1.847	0.118	5.599	0.357	5.417	0.523
Chest	0.644	0.050	5.050	0.378	5.997	0.426
Hip	0.830	0.071	4.820	0.424	6.182	0.388
Wrist	2.861	0.607	2.575	0.572	7.458	0.111

Using these estimates, if we are given a new predictor value x, then we predict the corresponding BFP (y) to be

$$\hat{y} = \hat{\beta}_0 + \hat{\beta}_1 x. \tag{7.4}$$

Modelling the response variable as a linear function of one predictor is known as *simple linear regression*. Table 7.1 lists $\hat{\beta}_1$ for models based on the four predictors most highly correlated with BFP, and also on a predictor with a rather weak correlation (wrist circumference).

■ ***Exercise*** 7.2 The aim of this exercise is to derive the maximum likelihood estimates of β_0 and β_1 given by equations (7.2) and (7.3).

(a) Write down the likelihood function $\mathcal{L}(\beta_0, \beta_1, \sigma)$ corresponding to the linear model $Y = \beta_0 + \beta_1 X + \mathcal{E}$.

(b) Determine the log-likelihood and hence explain why maximum likelihood estimates of β_0 and β_1 minimise the RSS

$$\text{RSS}(\beta_0, \beta_1) = \sum_{i=1}^{n} (\beta_0 + \beta_1 x_i - y_i)^2.$$

(c) By solving the equations

$$\frac{\partial \text{RSS}}{\partial \beta_0} = 0 \quad \text{and} \quad \frac{\partial \text{RSS}}{\partial \beta_1} = 0,$$

find $\hat{\beta}_0$ and $\hat{\beta}_1$.

To estimate the standard errors in our coefficient estimates, we'll use non-parametric bootstrap resampling (Section 6.6.2). Approximate 95% confidence intervals are then given by $\hat{\beta}_1 \pm 2\hat{se}(\hat{\beta}_1)$. For example, from Table 7.1, if we use abdomen circumference as our predictor, then an approximate 95% confidence interval for β_1 is $[0.57, 0.70]$. We may conclude that, insofar as our simple model is an accurate description of the probability density of BFP given abdomen circumference, an extra centimetre around the abdomen corresponds to an increase of between 0.57% and 0.71% BFP for an 'average' man. More formally, according to our model, the average BFP of men with abdomen circumference $(x + 1)$ cm is somewhere between 0.57% and 0.71% greater than the average BFP of men with an abdomen circumference x cm.

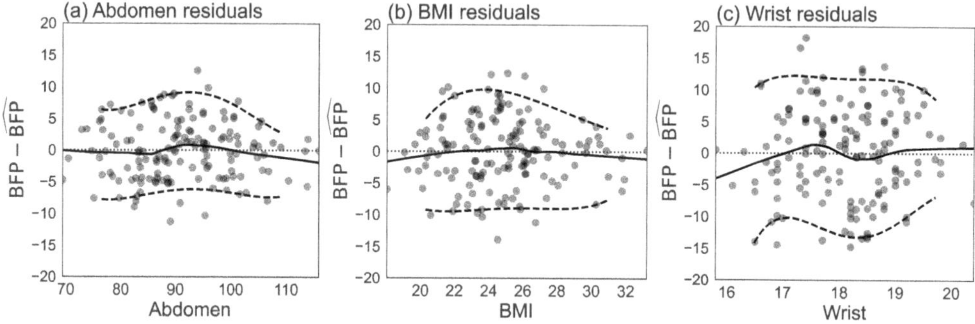

Figure 7.4 (a–c) Scatter plots of the residuals $e_i = y_i - \hat{y}_i$ (or BFP − $\widehat{\text{BFP}}$) against the predictors x_i for single-predictor models. The solid lines give the approximate relationship between the mean residual and the predictor. The dashed lines show approximate 5% and 95% quantiles of the residual distribution.

7.5.1 Checking Assumptions

Having fitted a conditional probabilistic model of our data-generating mechanism, we can use it to retrospectively check the modelling assumption on which it was based. After estimating the coefficients, our approximation to the true relationship between predictor and response is

$$Y = \hat{\beta}_0 + \hat{\beta}_1 X + \mathcal{E},$$

where $\mathcal{E} \sim \mathcal{N}(0, \hat{\sigma}^2)$ and $\mathcal{E} \perp\!\!\!\perp X$. Here $\hat{\sigma}$ is the maximum likelihood estimate of σ. Letting our prediction of the response based on X define a random variable,

$$\hat{Y} = \hat{\beta}_0 + \hat{\beta}_1 X,$$

our fitted model can be written

$$Y - \hat{Y} = \mathcal{E}.$$

According to our assumptions about $\mathcal{E}$, the random variable $Y - \hat{Y}$ should be normal and independent of X. Realisations of $Y - \hat{Y}$ are the *residuals*, $e_i = y_i - \hat{y}_i$.

Figure 7.4 shows scatter plots of the points (x_i, e_i) for three single-predictor models. These plots provide a visual means to check our approximation that $\mathcal{E}$ is independent of X, and that our (linear) model of the systematic component of the relationship between Y and X is appropriate. If $\mathcal{E}$ is independent of X, then (in particular) its variance should not depend on X. This constant variance property is called *homoscedasticity*. The dashed lines superimposed on the residual distributions in Figure 7.4 provide estimates of the 5% and 95% quantiles of the distribution as a function of the predictor. These quantiles vary with the predictor, but the variations are small. We usually only worry about heteroscedasticity (departures from homoscedasticity) when there is a clear 'funnel' in the residual plot, with the width of the scatter increasing or decreasing monotonically with the predictor over a substantial range. The solid lines in Figure 7.4 approximate variations in the mean of the residual distribution with the predictor. According to our assumptions, $\mathbb{E}(\mathcal{E}) = \mathbb{E}(Y - \hat{Y}) = 0$, so strong and systematic deviations from the line $e = 0$ would be a sign that our model of the relationship between Y and X was inappropriate. The deviations in the figure are not particularly alarming. We

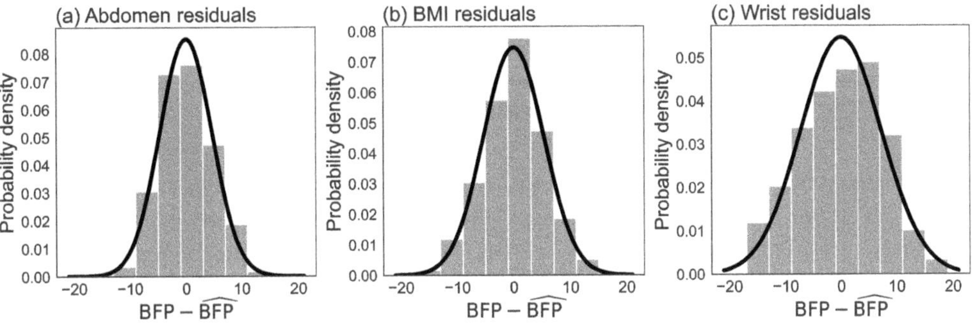

Figure 7.5 (a–c) Histograms approximating the distributions of residuals for single-predictor linear models. The solid lines show maximum likelihood normal probability densities.

can also check our assumption that the residuals are normally distributed. Figure 7.5 shows histograms of the residuals plotted in Figure 7.4, together with maximum likelihood normal probability densities estimated from the residuals. The distributions of the residuals do not differ substantially from normal distributions.

7.5.2 Coefficients and Correlations

We can quantify how much of the variation in Y is predicted by a particular model by comparing the residual sum of squares

$$\text{RSS} = \sum_{i=1}^{n}(y_i - \hat{y}_i)^2$$

with the *total sum of squares*

$$\text{TSS} = \sum_{i=1}^{n}(y_i - \bar{y})^2.$$

Here, as usual, $\hat{y}_i$ is the response predicted by the model for the ith example, and $\bar{y}$ is the empirical average of the actual responses. The sums may be taken over the training data or over a dataset that was not used for training (e.g. a fold or a test set). The TSS is a measure of total variation, while the RSS is a measure of the variation left unaccounted for by the model's predictions. We would like the RSS to be small relative to the TSS. Accordingly, we define the *coefficient of determination*, denoted R^2, by

$$R^2 = 1 - \frac{\text{RSS}}{\text{TSS}}. \tag{7.5}$$

This measures the fraction of the variation in Y which is predicted by the model (we expand on this in Box 7.2). If the same dataset is used to estimate model coefficients *and* to evaluate the RSS and TSS, then we will see that R^2 has a number of attractive properties, including a close relationship to the Pearson correlation. While it is most often used as a way to characterise linear models, we can also calculate R^2 for nonlinear models; the TSS and RSS may be computed for any family of prediction function.

★ Box 7.2 Interpretation of R^2 for general models

If we attempt to predict our response variable without reference to any predictor variables, then the prediction that minimises the residual sum of squares is $\hat{y}_i = \bar{y}$ (see Exercise 2.1). Thus, the TSS may be seen as the RSS we would obtain *without a model* (or with a *null model*). When our prediction is based on knowledge about the x_is, then we expect deviations of $\hat{y}_i$ from y_i to be smaller, meaning that RSS $\leq$ TSS. Thus, the amount of variation that is explained *by using a model* is TSS − RSS, and the fraction of variance explained is

$$\frac{\text{TSS} - \text{RSS}}{\text{TSS}} = R^2.$$

Moreover, $0 \leq R^2 \leq 1$. If R^2 is computed using a model with parameters estimated from a dataset different to that used to compute the RSS and TSS, then it may lead to *worse* results than not using a model. In that case, we would find $R^2 < 0$.

Let us assume that the sum in the RSS is taken over the same data that was used to train a single-predictor linear model. In this case, we can make the arguments of Box 7.2 more precise. We achieve this by decomposing the TSS as follows:

$$\begin{aligned}\sum_{i=1}^{n}(y_i - \bar{y})^2 &= \sum_{i=1}^{n}(y_i - \hat{y}_i + \hat{y}_i - \bar{y})^2 \\ &= \sum_{i=1}^{n}(\hat{y}_i - \bar{y})^2 + \sum_{i=1}^{n}(\hat{y}_i - y_i)^2 + 2\sum_{i=1}^{n}(y_i - \hat{y}_i)(\hat{y}_i - \bar{y}).\end{aligned}$$

It follows from the stationary point conditions used to estimate β_0 and β_1 that the final summation on the right-hand side above is zero (see Exercise 7.3). We now define the sum of the squared deviations of the predictions from the mean response as the *explained sum of squares*

$$\text{ESS} = \sum_{i=1}^{n}(\hat{y}_i - \bar{y})^2.$$

Here 'explained' just means *predicted by the model*. (This mild abuse of language is unfortunately common in the statistics and machine learning literature.) So, for a linear model,

$$\text{TSS} = \text{ESS} + \text{RSS}. \tag{7.6}$$

This equation also holds for linear models with multiple predictors. Since all of these quantities are positive, we have that $0 \leq \text{RSS} \leq \text{TSS}$, so $R^2 \in [0, 1]$. The decomposition of the TSS given by (7.6) has a useful interpretation. The TSS is a measure of the total amount of variation that needs to be accounted for. The ESS is the amount of variation which is 'explained' – that is, predicted – by our model. The RSS is the amount of variation which is left 'unexplained', either because our model is too simple to describe it, or because the data contains inherently unpredictable noise, or some combination of these two factors. We emphasise that the relationship (7.6) holds *only* when the model parameters are estimated using the same data used to compute the sums of squares. Using the *same data* to fit a model and to measure its performance might set alarm bells ringing in the minds of alert readers. We will return to this point in Section 7.6 where we analyse R^2 for models with multiple predictors.

■ ***Exercise*** 7.3 The RSS for a linear model, viewed as a function of β_0 and β_1, is

$$\mathrm{RSS}(\beta_0, \beta_1) = \sum_{i=1}^{n} (\beta_0 + \beta_1 x_i - y_i)^2.$$

(a) The maximum likelihood estimates $\hat{\beta}_0$ and $\hat{\beta}_1$ satisfy the stationary point conditions $\partial_{\beta_0}\mathrm{RSS}(\hat{\beta}_0, \hat{\beta}_1) = 0$ and $\partial_{\beta_1}\mathrm{RSS}(\hat{\beta}_0, \hat{\beta}_1) = 0$. Using these, show that

$$\sum_{i=1}^{n} (\hat{y}_i - y_i) = \sum_{i=1}^{n} x_i(\hat{y}_i - y_i) = 0,$$

where $\hat{y}_i = \hat{\beta}_0 + \hat{\beta}_1 x_i$. Hence, show that

$$\sum_{i=1}^{n} \hat{y}_i(y_i - \hat{y}_i) = 0.$$

(b) Use the results established in part (a) to demonstrate that for a linear model,

$$\sum_{i=1}^{n} (y_i - \hat{y}_i)(\hat{y}_i - \bar{y}) = 0.$$

When using a linear model, the coefficient of determination is closely related to the correlation between Y and X. If the model makes perfect predictions, then RSS $= 0$ and $R^2 = 1$. This situation corresponds to a data-generating process of the form

$$Y = \beta_0 + \beta_1 X$$

for some suitable distribution of X. *All* of the variation in Y is 'explained' by variations in X, so Y and X are *perfectly correlated* (meaning that $\rho(Y, X) = \pm 1$). Turning now to the other extreme, suppose that X tells us nothing about the value of Y. This would correspond to a data-generating process

$$Y = \beta_0 + \mathcal{E},$$

where $\beta_0 = \mathbb{E}(Y)$. Now Y and X are *perfectly uncorrelated* ($\rho(Y, X) = 0$). If our dataset consisted of samples generated by this process, then fitting a linear model would lead to parameter estimates

$$\hat{\beta}_1 \approx 0,$$
$$\hat{\beta}_0 \approx \bar{y},$$

corresponding to a prediction function $\hat{y} = \bar{y}$. We would find RSS $=$ TSS and $R^2 = 0$. The fact that R^2 is 1 when Y and X are perfectly correlated and 0 when they are perfectly uncorrelated is no coincidence. We can show that for a simple linear model, when R^2 is computed using the same dataset than was used to estimate coefficients, then it is equal to the square of Pearson correlation coefficient between Y and X for that dataset. Exercise 7.4 takes you through the calculation.

■ ***Exercise*** 7.4 The aim of this exercise is to confirm that for a simple linear model, the coefficient of determination is equal to the square of the sample correlation. Consider the model

$$Y = \beta_0 + \beta_1 X + \mathcal{E},$$

where β_0 and β_1 are to be estimated using a dataset $D_{\text{train}} = \{(x_i, y_i)\}_{i=1}^n$. Let $\hat{\rho}$ be the Pearson correlation between Y and X, and define variance estimates $\hat{\sigma}_y^2 = n^{-1}\sum_{i=1}^n (y_i - \bar{y})^2$ and $\hat{\sigma}_x^2 = n^{-1}\sum_{i=1}^n (x_i - \bar{x})^2$.

(a) Making use of equation (7.2), show that $\hat{\beta}_1 = \frac{\hat{\rho}\hat{\sigma}_y}{\hat{\sigma}_x}$.
(b) Show that the RSS computed from D_{train} may be written

$$\text{RSS} = \sum_{i=1}^n \left(\hat{\beta}_1(x_i - \bar{x}) - (y_i - \bar{y})\right)^2.$$

(c) Use your answers to (a) and (b) to show that $\text{RSS} = n\hat{\sigma}_y^2(1 - \hat{\rho}^2)$.
(d) Using the fact that $\text{TSS} = n\hat{\sigma}_y^2$, show that $R^2 = 1 - \text{RSS}/\text{TSS} = \hat{\rho}^2$.

7.5.3 Standardising Predictors

It is often useful to *standardise* predictors so that their ranges of variation within the training data are similar. This makes the values of $\hat{\beta}_1$ for different single-predictor models – each measuring the sensitivity of Y to a different predictor – directly comparable. We can quantify the magnitude of variations in predictor X_k using the standard deviation estimate

$$\hat{\sigma}_k = \sqrt{\frac{1}{n}\sum_{i=1}^n (x_{ik} - \bar{x}_k)^2}, \tag{7.7}$$

where the sum is taken over the training data and $\bar{x}_k$ is the (training) sample mean of X_k. An alternative to the estimate (7.7) is the unbiased 'sample standard deviation', s, defined in Box 7.3. For large n, the differences are minimal.

★ Box 7.3 Sample standard deviation

Let $\underline{x} = (x_1, \ldots, x_n)$ be a vector of n observations of a random variable X. We view $\underline{x}$ as a single observation of the random vector $\underline{X} = (X_1, \ldots, X_n)$, where the X_i are i.i.d. copies of X. Define the true mean and variance of X to be $\mu = \mathbb{E}(X)$ and $\sigma^2 = \text{Var}(X)$. Consider the estimator function

$$t(\underline{X}) = c\sum_{i=1}^n \left(X_i - \frac{1}{n}\sum_{j=1}^n X_j\right)^2 = c\left(\sum_{i=1}^n X_i^2 - \frac{1}{n}\left(\sum_{i=1}^n X_i\right)^2\right).$$

If $\mathbb{E}(t(\underline{X})) = \sigma^2$, then t is an *unbiased* estimator of σ^2. Using the decomposition $\mathbb{E}(Y^2) = \text{Var}(Y) + \mathbb{E}(Y)^2$ and the fact that if $Y \perp\!\!\!\perp Z$ then $\text{Var}(Y+Z) = \text{Var}(Y) + \text{Var}(Z)$, we have

$$\mathbb{E}\left(\sum_{i=1}^{n} X_i^2\right) = n(\sigma^2 + \mu^2), \quad \mathbb{E}\left(\left(\sum_{i=1}^{n} X_i\right)^2\right) = n\sigma^2 + (n\mu)^2,$$

so $\mathbb{E}(t(\underline{X})) = c(n-1)\sigma^2$. Therefore, if $c = 1/(n-1)$, then t is an unbiased estimator of σ^2. The estimate $t(\underline{x})$ is called the *sample variance*,

$$s^2 = \frac{1}{n-1}\sum_{i=1}^{n}(x_i - \bar{x})^2,$$

where $\bar{x} = n^{-1}\sum_{i=1}^{n} x_i$ is the *sample mean*. The sample standard deviation is $s = \sqrt{s^2}$.

■ ***Exercise*** 7.5 Write out a detailed derivation of the result $\mathbb{E}(t(\underline{X})) = c(n-1)\sigma^2$, using the sketch in Box 7.3 to guide you.

Having defined our standard deviation estimates, we define *standardised* predictors

$$z_{ik} = \frac{x_{ik} - \bar{x}_k}{\hat{\sigma}_k}. \tag{7.8}$$

These predictors can be used to fit models exactly as we would using 'raw' predictors. Coefficient estimates based on standardised predictors can be found in Table 7.1 and are denoted $\hat{\beta}_1^*$ rather than $\hat{\beta}_1$. Comparing different predictors in Table 7.1, we see that as the R^2 value reduces, indicating a lower magnitude of correlation between the predictor and the response variable, so does the coefficient of the predictor in the model. This reflects the fact that for standardised predictor models, we have

$$\hat{\beta}_1^* = \hat{\rho}(Y, X_k)\hat{\sigma}_y,$$

where $\hat{\rho}(Y, X_k)$ denotes the Pearson correlation between Y and X_k, and $\hat{\sigma}_y$ the estimated standard deviation of Y. We also see from Table 7.1 that the standard errors are larger for smaller coefficients. This reflects the fact that, in a linear model, correlations of smaller magnitude between Y and X_k mean that a greater proportion of the variability in Y is attributable to noise. Hence, rerunning the dataset generation process is likely to produce more varied datasets and greater variations in parameter estimates.

The standardised predictors produced by transformation (7.8) have unit sample variance *and* zero sample mean within the training data. Strictly speaking, we only needed to standardise the variance in order to make sensitivities comparable. Standardising the mean ensures that transformed predictors vary over a similar interval and allows us to more easily compare their empirical distributions, when this is of interest.

7.6 Multiple-Predictor Models

If we are given a choice between two or more simple linear models based on different predictors, then (provided the predictor–response relationships are linear) the predictor whose correlation with the response has greatest magnitude will produce the most accurate predictive model. Of course, in practice we can only estimate correlations, so selecting the best model is a matter of risky inference. One option is to compare R^2 values; another is to compare model errors computed by cross-validation. There is an important conceptual difference here. R^2 (as it is usually calculated) is a measure of the ability of a linear model to describe the variation in the data it was trained on. By contrast, cross-validated errors measure performance on unseen data. When we are comparing simple linear models, this difference will usually not matter much, because unless our dataset is very small or very noisy, overfitting is unlikely to be a problem. When we start building models with multiple predictors, the situation changes, and some new questions emerge. Will adding more predictors necessarily produce more accurate predictions? If not, then which predictors should we use?

7.6.1 Performance in Multiple Predictor Models

We have labelled our predictors $X_1, X_2, \ldots, X_p$ (where $p = 13$ for the body fat data). Suppose we decide to use only a subset of them. That is, we assume a model of the form

$$Y = \beta_0 + \sum_{k \in P} \beta_k X_k + \mathcal{E},$$

where $P \subseteq \{1, \ldots, p\}$. This is the *multiple linear regression* model. After we have estimated our model coefficients from the training data, we can view the predictions of our model as realisations of a random variable

$$\hat{Y} = \hat{\beta}_0 + \sum_{k \in P} \hat{\beta}_k X_k.$$

The coefficient of determination is equal to the square of the Pearson correlation between Y and $\hat{Y}$ computed using the training data. That is,

$$R^2 = \hat{\rho}(Y, \hat{Y})^2.$$

The R^2 statistic is a measure of how much of the variation in our training data is described by the model. But as noted earlier, we should be cautious about interpreting it as a measure of predictive performance. Using the same data to fit a model and to calculate performance measures risks overfitting. Adding more predictors to a model, even if they are only weakly correlated with the response, will always decrease the RSS and therefore increase R^2. The overfitting problem recedes as the number of datapoints, n, becomes large compared to the number of predictors, p. Intuitively, when $n \gg p$ there is sufficient data for the model to distinguish the systematic relationships from noise. Overfitting can be detected by applying cross-validation to calculate a 'test' value of R^2. We compute the RSS and TSS and R^2 for each fold based on the model obtained by fitting to the remaining training data. Averaging the R^2 values over the folds we obtain R^2_{test}. This is a measure of the amount of variation our model is capable of describing in *unseen* data. If we find that

$$R^2_{\text{test}} \ll R^2,$$

then this is a warning that overfitting has occurred.

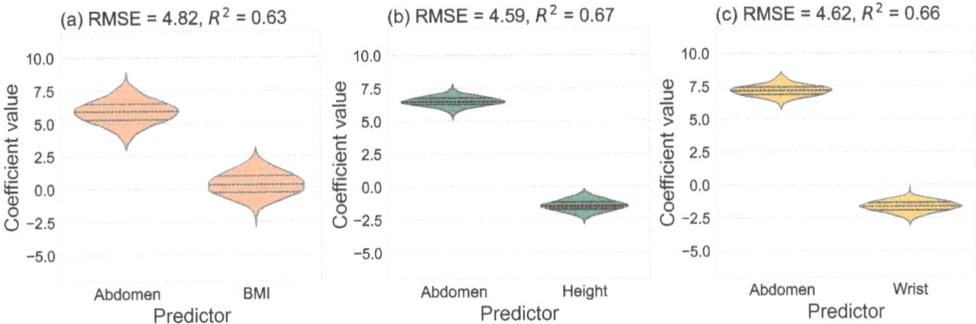

Figure 7.6 (a–c) Bootstrapped sampling distributions of standardised predictor coefficients for two-predictor models. Plot titles give cross-validated RMSE estimates and R^2 values. For reference, the cross-validated RMSE for a linear model using only abdomen is 4.79.

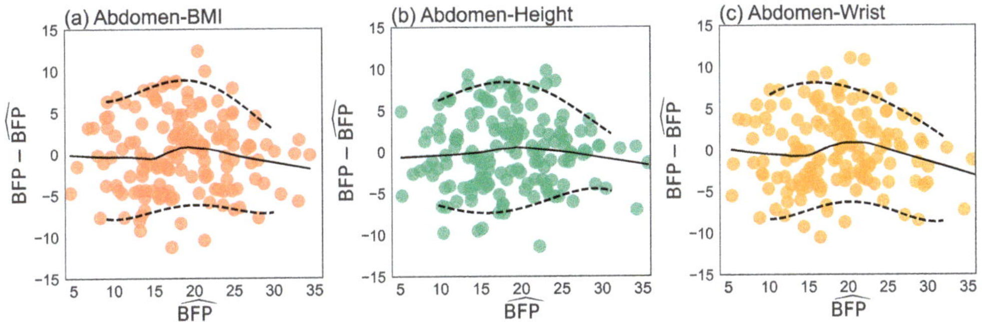

Figure 7.7 (a–c) Scatter plots of the residuals $e_i = y_i - \hat{y}_i$ (or $\mathrm{BFP} - \widehat{\mathrm{BFP}}$) against the predictions $\hat{y}_i$ for three two-predictor models. The solid lines give the approximate relationship between the mean residual and the predictor. The dashed lines show approximate 5% and 95% quantiles of the residual distribution.

7.6.2 A Two-Predictor Model

Let us start by fitting two-predictor models which all use abdomen (the best single predictor) plus one extra measurement. We will use standardised predictors so that the magnitudes of their coefficients may be easily compared. Consider the model which uses BMI as its extra predictor. Using the full training dataset to estimate parameter values, we obtain the prediction function

$$\hat{y} = 18.54 + 5.86 z_{\mathrm{Abd}} + 0.33 z_{\mathrm{BMI}}, \tag{7.9}$$

where z_{Abd} and z_{BMI} are standardised abdomen and BMI measurements. When using standardised predictors, the 'intercept' of our prediction function (where the regression plane intersects the y axis) is the mean BFP in our training data. The standardised predictors represent deviations from the mean abdomen and BMI values. For now we'll interpret the predictor coefficients as providing a measure – not yet precisely defined – of how sensitive BFP is to these deviations. Figure 7.6a is a 'violin plot' of the bootstrap sampling distributions of the coefficient estimates for this model. The horizontal widths of the violins represent sampling distribution densities for each model coefficient. The solid and dashed horizontal lines within each violin give the mean and quartiles of the distribution.

Figure 7.7 shows residual plots for the abdomen–BMI model (a) and two others that we will also examine. When we have more than one predictor, we plot the residuals $e_i = y_i - \hat{y}_i$, against the predictions $\hat{y}_i$, so each point is of the form $(\hat{y}_i, e_i)$. If a linear model is sufficient to capture systematic variations of Y with $\boldsymbol{X}$, then we should have $\mathbb{E}(Y - \hat{Y}|\boldsymbol{X} = \boldsymbol{x}) = 0$, for all $\boldsymbol{x}$. That is, the mean of the residual distributions should be close to zero, whatever the value of $\hat{y}$. If our noise term $\mathcal{E}$ is independent of the predictors $\boldsymbol{X}$, then the width of the residual distribution should remain approximately constant. There is a hint in the residual distributions in Figure 7.7 of declining width for larger predicted BFP, but the effect is not strong enough for us to rethink our assumptions.

Let us examine the properties of our two-predictor model (7.9). The first point of interest is that adding BMI to the model hasn't improved its predictive abilities – in fact, according to our cross-validation RMSE estimate it is slightly worse (two-predictor RMSE = 4.82 vs. single-predictor RMSE = 4.79). On its own BMI was the second best predictor of BFP, so it might seem strange that combining it with abdomen doesn't improve performance. Let us try to understand why. Looking at the coefficients of our fitted model (7.9), we see that the abdomen coefficient is ≈ 18 times larger in magnitude than the BMI coefficient. That is, in our new prediction function abdomen circumference is doing the heavy lifting and very little use is made of BMI, even though they are both highly correlated with BFP. Looking at the sampling distribution of $\hat{\beta}^*_{\text{BMI}}$ (Figure 7.6a), we also see that even the *sign* of our estimate is a matter of roughly equal chance depending on noise in the dataset-generation process. In a simple linear model, the coefficient of the predictor is proportional to the correlation coefficient between the response and the predictor. This is no longer the case when we have multiple predictors. To understand what has changed, it is useful to take a step back from the real world and think about models purely as relationships between sets of random variables.

7.6.3 Analysing Models Probabilistically

Setting aside any real-world interpretation, the model

$$\begin{aligned} Y &= \beta_0 + \beta_1 X_1 + \beta_2 X_2 + \mathcal{E}, \\ \mathcal{E} &\sim \mathcal{N}(0, \sigma^2) \end{aligned}$$

with $\mathcal{E} \perp\!\!\!\perp \{X_1, X_2\}$ specifies a relationship between random variables. We can think of the model as a partial description of a data-generating process that spits out random variable triples (X_1, X_2, Y). The model is partial because it says nothing about the individual *marginal* distributions of these variables, only about the relationship between them.

Let's suppose we've observed that $X_2 = x_2$, but we haven't yet observed X_1 and Y. We have

$$Y = \underbrace{\beta_0 + \beta_2 x_2}_{\text{adjusted 'intercept'}} + \beta_1 X_1 + \mathcal{E}.$$

This is a just a one-predictor model with an adjusted intercept. Thus, provided we observe X_2, the coefficient of X_1 can be interpreted in the same way it would be in a one-predictor model, with the caveat that this new model only describes outputs of our data-generating process which have $X_2 = x_2$. We can relate β_1 to the *conditional covariance* between Y and X_1 given that $X_2 = x_2$, defined as

$$\text{Cov}(Y, X_1|X_2 = x_2) = \mathbb{E}((Y - \mathbb{E}(Y|X_2 = x_2))(X_1 - \mathbb{E}(X_1|X_2 = x_2))|X_2 = x_2).$$

We can write this in a more compact form by defining $\mu_{Y|X_i} = \mathbb{E}(Y|X_i)$ etc. Then

$$\text{Cov}(Y, X_1|X_2 = x_2) = \mathbb{E}((Y - \mu_{Y|X_2})(X_1 - \mu_{X_1|X_2})|X_2 = x_2).$$

Computing this covariance using our model, we find (see Exercise 7.9) that

$$\beta_1 = \frac{\text{Cov}(Y, X_1|X_2 = x_2)}{\text{Var}(X_1|X_2 = x_2)},$$

where $\text{Var}(X_1|X_2 = x_2) = \text{Cov}(X_1, X_1|X_2 = x_2)$. From this we see that β_1 measures the strength of the linear relationship between Y and X_1, given the value of X_2, *whatever* that value is.

Now let's take a step back towards the real world and imagine that our data-generating process is sampling from a population of triples $\{(x_{i1}, x_{i2}, y_i)\}_{i=1}^N$. We'll assume that the population is effectively infinite ($N \approx \infty$), so that it makes sense to consider the *subpopulation* with a specific x_2 value. The coefficient β_1 measures the strength of the linear relationship between Y and X_1, or equivalently the expected change in Y per unit increase in X_1, within this subpopulation.

Armed with this theoretical perspective, let's return to our body fat predictions, assuming our model is a good description of reality. We'll work with standardised versions of our predictors so the magnitudes of their coefficients can be compared. That is,

$$Y = \beta_0^* + \beta_1^* Z_1 + \beta_2^* Z_2 + \mathcal{E},$$

where $\mathbb{E}(Z_1) = \mathbb{E}(Z_2) = 0$ and $\text{Var}(Z_1) = \text{Var}(Z_2) = 1$. Suppose that Z_1 is BMI and Z_2 is abdomen circumference. If we fix Z_2 then β_1^* measures the strength of the linear relationship between BFP and standardised BMI within the subpopulation with the given standardised abdomen circumference, or equivalently the expected increase in BFP per unit increase in standardised BMI within this subpopulation. The small magnitude of our estimate of β_1^* now has a concrete interpretation – amongst the subset of the population who have a given abdomen circumference there will be variations in BMI, but they don't provide a great deal of information about BFP. Of course we can reverse this interpretation and consider a subset of the population who all have the same BMI. Our results suggest that within this subset, abdomen size remains a good predictor of BFP. We might speculate that within this group, there will be individuals who carry a lot of their weight in the form of muscles, and individuals who carry a lot of their weight as fat. In males this fat is often concentrated around the abdomen meaning that abdomen circumference remains a strong predictor of BFP even if we already know the BMI. The key point here is that in multiple linear regression, each predictor coefficient is measuring the strength of its correlation with the response, *conditional* on the value of the other predictors.

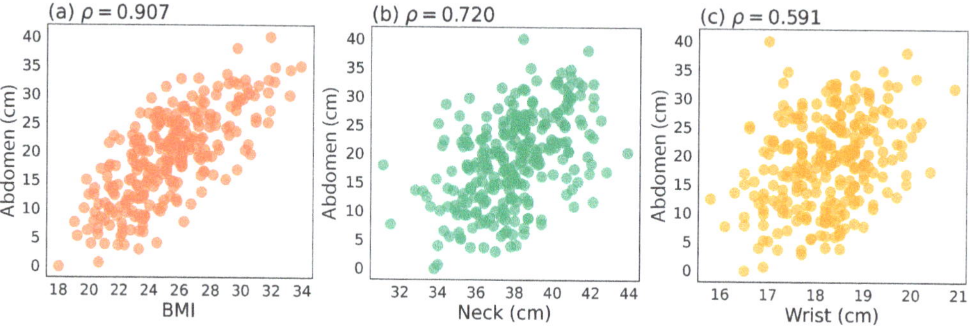

Figure 7.8 (a–c) Scatter plots of the predictor pairs used for the two-predictor models in Figure 7.6. BMI and Abdomen show high collinearity.

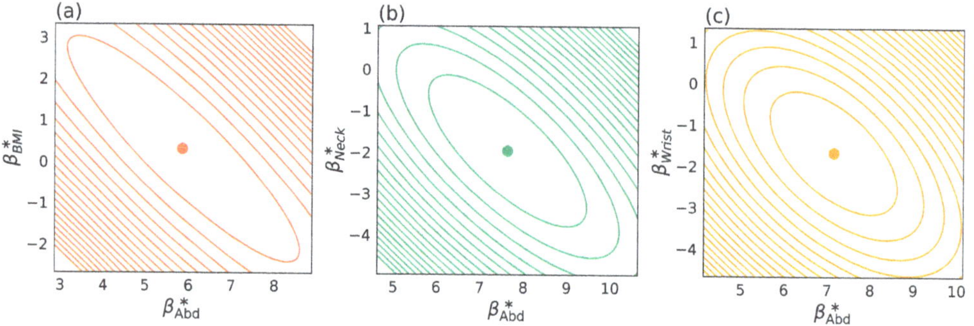

Figure 7.9 (a–c) Contour plots of the RSS with equally spaced contour levels for the two-predictor models in Figure 7.6. The central points mark the minima of the RSS, corresponding to maximum likelihood coefficient estimates.

7.6.4 Combining Predictors

Let us now examine the coefficient sampling distribution of our abdomen- and BMI-based prediction model shown in Figure 7.6a. Compared to the two alternative two-predictor models in Figure 7.6b and c, which replace BMI with neck and wrist, standard errors in the abdomen-BMI model are large. This reflects the fact that while abdomen and BMI are individually good predictors of BFP, they also exhibit a comparatively high level of *collinearity*. That is, they have a strong linear relationship (and a high correlation), as can be seen from Figure 7.8. Strongly collinear predictors tend to increase or decrease together, so using them in combination typically yields minimal improvements in predictive performance. It is also difficult to determine their respective contribution to variations in the response. We can explain this using Figure 7.9a, which shows the dependence of the RSS on β^*_{Abd} and β^*_{BMI} when $\beta^*_0 = \bar{y}$. The maximum likelihood estimate $(\hat{\beta}^*_{\text{Abd}}, \hat{\beta}^*_{\text{BMI}})$ lies at the centre of an elongated region within which the RSS takes similar values. Points within this region represent alternative coefficient estimates all of which are similarly plausible. If we now look at the RSS plots in Figures 7.9b and c, obtained by replacing BMI with (in turn) neck

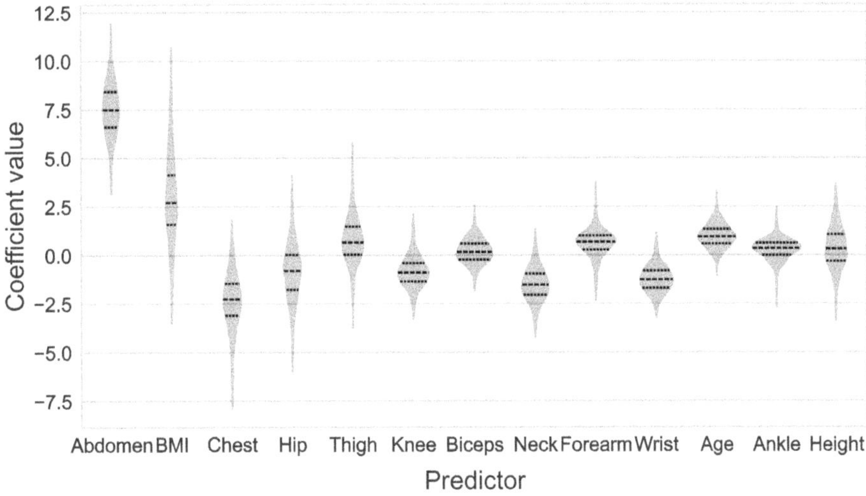

Figure 7.10 Bootstrapped sampling distributions of standardised predictor coefficients for the linear model using all predictors. The cross-validated error estimate for this model is RMSE = 4.822.

and wrist, we see that the central regions are less elongated, implying a narrower range of plausible coefficient estimates. This is due to lower levels of collinearity between the predictors.

It is clear that simply combining the most effective individual predictors is not always the best way to improve predictive accuracy. We need predictors that *complement* each other in the sense that their individual associations with the response variable are as strong as possible, while their associations with each other are weak (there is minimal collinearity). In Chapter 12, we will show how to construct linear combinations of variables whose sample correlations with one another are zero, eliminating collinearity altogether. Here we'll stick to models which use a subset of the original predictors. Consider the bootstrap sampling distributions shown in Figure 7.6b and c for the abdomen–neck and abdomen–wrist models. Despite the fact that neck and wrist have substantially lower individual correlations with BFP than does BMI, in combination with abdomen they not only achieve better predictive performance but also have smaller standard errors in their coefficients. This increased precision is a consequence of reduced collinearity. Interestingly, for both models the additional predictor has a *negative* relationship with the response. That is, once we have measured the abdomen, then a larger neck or wrist is associated with a lower BFP. We might speculate that amongst men with the same abdomen size, larger wrists or neck is associated with a larger or more muscular frame, and therefore a lower proportion of fat.

In Section 7.6.3, we will take a first look at the problem of *systematically* selecting optimal sets of predictors. Before we do so, let us first explore what happens when we use *all* the predictors available to us. Figure 7.10 is a violin plot of the predictor coefficient sampling distributions in a model based on all 13 predictors. We highlight two points of interest. First, the cross-validated RMSE = 4.822 is the highest of all the models we have considered so far. From this we can conclude if we want to optimise predictive performance we should not use all the predictors. The second point of interest is that the bootstrap standard errors in this all-predictor model are also the largest we have seen so far. This reflects the fact that when more predictors are available, there are more ways to explain the pattern of responses in terms of them, and therefore less certainty about each possible explanation. The R^2 values for this all-predictor model are

$$R^2 = 0.702,$$
$$R^2_{\text{test}} = 0.525,$$

where R^2_{test} was computed using the RSS and TSS evaluated on held-out folds during cross-validation. As expected, R^2 for this model is larger than the R^2 values for all previous models, because adding more predictors can only decrease the RSS evaluated on the training data. However, R^2_{test} is substantially smaller, providing further evidence of overfitting.

7.7 Best Subset Selection

So far we have experimented with various combinations of predictors, and we have established that the optimal model – as determined by cross-validation – must use a strict subset of them. With 13 predictors in total, there are $2^{13} = 8192$ possible subsets (of which 8,191 are strict). We could test them all and pick the one with the lowest cross-validation error. This is feasible using the current dataset, because fitting linear models can be done very efficiently. But if we had (say) 50 predictors, exhaustive search would be a non-starter.

An efficient alternative to exhaustive search is *forward stepwise selection*. Let p be the number of available predictors, where each predictor is labelled using an integer from the set $S = \{1, 2, \ldots, p\}$. The aim is to construct a sequence of predictor sets $P_0, P_1, P_2, \ldots, P_p$ where P_k is our guess at the best set of k predictors. The zeroth predictor set $P_0 = \{\}$ produces a 'null model' which contains no predictors. To find P_1 we test all possible single-predictor models and pick the one with the lowest cross-validation error. To find P_2, rather than testing *all* two-predictor models, we test all predictor sets which result from adding one more predictor to P_1, subject to the condition that the new predictor is not already in P_1. These sets are of the form

$$P_1 \cup \{i\}, \quad \text{where} \quad i \in S \setminus P_1.$$

Here $S \setminus P_1$ is the set of predictors which are not in P_1[6]. We choose P_2 to be the set that produces the lowest cross-validation error in these tests. We can construct P_{k+1} from P_k in a similar way. Let $\text{MSE}(P)$ be the cross-validated MSE of the model with predictor set P. Given P_k, we find the extra predictor that when added to P_k produces the smallest error:

$$\hat{i} = \underset{i \in S \setminus P_k}{\arg\min}\, \text{MSE}\left(P_k \cup \{i\}\right).$$

We then add $\hat{i}$ to P_k, to form P_{k+1}

$$P_{k+1} = P_k \cup \{\hat{i}\}.$$

Having constructed the sequence of best predictor sets of each size, $P_0, P_1, P_2, \ldots, P_p$, we

[6] Given two sets A and B, the 'set difference' $A \setminus B$ is the set of all elements of A which are not in B.

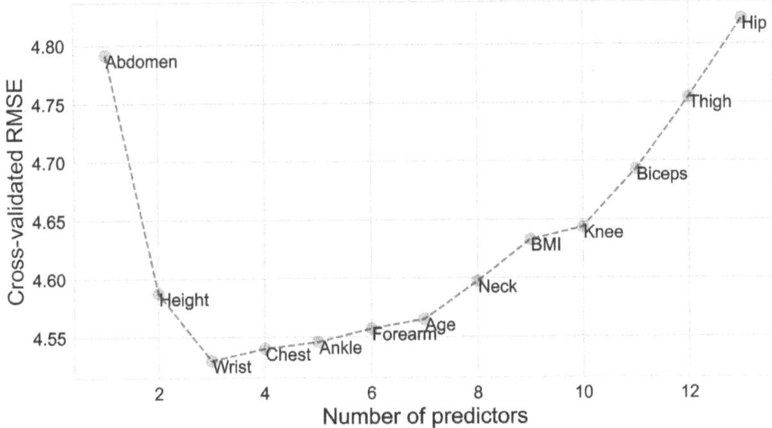

Figure 7.11 Cross-validated RMSEs for models using predictor sets determined by forward stepwise selection. Points are labelled with the additional predictor added to the previous subset.

find the best overall predictor set

$$\hat{P} = \arg\min_{P_k} \text{MSE}(P_k).$$

Without exhaustive search, we cannot *guarantee* to find the best subset. The idea of stepwise selection is to *efficiently* find a model which, even if it is not strictly optimal, still comes close. If we have p predictors, then the number of models we need to test using stepwise selection is $1 + p(p+1)/2$ versus 2^p for exhaustive search. For example, when $p = 13$ this means performing 92 tests versus 8192.

Figure 7.11 shows the cross-validated RMSEs obtained using the predictor subset sequence $P_1, P_2, \ldots$ determined by applying forward stepwise selection to our body fat training data. The relationship between cross-validation error and model complexity (as measured by the number of predictors used) follows the characteristic U-shape we saw in Chapters 2 and 5. In the current context, using too few predictors causes underfitting – we are assuming an excessively simple relationship between predictors and response and so failing fully to exploit systematic dependencies to improve predictions. Using too many predictors causes overfitting – by allowing an excessively complex relationship, we end up fitting to noise in the training data rather than systematic dependencies. From Figure 7.11, we see that the optimal subset contains three predictors,

$$P_3 = \{\text{abdomen}, \text{height}, \text{wrist}\}.$$

Now that we have selected an optimal model, we can break out our test set. Using our optimal predictor set, we fit to the full training set, obtaining the prediction function

$$\hat{y} = 18.54 + 7.12 z_{\text{Abd}} - 1.10 z_{\text{Height}} - 1.17 z_{\text{Wrist}}. \tag{7.10}$$

Applying this function to the predictors in the test set and comparing to the test responses, we obtain

$$\mathrm{RMSE}(P_3) = 4.17.$$

For reference, using abdomen alone we obtain RMSE = 4.67 on the test set, satisfyingly larger than our optimal model.

Let us pause briefly to interpret our final fitted model (7.10). As with previous models, abdomen plays the most important role. The height coefficient is negative, indicating that for males with a given abdomen and wrist size, being taller is associated with a lower BFP. This makes sense; we would expect taller people to have generally larger body measurements, so a taller man with the same abdomen and wrist size as a shorter man typically carries a smaller proportion of fat. The wrist coefficient is also negative: for males with a given abdomen circumference and height, a larger wrist size is associated with a lower BFP. Again, it's not hard to see why. With height and abdomen circumference held fixed, a larger wrist size would likely indicate a larger and broader skeleton. A large-boned man who has the same height and abdomen circumference as a finely built man will tend to have a greater mass of muscle and bone, and therefore a smaller percentage of fat.

7.8 The Bias–Variance Trade-Off

When building a statistical model, decisions must be made about how complex it should be. These decisions take many different forms. In Chapters 2 and 5, where we estimated one-variable regression functions, we controlled complexity via the polynomial degree of our prediction function. In Chapter 3, where we modelled a density as a sum of Gaussian blobs, we adjusted the number of blobs. In this chapter, we controlled the number of predictors. These are all examples of changing complexity in discrete steps, but in Chapter 9 we will see that it can be sometimes be varied continuously. Although complexity can be adjusted in many ways, when we examine its impact on the ability of a model to predict patterns in unseen data, we see a recurring theme. Graphs of performance against complexity tend to be U-shaped: too little complexity leads to underfitting; too much, to overfitting. Both hurt performance. Finding the sweet spot between underfitting and overfitting may be understood in a precise mathematical way as balancing two distinct sources of error: *bias* and *variance*.

To keep our calculations concrete, let us consider fitting the following conditional probabilistic model of Y given $\boldsymbol{X}$:

$$Y = g(\boldsymbol{X}; \boldsymbol{\theta}) + \mathcal{E},$$
$$\mathcal{E} \sim \mathcal{N}(0, \sigma^2).$$

We will suppose that the true relationship between Y and $\boldsymbol{X}$ is given by

$$Y = \mathfrak{g}(\boldsymbol{X}) + \mathcal{E},$$

where $\mathfrak{g}$ is an unknown function that we aim to approximate with $g(\boldsymbol{x}; \boldsymbol{\theta})$ for some suitably chosen value of $\boldsymbol{\theta}$. We assume that the true noise variable is the same as the noise variable in our model and that $\mathcal{E}$ is independent of $\boldsymbol{X}$. The variance σ^2 is an unknown constant. Given a dataset $D = \{(\boldsymbol{x}_i, y_i)\}_{i=1}^n$, we estimate $\boldsymbol{\theta}$ by minimising the RSS, which we here denote l for loss:

$$l(\boldsymbol{\theta}; D) = \sum_{i=1}^{n} (g(\boldsymbol{x}_i, \boldsymbol{\theta}) - y_i)^2.$$

Our prediction function is

$$\hat{y} = \hat{g}(\boldsymbol{x}) = g(\boldsymbol{x}; \hat{\boldsymbol{\theta}}) = g(\boldsymbol{x}; \boldsymbol{t}(D)),$$

where

$$\boldsymbol{t}(D) = \arg\min_{\boldsymbol{\theta}} l(\boldsymbol{\theta}; D).$$

Our particular D may be viewed as a single realisation of a random dataset $\mathcal{D} = \{(\boldsymbol{X}_i, Y_i)\}_{i=1}^{n}$, where the pairs $(\boldsymbol{X}_i, Y_i)$ are independent and identically distributed. If we were to repeat the dataset-generating process many times, we would obtain a sequence of realisations of the random vector $\hat{\boldsymbol{\Theta}} = \boldsymbol{t}(\mathcal{D})$, representing samples from the estimator's sampling distribution. Each of the parameter vector estimates would correspond to a different prediction function.

When training and testing a model, there are two sources of uncertainty: the training data and the test data. We assume that both are drawn from the same distribution (i.e. that all the datapoints are independent realisations of $(\boldsymbol{X}, Y)$). It is useful to distinguish between the performance of a single fitted model and the performance of the overall *procedure* of fitting a model and then using it to make predictions. The procedure specifies both the mathematical form of the model to be fitted (which determines its complexity) and the fitting method, but *not* the specific values of the fitted parameters. These are determined by the training data. The performance of a procedure can be measured by averaging the average test error over all realisations of the training data. Notice that this is a *double* average.

Let's fix a particular predictor value $\boldsymbol{x}$ and compute this double average. We first define a 'predicted response' random variable which depends on the training dataset,

$$\hat{Y} = g(\boldsymbol{x}, \boldsymbol{t}(\mathcal{D})).$$

Now consider a response from the true data-generating process with $\boldsymbol{X} = \boldsymbol{x}$,

$$Y = \mathfrak{g}(\boldsymbol{x}) + \mathcal{E}.$$

The squared test error is

$$(\hat{Y} - Y)^2 = (g(\boldsymbol{x}; \boldsymbol{t}(\mathcal{D})) - \mathfrak{g}(\boldsymbol{x}) - \mathcal{E})^2.$$

Notice that there are two independent sources of randomness here: the training dataset $\mathcal{D}$ and the test noise variable $\mathcal{E}$. To measure our procedure's performance, we average the squared error over $\mathcal{D}$ and $\mathcal{E}$. This gives what is known as the procedure's *risk*,

$$\begin{aligned} \text{RISK}(\boldsymbol{x}) &= \mathbb{E}\left((\hat{Y} - Y)^2\right) \\ &= \mathbb{E}\left((g(\boldsymbol{x}; \boldsymbol{t}(\mathcal{D})) - \mathfrak{g}(\boldsymbol{x}) - \mathcal{E})^2\right). \end{aligned}$$

We will simplify this expectation in two stages. First, we – or rather you, in the exercise below – will calculate the expected squared error conditional on $\mathcal{D}$,

$$\mathbb{E}\left((g(\boldsymbol{x}; \boldsymbol{t}(\mathcal{D})) - \mathfrak{g}(\boldsymbol{x}) - \mathcal{E})^2 | \mathcal{D}\right) = (g(\boldsymbol{x}; \boldsymbol{t}(\mathcal{D})) - \mathfrak{g}(\boldsymbol{x}))^2 + \sigma^2. \tag{7.11}$$

This is a random variable whose randomness is derived entirely from the training data.

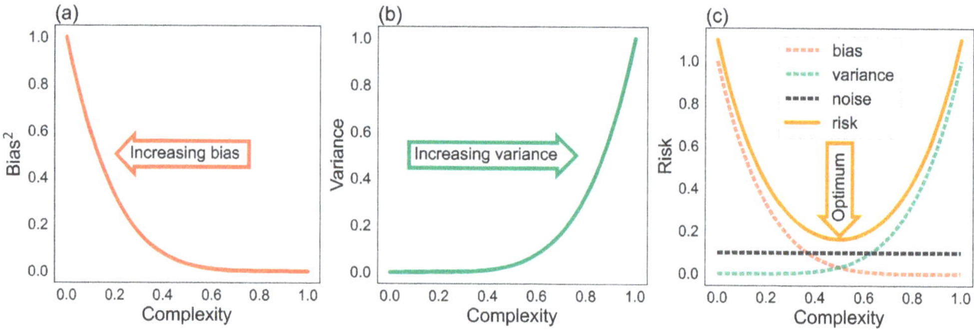

Figure 7.12 (a–c) The bias variance trade-off.

■ ***Exercise*** 7.6 Making use of the fact that $\mathbb{E}(\mathcal{E}) = 0$ and $\mathbb{E}(\mathcal{E}^2) = \sigma^2$, and that $\mathcal{E}$ is independent of the training data, verify equation (7.11).

We can now compute the risk by averaging over $\mathcal{D}$:

$$\text{RISK}(\boldsymbol{x}) = \underbrace{\mathbb{E}\left((g(\boldsymbol{x};\boldsymbol{t}(\mathcal{D})) - \mathfrak{g}(\boldsymbol{x}))^2\right)}_{\text{Reducible}} + \underbrace{\sigma^2}_{\text{Irreducible}}. \tag{7.12}$$

We have labelled the first term on the right-hand side 'reducible' because its magnitude is affected by the mathematical form of $g(\boldsymbol{x},\boldsymbol{\theta})$, and therefore it is possible to reduce it using good modelling choices. The second term is labelled 'irreducible' because it derives from the noise $\mathcal{E}$, which we cannot control. The first term can be decomposed into two further terms, each of which captures a distinctive source of reducible error. We first define the average value of the prediction function over all possible training sets,

$$\bar{g}(\boldsymbol{x}) = \mathbb{E}\left(g(\boldsymbol{x};\boldsymbol{t}(\mathcal{D}))\right) = \mathbb{E}\left(g(\boldsymbol{x};\hat{\boldsymbol{\Theta}})\right).$$

Here we have made use of the fact that averaging over the training data is equivalent to averaging over the sampling distribution of the parameter vector. We then have

$$\text{RISK}(\boldsymbol{x}) = \underbrace{\mathbb{E}\left((g(\boldsymbol{x};\hat{\boldsymbol{\Theta}}) - \bar{g}(\boldsymbol{x}))^2\right)}_{\text{Variance}} + \underbrace{(\bar{g}(\boldsymbol{x}) - \mathfrak{g}(\boldsymbol{x}))^2}_{\text{Bias}^2} + \underbrace{\sigma^2}_{\text{Noise}}. \tag{7.13}$$

We prove this identity in Box 7.4. Let us examine each of the three terms on the right-hand side of (7.13). The first term – the variance – is the mean squared discrepancy between the average prediction and the predictions produced using the various different parameter estimates drawn from the sampling distribution. It is not a measure of the ability of the prediction function to make accurate predictions; rather, it is a measure of the prediction function's sensitivity to variations in the training data. A complex model with many parameters will have high variance because its flexibility will allow it to adapt to noise in the training data. (We saw this effect in the all-predictor model fitted to our body fat data.) High variance is therefore synonymous with overfitting. The second term – the squared bias – measures the extent to which the model *averaged over the sampling distribution* matches $\mathfrak{g}(\boldsymbol{x})$. Here, flexibility is an asset. If for *some* $\boldsymbol{\theta}$ the function $g(-;\boldsymbol{\theta})$ approximates $\mathfrak{g}$ well, then it tends to be the case that $\bar{g}(\boldsymbol{x})$ lies close to $\mathfrak{g}(\boldsymbol{x})$. Conversely, if a model is incapable of approximating $\mathfrak{g}$ well (however one tunes its parameters), then it usually will *not* be the case that averaging over the sampling distribution yields a function close to $\mathfrak{g}$. The bias term therefore tends to be large for simpler models and small for complex models – the opposite behaviour to the variance. The net effect is that the risk is U-shaped (see Figure 7.12). Excessively low complexity leads to high bias, low variance and high risk. Excessively high complexity leads to high variance, low bias and high risk. To minimise risk, we must adjust complexity to optimally balance bias and variance.

★ Box 7.4 The bias-variance trade-off identity

Equation (7.12) for the risk of a prediction model may be written

$$\mathrm{RISK}(\boldsymbol{x}) = \mathbb{E}\left((g(\boldsymbol{x};\hat{\boldsymbol{\Theta}}) - \mathfrak{g}(\boldsymbol{x}))^2\right) + \sigma^2.$$

We can decompose the expectation on the right-hand side into a variance term and a bias term. To simplify intermediate steps in our calculation we will use the abbreviations:

$$\begin{aligned}\hat{g} &= g(\boldsymbol{x};\hat{\boldsymbol{\Theta}}),\\ \bar{g} &= \bar{g}(\boldsymbol{x}),\\ \mathfrak{g} &= \mathfrak{g}(\boldsymbol{x}).\end{aligned}$$

Notice that $\hat{g}$ is a random variable satisfying $\mathbb{E}(\hat{g}) = \bar{g}$, whereas $\bar{g}$ and $\mathfrak{g}$ are non-random. Thus,

$$\begin{aligned}\mathbb{E}\left((g(\boldsymbol{x};\hat{\boldsymbol{\Theta}}) - \mathfrak{g}(\boldsymbol{x}))^2\right) &= \mathbb{E}\left(((\hat{g} - \bar{g}) + (\bar{g} - \mathfrak{g}))^2\right)\\ &= \mathbb{E}\left((\hat{g} - \bar{g})^2 + (\bar{g} - \mathfrak{g})^2 + 2(\hat{g} - \bar{g})(\bar{g} - \mathfrak{g})\right)\\ &= \mathbb{E}\left((\hat{g} - \bar{g})^2\right) + (\bar{g} - \mathfrak{g})^2 + 2(\bar{g} - \mathfrak{g})\underbrace{\mathbb{E}(\hat{g} - \bar{g})}_{=\bar{g}-\bar{g}=0}\\ &= \mathbb{E}\left((g(\boldsymbol{x};\hat{\boldsymbol{\Theta}}) - \bar{g}(\boldsymbol{x}))^2\right) + (\bar{g}(\boldsymbol{x}) - \mathfrak{g}(\boldsymbol{x}))^2.\end{aligned}$$

7.9 Chapter Summary

In this chapter we have explored the simple and multiple linear regression models. They are workhorses of predictive modelling due to their simplicity and ease of interpretation. We have also introduced a number of core concepts of predictive modelling and inference.

- The **simple** and **multiple linear regression** models assume that the response variable is a linear function of the predictor(s), plus a noise term.
- The **regression function** is the expected value of the response given the predictor(s).
- The **coefficient of determination**, R^2, measures the proportion of response variance explained by the regression function.
- R^2 is equal to the square of the Pearson correlation coefficient between the response, Y,

and the predictor, X in simple linear regression and between the response, Y, and the prediction, $\hat{Y}$ in multiple linear regression.

- In the multiple linear regression model, the coefficient of predictor X_i is the expected change in Y associated with a one-unit increase in X_i keeping the other predictors fixed.
- **Standardising** predictors allows us to measure the sensitivity of model predictions to changes in predictor variables by comparing the sizes of their regression coefficients.
- Strong **collinearity** between predictors increases uncertainty in their fitted coefficient values.
- Models that use a subset of the available predictors may perform better on unseen data than models which use all the predictors and consequently overfit.
- The squared error **risk** of a procedure for constructing prediction functions is the squared prediction error averaged over the training data-generation process, and over noise in the test data. Risk can be decomposed into **bias**, **variance** and **noise**.

7.10 Further Exercises

■ ***Exercise*** 7.7 Suppose that $Y = \beta_0 + \beta_1 X + \mathcal{E}$, where $\mathcal{E} \sim \mathcal{N}(0, \sigma^2)$ and $\mathcal{E} \perp\!\!\!\perp X$. Suppose also that $\mathbb{E}(X) = \mu_X$ and $\mathrm{Var}(X) = \sigma_X^2$.

(a) Define $\mu_Y = \mathbb{E}(Y)$ and $\sigma_Y^2 = \mathrm{Var}(Y)$. Show that $\mu_Y = \beta_0 + \beta_1 \mu_X$, $\sigma_Y^2 = \beta_1^2 \sigma_X^2 + \sigma^2$ and $\mathrm{Cov}(X, Y) = \beta_1 \sigma_X^2$.

(b) Let ρ_{XY} be the correlation between X and Y. Show that

$$\rho_{XY} = \frac{\beta_1 \sigma_X}{\sqrt{\beta_1^2 \sigma_X^2 + \sigma^2}}.$$

(c) The random variables X_1 and X_2 are said to follow the *bivariate normal distribution* if they have joint density

$$f_{X_1,X_2}(x_1, x_2) = \frac{\exp\left(-\frac{1}{2(1-\rho_{12}^2)}\left(\frac{(x_1-\mu_1)^2}{\sigma_1^2} - \frac{2\rho_{12}(x_1-\mu_1)(x_2-\mu_2)}{\sigma_1\sigma_2} + \frac{(x_2-\mu_2)^2}{\sigma_2^2}\right)\right)}{2\pi\sigma_1\sigma_2\sqrt{1-\rho_{12}^2}}, \quad (7.14)$$

where $\mu_i = \mathbb{E}(X_i)$, $\mathrm{Var}(X_i) = \sigma_i^2$ and $\rho_{12} = \mathrm{Cov}(X_1, X_2)/(\sigma_1\sigma_2)$. Let $f_{X,Y}(x, y)$ be the joint density of X and Y (the random variables considered in (a) and (b)). Suppose that $X \sim \mathcal{N}(\mu_X, \sigma_X^2)$. Starting from the factorisation $f_{X,Y}(x, y) = f_{Y|X}(y|x) f_X(x)$, show that $f_{X,Y}(x, y)$ can be expressed in the form (7.14) where $\sigma_1 = \sigma_X, \sigma_2 = \sigma_Y, \rho_{12} = \rho_{XY}, \mu_1 = \mu_X$ and $\mu_2 = \mu_Y$.

■ ***Exercise*** 7.8 Suppose that $Y = \beta_0 + \beta_1 X_1 + \beta_2 X_2 + \mathcal{E}$, where $\mathcal{E} \sim \mathcal{N}(0, \sigma^2)$ and $\mathcal{E} \perp\!\!\!\perp \{X_1, X_2\}$. Define $\sigma_{Yk} = \mathrm{Cov}(Y, X_k)$, where $k \in \{1, 2\}$ and also $\sigma_{12} = \mathrm{Cov}(X_1, X_2)$, $\sigma_1^2 =$

$\text{Var}(X_1)$ and $\sigma_2^2 = \text{Var}(X_2)$. Show that

$$\beta_1 = \frac{\sigma_{Y1}\sigma_2^2 - \sigma_{12}\sigma_{Y2}}{\sigma_1^2\sigma_2^2 - \sigma_{12}^2}.$$

Find a similar expression for β_2.

■ ***Exercise*** 7.9 Suppose that $Y = \beta_0 + \beta_1 X_1 + \beta_2 X_2 + \mathcal{E}$, where $\mathcal{E} \sim \mathcal{N}(0, \sigma^2)$ and $\mathcal{E} \perp\!\!\!\perp \{X_1, X_2\}$. Show that

$$\beta_1 = \frac{\text{Cov}(Y, X_1|X_2)}{\text{Var}(X_1|X_2)},$$

where the conditional covariance is defined

$$\text{Cov}(Y, X_1|X_2) = \mathbb{E}((Y - \mathbb{E}(Y|X_2))(X_1 - \mathbb{E}(X_1|X_2))|X_2)$$

and $\text{Var}(X_1|X_2) = \text{Cov}(X_1, X_1|X_2)$. Find a similar expression for β_2.

■ ***Exercise*** 7.10 Life as a jazz trumpeter can be rewarding, but also financially risky. Let Y be the net profit of a newly trained trumpeter in their first year of performing, and let X be their score in the improvisation exam in the final year of music college. Suppose we have a dataset of scores and profits $D = \{(x_i, y_i)\}_{i=1}^n$.

(a) Suppose you want to estimate the profit from the score. We'll use the model

$$Y = \alpha X + \mathcal{E},$$

where $\mathcal{E} \sim \mathcal{N}(0, \sigma_1^2)$ and $\mathcal{E} \perp\!\!\!\perp X$. Find an expression for the maximum likelihood estimate of α in terms of D.

(b) Now suppose you want to estimate the score from the profit, using the model

$$X = \beta Y + \mathcal{E},$$

where $\mathcal{E} \sim \mathcal{N}(0, \sigma_2^2)$ and $\mathcal{E} \perp\!\!\!\perp Y$. Find an expression for the maximum likelihood estimate of β in terms of D.

(c) Suppose that the true marginal distribution of X is $X \sim \mathcal{N}(20, 5^2)$ and that in reality Y is related to X by $Y = X + \mathcal{E}$, where $\mathcal{E} \perp\!\!\!\perp X$ and $\mathcal{E} \sim \mathcal{N}(0, 10^2)$. Show that the correlation between Y and $\mathcal{E}$ is $\rho(Y, \mathcal{E}) = 2\sqrt{5}/5$. In light of this, are the assumptions in part (b) correct?

(d) Show that, according to the true data-generating process, as $n \to \infty$, $\hat{\alpha} \to 1$ and $\hat{\beta} \to 17/21$. Explain why $\hat{\beta} \neq 1$ in this limit.

8

Directed Graphical Models

As we dig more deeply into Bayesian ideas in Chapters 9 and 10, we'll often need to reason carefully about conditional probabilistic relationships between random variables. When these relationships get complicated, it can be helpful to represent them diagrammatically. This chapter introduces *directed acyclic graphs* (DAGs) as a tool for doing just that. DAGs are sometimes referred to as *Bayesian networks*, but in fact they are useful more broadly. They can describe statistical models in general, clarify dependencies and support causal reasoning.

8.1 Directed Acyclic Graph Models

In the current context, a *graph* is a set of points or *nodes* and a set of lines or *edges* connecting pairs of nodes. Connected nodes are said to be *adjacent*, and a *path* through the graph is a sequence of adjacent nodes. If edges have a direction – if they are 'arrows' – then the graph is *directed*. A path that follows arrow directions is a *directed path*. A directed path that starts and ends in the same place is a *directed cycle*. A directed graph with no directed cycles is a DAG.

We can use DAGs to describe multivariate probability distributions in terms of conditional distributions. For simplicity we will work with continuous random variables and density functions, but note that the results presented in this chapter also hold for discrete random variables if we replace densities with mass functions. Consider the case of three continuous random variables X, Y, Z with joint density $f_{X,Y,Z}(x, y, z)$. According to the definition of conditional probability density (Appendix A), we have

$$\begin{aligned} f_{X|Y,Z}(x|y, z) &= \frac{f_{X,Y,Z}(x, y, z)}{f_{Y,Z}(y, z)}, \\ f_{Y|Z}(y|z) &= \frac{f_{Y,Z}(y, z)}{f_Z(z)}. \end{aligned}$$

Using these relationships, we can factorise the full multivariate density into a product of univariate conditional densities as follows:

$$\begin{aligned} f_{X,Y,Z}(x, y, z) &= f_{X|Y,Z}(x|y, z) f_{Y,Z}(y, z) \\ &= f_{X|Y,Z}(x|y, z) f_{Y|Z}(y|z) f_Z(z). \end{aligned} \tag{8.1}$$

Each of the factors in this expression is a density of *one* variable, conditional on the values of zero or more other variables, which we call its 'parents'. Such a factorisation can be represented graphically using a DAG. We first create a node for each random variable. Then, for each node, we add an incoming directed edge from each of its parents. According to these rules, factorisation (8.1) corresponds to the graph

(8.2)

Notice that the density of Z is not conditional on any other variable so the Z node has no parents. The factorisation (8.1) is not unique. Another possibility is

$$f_{X,Y,Z}(x, y, z) = f_{Z|Y,X}(z|y, x) f_{Y|X}(y|x) f_X(x), \tag{8.3}$$

corresponding to the DAG

(8.4)

Each of these two possibilities corresponds to a different ordering of our variables: (X, Y, Z) in the first case and (Z, Y, X) in the second.

■ ***Exercise*** 8.1 Find one further factorisation of $f_{X,Y,Z}(x, y, z)$ into a product of univariate conditional densities and draw the corresponding DAG.

In performing the above factorisations, we placed no restrictions on the dependencies among the random variables. As a result, all the above factorisations are entirely general: they can describe any multivariate distribution. What happens when we do introduce assumptions about the dependency structure? Consider the factorisation (8.1). It contains the factor $f_{X|Y,Z}(x|y, z)$, the conditional density of X given Y and Z. Suppose that this density does *not* depend on the value of Y, so that

$$f_{X|Y,Z}(x|y, z) = f_{X|Z}(x|z), \tag{8.5}$$

leading to the simplified factorisation

$$f_{X,Y,Z}(x, y, z) = f_{X|Z}(x|z) f_{Y|Z}(y|z) f_Z(z).$$

This change means that Y is no longer a parent of X, so we remove the directed edge $Y \to X$ from DAG (8.2). The resulting structure is

(8.6)

This DAG encodes the assumption that X and Y are *conditionally independent* given Z, written $X \perp\!\!\!\perp Y|Z$. That is, the joint distribution of X, Y and Z factorises as though all statistical dependence between X and Y is attributable to their individual relationships with Z. As a practical example of conditional independence, suppose that Hilda and Godiva share a flat. They always leave for work at the same time, but work in different offices which they walk to via different routes. If X and Y are the times Hilda and Godiva (respectively) arrive at work and Z is the time that they leave the flat, then we would not expect X and Y to be independent, but we might reasonably expect them to be *conditionally* independent *given* Z.

■ ***Exercise*** 8.2 Let W, X, Y and Z be random variables. Suppose that their joint density admits the factorisation

$$f_{W,X,Y,Z}(w, x, y, z) = f_{W|Y}(w|y) f_{X|Y}(x|y) f_{Y|Z}(y|z) f_Z(z).$$

Draw the corresponding DAG.

■ ***Exercise*** 8.3 Suppose that X, Y and Z are mutually independent, so that $f_{X,Y,Z}(x, y, z) = f_X(x) f_Y(y) f_Z(z)$. Draw the corresponding DAG.

Given a factorisation of a multivariate density into univariate conditional densities, there is a unique DAG representation. The reverse is also true: for each DAG, there corresponds a unique factorisation. Suppose our DAG has nodes $X_1, \ldots, X_n$, and let $\boldsymbol{PA}_k$ be the set of parent nodes of X_k. To find the factorisation, for every node X_k we include a factor $f_{X_k|\boldsymbol{PA}_k}(x_k|\boldsymbol{pa}_k)$, giving the overall factorisation

$$f_{X_1,\ldots,X_n}(x_1, \ldots, x_n) = \prod_{k=1}^{n} f_{X_k|\boldsymbol{PA}_k}(x_k|\boldsymbol{pa}_k).$$

For example, the DAG

X_2

X_1 X_4

X_3

(8.7)

corresponds to the factorisation

$$f_{X_1,X_2,X_3,X_4}(x_1, x_2, x_3, x_4) = f_{X_1}(x_1) f_{X_2|X_1}(x_2|x_1) f_{X_3|X_1,X_2}(x_3|x_1, x_2) f_{X_4|X_3}(x_4|x_3).$$

■ ***Exercise*** 8.4 Let W, X, Y and Z be random variables. Write down the factorisation of their multivariate density which corresponds to the DAG

(8.8)

8.2 Conditional Independence

A DAG in which every pair of nodes is connected is said to be *fully connected*. We can think of a fully connected DAG as a model for a completely arbitrary multivariate distribution. In this sense, a fully connected DAG tells us nothing about the probability distribution it represents. In Section 8.1, we discovered, informally, that removing an edge produced a model in which two variables were conditionally independent. The more edges we remove from a DAG, the more constraints we place on the dependencies between variables. An important question is therefore: given a DAG, how do we determine all of the conditional independence relationships that its structure implies? We start by formally defining conditional independence and its connection to independence in Box 8.1.

★ Box 8.1 Conditional independence

Two continuous random variables X and Y are *marginally independent*, written $X \perp\!\!\!\perp Y$, if their joint density factorises into the product of their univariate marginals

$$f_{X,Y}(x, y) = f_X(x) f_Y(y).$$

Often, marginal independence is simply referred to as *independence*. Conditional independence is defined using a similar factorisation requirement, except that each of the densities is conditional on one or more additional variables. We say that X and Y are *conditionally independent* given Z, written $X \perp\!\!\!\perp Y|Z$, if

$$f_{X,Y|Z}(x, y|z) = f_{X|Z}(x|z) f_{Y|Z}(y|z). \tag{8.9}$$

More generally, if $\boldsymbol{A}$, $\boldsymbol{B}$ and $\boldsymbol{C}$ are disjoint sets of random variables – we may think of them as random vectors – then $\boldsymbol{A} \perp\!\!\!\perp \boldsymbol{B}|\boldsymbol{C}$ if

$$f_{\boldsymbol{A},\boldsymbol{B}|\boldsymbol{C}}(\boldsymbol{a}, \boldsymbol{b}|\boldsymbol{c}) = f_{\boldsymbol{A}|\boldsymbol{C}}(\boldsymbol{a}|\boldsymbol{c}) f_{\boldsymbol{B}|\boldsymbol{C}}(\boldsymbol{b}|\boldsymbol{c}).$$

Independence is a special case of conditional independence in the following sense:

$$X \perp\!\!\!\perp Y|\emptyset \equiv X \perp\!\!\!\perp Y,$$

where $\emptyset = \{\}$ is the empty set.

Exercise 8.5 invites you to demonstrate that the relationship $f_{X|Y,Z}(x|y, z) = f_{X|Z}(x|z)$, from which we informally derived the conditional independence relation $X \perp\!\!\!\perp Y|Z$, is equivalent to equation (8.9) in Box 8.1.

■ ***Exercise*** 8.5 The aim of this exercise is to establish formally that $f_{X|Y,Z}(x|y, z) = f_{X|Z}(x|z)$ implies $X \perp\!\!\!\perp Y|Z$, and vice versa.

(a) Show that

$$f_{X|Y,Z}(x|y, z) = \frac{f_{X,Y|Z}(x, y|z)}{f_{Y|Z}(y|z)}.$$

Hint: you may find it useful to start from the definition of the conditional probability density, $f_{X|Y,Z}(x|y, z) = f_{X,Y,Z}(x, y, z)/f_{Y,Z}(y, z)$.

(b) Use your previous result to show that $f_{X|Y,Z}(x|y,z) = f_{X|Z}(x|z)$ implies that

$$f_{X,Y|Z}(x,y|z) = f_{X|Z}(x|z) f_{Y|Z}(y|z).$$

(c) Show that the converse implication also holds.

Given a DAG we can systematically discover conditional independence relationships using a concept called *d-separation*. To motivate this concept, we consider three simple graphs, each with three nodes. The first is the *pipe*,

$$X \rightarrow Y \rightarrow Z, \tag{8.10}$$

which corresponds to the factorisation

$$f_{X,Y,Z}(x,y,z) = f_X(x) f_{Y|X}(y|x) f_{Z|Y}(z|y).$$

Using this factorisation, we can write the joint distribution of X and Z conditional on Y as

$$f_{X,Z|Y}(x,z|y) = \frac{f_{X,Y,Z}(x,y,z)}{f_Y(y)} = \frac{f_{X,Y}(x,y) f_{Z|Y}(z|y)}{f_Y(y)} = f_{X|Y}(x|y) f_{Z|Y}(z|y).$$

Therefore $X \perp\!\!\!\perp Z|Y$. As practical example of the pipe, suppose that Hrothgar sings a single musical note to Hilda, out of earshot of Godiva. Some time later, Hilda attempts to sing the same note to Godiva, who attempts to copy it. If X, Y and Z are the pitches of the notes sung by Hrothgar, Hilda and Godiva, then we might reasonably expect that $X \perp\!\!\!\perp Z|Y$.

Now consider the *fork*,

$$X \leftarrow Y \rightarrow Z. \tag{8.11}$$

This is equivalent to the first non-fully connected DAG we considered (8.6), but with variables Y and Z exchanged. As in the case of the pipe, we have $X \perp\!\!\!\perp Z|Y$: conditioning on the middle node Y renders X and Z independent. In both the pipe and the fork, Y is said to *block* the path between X and Z. Informally, conditioning on Y breaks any statistical dependence that might otherwise be present between X and Z due to their mutual connection with Y.

Finally, consider the *collider*,

$$X \rightarrow Y \leftarrow Z, \tag{8.12}$$

which corresponds to the factorisation

$$f_{X,Y,Z}(x,y,z) = f_X(x) f_Z(z) f_{Y|X,Z}(y|x,z). \tag{8.13}$$

From this we can compute the marginal bivariate density of X and Z:

$$f_{X,Z}(x,z) = \int f_{X,Y,Z}(x,y,z)dy = f_X(x) f_Z(z) \int f_{Y|X,Z}(y|x,z)dy = f_X(x) f_Z(z),$$

which shows that X and Z are marginally independent ($X \perp\!\!\!\perp Z$). Now consider their joint distribution conditional on Y:

$$f_{X,Z|Y}(x,z|y) = \frac{f_{X,Y,Z}(x,y,z)}{f_Y(y)} = \frac{f_X(x) f_Z(z) f_{Y|X,Z}(y|x,z)}{f_Y(y)}.$$

Since this expression cannot in general be written as $f_{X|Y}(x|y)f_{Z|Y}(z|y)$, the generic collider case is that X and Z are conditionally dependent given Y. (We phrase this cautiously because the factorisation in (8.13) does not *rule out* the possibility that Y is independent of X (in which case $f_{Y|X,Z} = f_{Y|Z}$) or of Z (in which case $f_{Y|X,Z} = f_{Y|X}$), or of both (in which case $f_{Y|X,Z} = f_Y$). Usually, though, when people talk about collider structures they take for granted that the middle node depends non-trivially on the other two.) Notice the contrast with the pipe and the fork structures: in the collider, the middle node Y does *not* transmit statistical dependence between X and Z when left unconditioned, but conditioning on Y introduces dependence. We say that Y blocks the path between X and Z when unconditioned, but opens the path when conditioned on. Here is an example. Suppose that the sales revenue, Y, of a musical artist is the sum of their talent, X, their beauty, Z, and luck, $\mathcal{E}$, so

$$Y = X + Z + \mathcal{E}.$$

Suppose also that beauty, talent and luck are bestowed upon musicians independently. From the above relationship, we see that for a given size of revenue, a more beautiful artist will tend to be less talented and vice versa. Beauty and talent are therefore dependent (and negatively correlated) given revenue.

So far we have considered isolated pipes, forks and colliders. However, these structures can also occur as parts (subgraphs) of larger DAGs – and then the story is a bit more complicated. When a pipe or a fork is part of a larger DAG, conditioning on its middle node might *not* remove all dependence between the end nodes, because there could be other unblocked paths connecting those nodes. When a collider is part of a larger DAG, its middle node Y may be the parent of one or more nodes, which may be the parents of further nodes and so on, forming a set of *descendants* of Y, written $\boldsymbol{DESC}(Y)$. It can be shown that colliders become unblocked if Y *or any of its descendants* are conditioned on. The concept of d-separation is a generalisation of these examples which provides a method for establishing conditional independence relationships in any DAG. It is described in Box 8.2.

★ Box 8.2 d-separation and conditional independence

Suppose that $\boldsymbol{A}$, $\boldsymbol{B}$ and $\boldsymbol{C}$ are three disjoint sets of nodes in a DAG. An undirected path, P, from a node in $\boldsymbol{A}$ to a node in $\boldsymbol{B}$ is said to be blocked by $\boldsymbol{C}$ if *at least one* of the following conditions are satisfied:

(a) P contains a pipe $X \to Y \to Z$ or $X \leftarrow Y \leftarrow Z$, where $Y \in \boldsymbol{C}$.
(b) P contains a fork $X \leftarrow Y \to Z$, where $Y \in \boldsymbol{C}$.
(c) P contains a collider $X \to Y \leftarrow Z$, where $Y \notin \boldsymbol{C}$ and $\boldsymbol{DESC}(Y) \cap \boldsymbol{C} = \emptyset$.

If all paths from $\boldsymbol{A}$ to $\boldsymbol{B}$ are blocked by $\boldsymbol{C}$, then $\boldsymbol{A}$ and $\boldsymbol{B}$ are said to be d-separated by $\boldsymbol{C}$; in a distribution that factorises according to the DAG, this implies $\boldsymbol{A} \perp\!\!\!\perp \boldsymbol{B} | \boldsymbol{C}$.

We will illustrate d-separation using the following DAG:

(8.14)

As a first example, let us demonstrate that $Y \perp\!\!\!\perp V|\{W, X\}$. There are two undirected paths between Y and V. First, we have

$$Y \rightarrow \underbrace{Z \leftarrow X \leftarrow V}_{\text{pipe}}.$$

The path contains a pipe whose central node is in the conditioning set $\{W, X\}$, so it is blocked. Notice that the path also contains a collider $Y \rightarrow Z \leftarrow X$, and since Z is not in $\{W, X\}$ and has no descendants, this would also be sufficient to demonstrate that the path is blocked. The second undirected path is

$$\underbrace{Y \leftarrow W \rightarrow X}_{\text{fork}} \leftarrow V.$$

This path contains a fork whose central node is in the conditioning set, so it too is blocked. Notice that there is also a collider $W \rightarrow X \leftarrow V$ whose central node *is* in the conditioning set, but this does not 'unblock' the path. For a path to be blocked it is sufficient for *any segment of the path* to be blocked. Since both paths between Y and V are blocked by $\{W, X\}$, Y and V are d-separated by $\{W, X\}$, and it follows that $Y \perp\!\!\!\perp V|\{W, X\}$.

As a second example, we have $V \perp\!\!\!\perp W|\emptyset$, since $V \rightarrow X \leftarrow W$ and $X \rightarrow Z \leftarrow Y$ are colliders, and neither X nor Z nor any descendent of X or Z is an element of $\emptyset$. Exercise 8.6 invites you to establish some further conditional independence relations for this DAG.

■ ***Exercise*** 8.6 With reference to DAG (8.14), use d-separation to establish the following conditional independence relations, explaining your reasoning in each case.

(a) $X \perp\!\!\!\perp Y|W$.
(b) $W \perp\!\!\!\perp Z|\{X, Y\}$.
(c) $V \perp\!\!\!\perp Y|\emptyset$.

8.3 Chapter Summary

In this chapter, we have introduced the DAG representation of a multivariate probability distribution. DAGs are useful tools for understanding the structure of probabilistic models and the dependencies between their variables. We will make considerable use of them in Chapters 9, 10, 12 and 14.

- Every DAG corresponds to a particular **factorisation** of a joint mass or density function into a product of conditional mass or density functions.
- The DAG representation of a given distribution implies **constraints** on the dependencies between variables. The DAG does not fully specify the distribution.
- Conditional independence relationships between variables can be inferred from DAGs using the concept of **d-separation**.

8.4 Further Exercises

■ ***Exercise*** 8.7 Let W, X, Y and Z be random variables. Write down the factorisation of their multivariate density which corresponds to the DAG

$$W \rightarrow X \rightarrow Y \leftarrow Z. \tag{8.15}$$

■ ***Exercise*** 8.8 Two sets of random variables $\{X_1, \ldots, X_n\}$ and $\{Y_1, \ldots, Y_n\}$ are *independent*, written $\{X_1, \ldots, X_n\} \perp\!\!\!\perp \{Y_1, \ldots, Y_n\}$, if and only if their joint density factorises as follows:

$$f_{X_1,\ldots,X_n,Y_1,\ldots,Y_n}(x_1, \ldots, x_n, y_1, \ldots, y_n) = f_{X_1,\ldots,X_n}(x_1, \ldots, x_n) f_{Y_1,\ldots,Y_n}(y_1, \ldots, y_n).$$

Equivalently, the random vectors $\boldsymbol{X} = (X_1, \ldots, X_n)$ and $\boldsymbol{Y} = (Y_1, \ldots, Y_n)$ are independent ($\boldsymbol{X} \perp\!\!\!\perp \boldsymbol{Y}$) if and only if $f_{\boldsymbol{X},\boldsymbol{Y}}(\boldsymbol{x}, \boldsymbol{y}) = f_{\boldsymbol{X}}(\boldsymbol{x}) f_{\boldsymbol{Y}}(\boldsymbol{y})$.

(a) Suppose that $X \perp\!\!\!\perp \{Y, Z\}$. Write down the corresponding factorisation of $f_{X,Y,Z}(x, y, z)$.
(b) Provide an expression for the marginal distribution of Y as an integral over $f_{Y,Z}(y, z)$.
(c) The joint density of $\{X, Y\}$ may be calculated from the joint density of $\{X, Y, Z\}$ as

$$f_{X,Y}(x, y) = \int f_{X,Y,Z}(x, y, z) dz.$$

Show that $X \perp\!\!\!\perp \{Y, Z\}$ implies that $f_{X,Y}(x, y) = f_X(x) f_Y(y)$.
(d) Establish the *decomposition property*

$$X \perp\!\!\!\perp \{Y, Z\} \implies (X \perp\!\!\!\perp Y) \wedge (X \perp\!\!\!\perp Z),$$

where $\wedge$ is the logical AND symbol.
(e) Show that the converse of the result in (d) is not true (i.e. that the right-hand side does not imply the left-hand side) by considering the following case: Y and Z are i.i.d. random variables with common distribution Uniform(0, 1), and $X = (Y + Z)$ mod 1.

■ ***Exercise*** 8.9 In this exercise, we assume that $X \perp\!\!\!\perp Y | Z$ and that $X \perp\!\!\!\perp Z$.

(a) Factorise $f_{X,Y|Z}(x, y|z)$ based on the fact that $X \perp\!\!\!\perp Y|Z$.
(b) Show that $f_{X|Y,Z}(x|y, z) = f_X(x)$.
(c) Write down the independence relationship implied by $f_{X|Y,Z}(x|y, z) = f_X(x)$.

■ ***Exercise*** 8.10 Harold's Hammer Company uses steel and wood to make its products. The market prices of wood and steel fluctuate independently. The price of Harold's hammers depends on these two prices and on some other random factors. Let S, W and H be random variables giving the prices of steel, wood and hammers, respectively. Draw a DAG representing their relationship, and also provide the corresponding factorisation of their joint density. State whether each of the following statements are true of false.

(a) $S \perp\!\!\!\perp W$
(b) $S \perp\!\!\!\perp W | \emptyset$
(c) $S \perp\!\!\!\perp W | H$

■ ***Exercise*** 8.11 Given the DAG

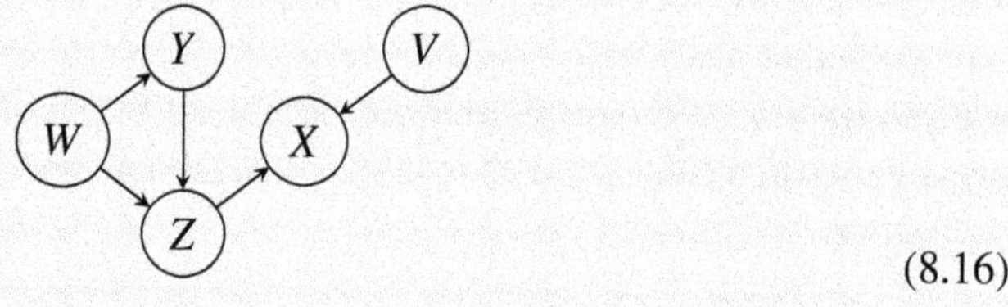

(8.16)

use d-separation to establish the following conditional independence relations, explaining your reasoning in each case.

(a) $V \perp\!\!\!\perp Y | Z$.
(b) $V \perp\!\!\!\perp Y | \emptyset$.

Explain why you cannot deduce $V \perp\!\!\!\perp Y | X$ from the DAG.

■ ***Exercise*** 8.12 Given the DAG

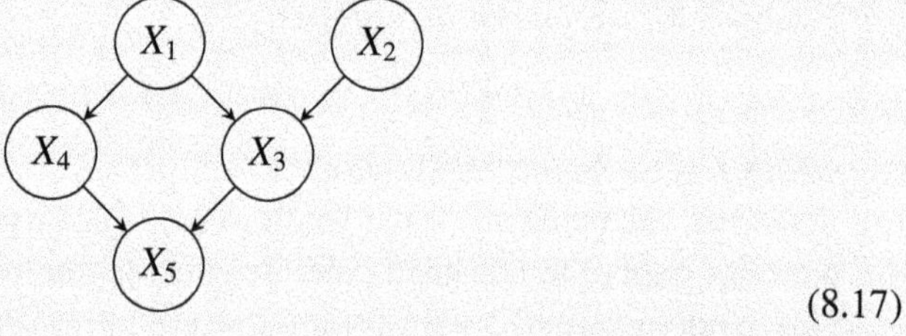

(8.17)

use d-separation to establish the following conditional independence relations, explaining your reasoning in each case.

(a) $X_1 \perp\!\!\!\perp X_2$
(b) $X_2 \perp\!\!\!\perp \{X_1, X_4\}$
(c) $X_3 \perp\!\!\!\perp X_4 | \{X_1, X_2\}$
(d) $X_5 \perp\!\!\!\perp \{X_1, X_2\} | \{X_3, X_4\}$
(e) $\{X_4, X_5\} \perp\!\!\!\perp X_2 | \{X_1, X_3\}$

9

Bayesian Linear Regression, Priors, and Regularisation

In Chapter 7, we introduced simple and multiple linear regression from a frequentist perspective, and we explored how varying the number of predictors in a model affected its predictive power. We also introduced the bias-variance trade-off as a universal way to understand how model complexity affects performance. In this chapter, we generalise linear regression to the Bayesian setting, requiring us to specify *priors* on model parameters. We show that by using *shrinkage priors* we can control the complexity of a model, and reduce overfitting, using a single continuous parameter. This is one example of a very broad class of methods called *regularisation*. The central idea of regularisation is to reduce overfitting by penalising model complexity. In the Bayesian setting, the complexity penalty can be seen as arising from a prior. However, complexity penalties are often added directly to loss functions – to the RSS or to negative log-likelihood, for example – rather than being formally incorporated in a probabilistic model. Shrinkage priors will appear again in Chapter 10, regularised loss functions will be used when we introduce classification in Chapter 11, and we will see examples of *implicit regularisation* when we come to study neural networks in Chapter 13.

9.1 Bayesian Linear Regression

To define the simple and multiple Bayesian linear regression models, we generalise the frequentist models introduced in Chapter 7 by promoting model parameters to be latent random variables. We then compute the posterior distribution of these variables given a dataset. This posterior is the Bayesian counterpart of the bootstrap sampling distributions we obtained in Chapter 7.

9.1.1 Simple Bayesian Linear Regression

Consider the simple linear regression problem where X is a predictor variable and Y a response. In the frequentist approach, we assume a linear relationship of the form

$$Y = \beta_0 + \beta_1 X + \mathcal{E}, \tag{9.1}$$
$$\mathcal{E} \sim \mathcal{N}(0, \sigma^2), \tag{9.2}$$

where $\mathcal{E}$ and X are independent and the parameter vector $\boldsymbol{\theta} = (\beta_0, \beta_1, \sigma)$ is assumed to be a fixed but unknown quantity. In the Bayesian approach, variables and parameters are treated on an equal footing, so the parameters form a random vector

$$\boldsymbol{\Theta} = (B_0, B_1, \Sigma).$$

According to this view, our model contains six random variables, X, Y, B_0, B_1, Σ and $\mathcal{E}$, where Y is defined in terms of the others as follows:

$$Y = B_0 + B_1 X + \mathcal{E}. \tag{9.3}$$

We can think of equation (9.1) as specifying the relationship between Y, X and $\mathcal{E}$ conditional on the event $\boldsymbol{\Theta} = (\beta_0, \beta_1, \sigma)$. We will find it useful to separate the coefficients B_0 and B_1 into their own vector $\boldsymbol{B} = (B_0, B_1)$. According to definition (9.3), the density of Y conditional on $\boldsymbol{B}$, Σ and X is

$$f_{Y|\boldsymbol{B},\Sigma,X}(y|\boldsymbol{\beta}, \sigma, x) = \mathcal{N}(y|\beta_0 + \beta_1 x, \sigma^2),$$

where $\mathcal{N}(y|\mu, \sigma^2)$ is the normal probability density function with mean μ and variance σ^2.

Our goal is to combine priors on $\boldsymbol{B}$ and Σ with a dataset of predictor–response observations $D = \{(x_i, y_i)\}_{i=1}^n$ to infer posterior parameter distributions. To achieve this we begin by factorising the joint distribution of our model variables as follows:

$$f_{Y,\boldsymbol{B},\Sigma,X}(y, \boldsymbol{\beta}, \sigma, x) = f_{Y|\boldsymbol{B},\Sigma,X}(y|\boldsymbol{\beta}, \sigma, x) f_{\boldsymbol{B},\Sigma|X}(\boldsymbol{\beta}, \sigma|x) f_X(x).$$

Here $f_{\boldsymbol{B},\Sigma|X}(\boldsymbol{\beta}, \sigma|x)$ is the prior parameter distribution given the predictor X. Let us assume that $\boldsymbol{B}$ and Σ are independent of X and of each other, so

$$f_{\boldsymbol{B},\Sigma|X}(\boldsymbol{\beta}, \sigma|x) = f_{\boldsymbol{B}}(\boldsymbol{\beta}) f_\Sigma(\sigma).$$

Hence,

$$f_{Y,\boldsymbol{B},\Sigma,X}(y, \boldsymbol{\beta}, \sigma, x) = f_{Y|\boldsymbol{B},\Sigma,X}(y|\boldsymbol{\beta}, \sigma, x) f_{\boldsymbol{B}}(\boldsymbol{\beta}) f_\Sigma(\sigma) f_X(x).$$

This factorisation may be represented by the DAG

B

X → Y ← Σ. (9.4)

Here we have refined our DAG notation by shading the nodes of variables that can be observed to distinguish them from *latent* variables – the model parameters – which cannot be observed.

In the context of regression, the predictor variable X will be observed (or otherwise fixed) both when collecting training data and when making predictions. The conditional density of the remaining variables will be

$$f_{Y,\boldsymbol{B},\Sigma|X}(y, \boldsymbol{\beta}, \sigma|x) = f_{Y|\boldsymbol{B},\Sigma,X}(y|\boldsymbol{\beta}, \sigma, x) f_{\boldsymbol{B}}(\boldsymbol{\beta}) f_\Sigma(\sigma).$$

To represent this factorisation graphically, we introduce a *constant node*, x, which is not circled. Such nodes represent *known* quantities. In the current context, these can be observations of random variables, or additional known parameters used to define the probability distributions of the latent or observable variables. The new DAG is

(9.5)

DAG (9.5) represents a conditional probabilistic model of the parameters and response variable given the predictor. Since the predictor and response variables are scalars, we can think of the sequences of predictor and response observations as random vectors $\underline{X} = (X_1, \ldots, X_n)$ and $\underline{Y} = (Y_1, \ldots, Y_n)$. Assuming that successive realisations of the noise variable are independent, the response variables will be conditionally independent given the predictors. Therefore,

$$\begin{aligned} f_{\underline{Y},\boldsymbol{B},\Sigma|\underline{X}}(\underline{y}, \boldsymbol{\beta}, \sigma|\underline{x}) &= f_{\boldsymbol{B}}(\boldsymbol{\beta}) f_{\Sigma}(\sigma) \prod_{i=1}^{n} f_{Y|\boldsymbol{B},\Sigma,X}(y_i|\boldsymbol{\beta}, \sigma, x_i) \\ &= \underbrace{f_{\boldsymbol{B}}(\boldsymbol{\beta}) f_{\Sigma}(\sigma)}_{\text{prior}} \underbrace{\mathcal{L}(\boldsymbol{\beta}, \sigma)}_{\text{likelihood}} . \end{aligned} \tag{9.6}$$

Here we have assumed that the model parameters are sampled *once* and the same parameters are used to generate all the data, so the prior densities only appear once in the factorisation. There are $n + 3$ random variables in joint density (9.6), making it infeasible to draw the corresponding DAG unless n is small. For example, the DAG representation when $n = 3$ is

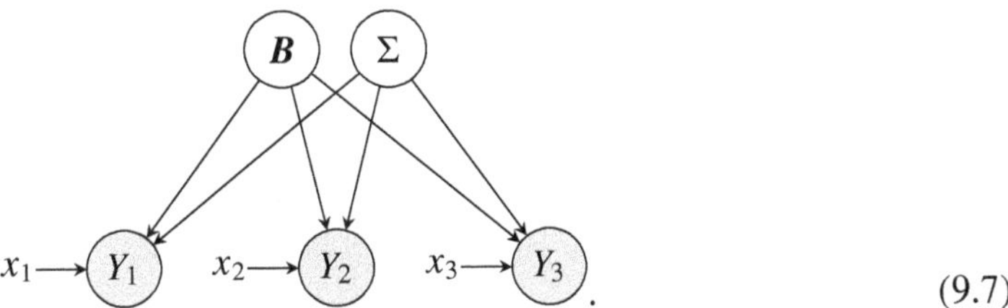

(9.7)

In order to represent the joint distribution of the parameters and the dataset for arbitrary n, we introduce a further notational refinement. Variables that lie inside a *plate* or box in a DAG are repeated n times. Using plate notation, the DAG for density (9.6) is

(9.8)

We are now able to derive an expression for the posterior parameter distribution:

$$f_{\boldsymbol{B},\Sigma|\underline{Y},\underline{X}}(\boldsymbol{\beta}, \sigma|\underline{y}, \underline{x}) = \frac{f_{\underline{Y},\boldsymbol{B},\Sigma|\underline{X}}(\underline{y}, \boldsymbol{\beta}, \sigma|\underline{x})}{f_{\underline{Y}|\underline{X}}(\underline{y}|\underline{x})} \propto \mathcal{L}(\boldsymbol{\beta}, \sigma) f_{\boldsymbol{B}}(\boldsymbol{\beta}) f_{\Sigma}(\sigma).$$

Here $f_{\underline{Y}|\underline{X}}(\underline{y}|\underline{x})$ is the marginal distribution of the responses given the predictors after averaging over the priors, which does not depend on the parameters. The above expression for the posterior fits the same pattern we saw in Chapter 6:

$$\text{Posterior} \propto \text{likelihood} \times \text{prior}.$$

Before turning to concrete examples, we extend this model to the case of multiple predictors.

9.1.2 Generalising to Multiple Predictors

If we have $p \geq 1$ predictor variables, our Bayesian linear model becomes

$$Y = B_0 + \sum_{k=1}^{p} B_k X_k + \mathcal{E},$$
$$\mathcal{E} \sim \mathcal{N}(0, \Sigma^2).$$

To write our model in compact form, we introduce an additional fictitious predictor X_0 which is always equal to 1. With this extra predictor included, our predictor vector is

$$\boldsymbol{X} = (1, X_1, \ldots, X_p)^T.$$

Here we have defined $\boldsymbol{X}$ as the transpose of a row vector meaning that it is a column vector. We may view a vector as a special case of a matrix which has either multiple rows and one column (a column vector – the default form) or multiple columns and one row (a row vector). Using this notation, the scalar ('dot') product of two vectors $\boldsymbol{u}$ and $\boldsymbol{v}$ of the same dimension may be written as a matrix product (see Appendix B),

$$\boldsymbol{u}^T \boldsymbol{v} = \sum_{i=1}^{n} u_i v_i.$$

Writing the model coefficients as a column vector,

$$\boldsymbol{B} = (B_0, B_1, \ldots, B_p)^T,$$

allows us to write our model in the compact form,

$$Y = \boldsymbol{X}^T \boldsymbol{B} + \mathcal{E}.$$

According to this model, the conditional distribution of Y is normal,

$$f_{Y|\boldsymbol{B},\Sigma,\boldsymbol{X}}(y|\boldsymbol{\beta}, \sigma, \boldsymbol{x}) = \mathcal{N}(y|\boldsymbol{x}^T\boldsymbol{\beta}, \sigma^2),$$

or equivalently,

$$Y|\{\boldsymbol{B} = \boldsymbol{\beta}, \Sigma = \sigma, \boldsymbol{X} = \boldsymbol{x}\} \sim \mathcal{N}(\boldsymbol{x}^T\boldsymbol{\beta}, \sigma^2).$$

To factorise the conditional density of the response and the parameters given the predictor vector, we follow the same steps as in the one-predictor case, but with the predictor replaced with a vector, yielding

$$f_{Y,\boldsymbol{B},\Sigma|\boldsymbol{X}}(y, \boldsymbol{\beta}, \sigma|\boldsymbol{x}) = f_{Y|\boldsymbol{B},\Sigma,\boldsymbol{X}}(y|\boldsymbol{\beta}, \sigma, \boldsymbol{x}) f_{\boldsymbol{B}}(\boldsymbol{\beta}) f_{\Sigma}(\sigma).$$

For multiple linear regression, the dataset is an observation of a sequence of predictor vectors $\boldsymbol{X}_1, \ldots, \boldsymbol{X}_n$ and responses $Y_1, \ldots, Y_n$. Under the assumption of i.i.d. realisations of the noise, the joint density of the responses and parameters given the predictors may be represented by the DAG

(9.9)

When each predictor is a vector, the set of all predictors in our dataset forms a matrix known as the *design matrix*:

$$\underline{\boldsymbol{X}} = (\boldsymbol{X}_1, \ldots, \boldsymbol{X}_n)^T.$$

The design matrix is an $n \times (p+1)$ matrix of random variables, each row of which is a single predictor vector. The element in row i, column j of the design matrix,

$$\left[\underline{\boldsymbol{X}}\right]_{ij} = X_{ij},$$

is the jth predictor for the ith example in the dataset (i.e. it is the jth component of predictor vector $\boldsymbol{X}_i$). Similarly, the set of all responses in our dataset forms a vector

$$\underline{Y} = (Y_1, \ldots, Y_n)^T.$$

Our dataset can be thought of as a single observation of the design matrix,

$$\underline{\boldsymbol{x}} = (\boldsymbol{x}_1, \ldots, \boldsymbol{x}_n)^T,$$

along with a single observation of the responses vector,

$$\underline{y} = (y_1, \ldots, y_n)^T.$$

The joint density of the observations and the parameters given the design matrix corresponding to DAG (9.9) is then

$$\begin{aligned} f_{\underline{Y},\boldsymbol{B},\Sigma|\underline{\boldsymbol{X}}}(\underline{y}, \boldsymbol{\beta}, \sigma|\underline{\boldsymbol{x}}) &= f_{\boldsymbol{B}}(\boldsymbol{\beta}) f_{\Sigma}(\sigma) \prod_{i=1}^{n} f_{Y|\boldsymbol{B},\Sigma,\boldsymbol{X}}(y_i|\boldsymbol{\beta}, \sigma, \boldsymbol{x}_i) \\ &= \underbrace{f_{\boldsymbol{B}}(\boldsymbol{\beta}) f_{\Sigma}(\sigma)}_{\text{prior}} \underbrace{\mathcal{L}(\boldsymbol{\beta}, \sigma)}_{\text{likelihood}}. \end{aligned}$$

Finally, we have the posterior

$$f_{\boldsymbol{B},\Sigma|\underline{Y},\underline{\boldsymbol{X}}}(\boldsymbol{\beta}, \sigma|\underline{y}, \underline{\boldsymbol{x}}) = \frac{f_{\underline{Y},\boldsymbol{B},\Sigma|\underline{\boldsymbol{X}}}(\underline{y}, \boldsymbol{\beta}, \sigma|\underline{\boldsymbol{x}})}{f_{\underline{Y}|\underline{\boldsymbol{X}}}(\underline{y}|\underline{\boldsymbol{x}})} \propto \mathcal{L}(\boldsymbol{\beta}, \sigma) f_{\boldsymbol{B}}(\boldsymbol{\beta}) f_{\Sigma}(\sigma).$$

Since the multiple linear model includes simple linear regression as a special case, we focus on the analysis of this form of the posterior in what follows.

9.1.3 Computing the Posterior

To obtain the posterior, we need to compute the likelihood and select our priors. Let us consider the likelihood first. We have

$$\mathcal{L}(\boldsymbol{\beta}, \sigma) = \prod_{i=1}^{n} \mathcal{N}(y_i | \boldsymbol{x}_i^T \boldsymbol{\beta}, \sigma^2)$$
$$= \frac{1}{\sigma^n (2\pi)^{n/2}} \exp\left(-\frac{1}{2\sigma^2}(\underline{y} - \underline{\boldsymbol{x}}\boldsymbol{\beta})^T(\underline{y} - \underline{\boldsymbol{x}}\boldsymbol{\beta})\right). \quad (9.10)$$

Although conceptually a likelihood function is not a probability density function of the model parameters, our calculations will proceed more smoothly if we can manipulate it into a form in which probability density functions of the parameters appear as factors. We begin by writing the product of univariate normal densities on the right-hand side of (9.10) as a single multivariate normal distribution, the general form of which is defined in Box 9.1.

★ Box 9.1 The multivariate normal distribution

Let $\mathbf{Z} = (Z_1, \ldots, Z_n)$ be a random vector with probability density function

$$f_Z(z; \boldsymbol{\mu}, \underline{\Sigma}) = \frac{1}{(2\pi)^{n/2} \det(\underline{\Sigma})^{1/2}} \exp\left(-\frac{1}{2}(z - \boldsymbol{\mu})^T \underline{\Sigma}^{-1}(z - \boldsymbol{\mu})\right),$$

where $\boldsymbol{\mu} = (\mu_1, \ldots, \mu_n)$ is an n-dimensional vector and $\underline{\Sigma}$ is a symmetric, positive definite $n \times n$ matrix having (matrix) inverse $\underline{\Sigma}^{-1}$. The random variables $Z_1, \ldots, Z_n$ are said to follow the *multivariate normal* distribution with means $\mu_1, \ldots, \mu_n$ and covariances

$$\mathrm{Cov}(Z_i, Z_j) = \mathbb{E}((Z_i - \mu_i)(Z_j - \mu_j)) = \left[\underline{\Sigma}\right]_{ij}.$$

We write $\mathbf{Z} \sim \mathcal{N}(\boldsymbol{\mu}, \underline{\Sigma})$. A shorthand for the multivariate normal density is $\mathcal{N}(z|\boldsymbol{\mu}, \underline{\Sigma})$. Do not confuse the covariance matrix $\underline{\Sigma}$ with the random variable Σ introduced in Section 9.1.1.

Using the definition in Box 9.1, we see that the right-hand side of equation (9.10) is equal to the density $\mathcal{N}(\underline{y}|\underline{\boldsymbol{x}}\boldsymbol{\beta}, \sigma^2\underline{\boldsymbol{I}})$, where $\underline{\boldsymbol{I}}$ is the identity matrix. Our aim now is to rearrange the likelihood so that a multivariate normal density for $\boldsymbol{B}$ appears as a factor. We do this using a multivariate version of 'completing the square', described in Box 9.2. If you wish to avoid the details of the calculations you may skip this box and also Exercise 9.1.

★ Box 9.2 Multivariate completing the square

Let z be a real-valued variable, and S and b real constants. Applying the identity

$$Sz^2 - 2bz = S\left(z - \frac{b}{S}\right)^2 - \frac{b^2}{S}$$

is known as 'completing the square'. We can generalise this identity to the case of vector variables as follows. Let z and $\boldsymbol{b}$ be n-dimensional vectors and $\underline{\boldsymbol{S}}$ a symmetric invertible matrix. We then have

$$z^T \underline{S} z - 2b^T z = (z - \underline{S}^{-1} b)^T \underline{S} (z - \underline{S}^{-1} b) - b^T \underline{S}^{-1} b.$$

The identity can be verified by expanding the quadratic form on the right-hand side.

By expanding the scalar product in equation (9.10) and completing the square, we obtain

$$(\underline{y} - \underline{x}\boldsymbol{\beta})^T (\underline{y} - \underline{x}\boldsymbol{\beta}) = (\boldsymbol{\beta} - \hat{\boldsymbol{\beta}})^T \underline{x}^T \underline{x} (\boldsymbol{\beta} - \hat{\boldsymbol{\beta}}) + (\underline{y} - \underline{x}\hat{\boldsymbol{\beta}})^T (\underline{y} - \underline{x}\hat{\boldsymbol{\beta}}), \tag{9.11}$$

where

$$\hat{\boldsymbol{\beta}} = (\underline{x}^T \underline{x})^{-1} \underline{x}^T \underline{y}. \tag{9.12}$$

■ ***Exercise* 9.1** This exercise invites you to derive the identity (9.11), where $\hat{\boldsymbol{\beta}}$ is given by equation (9.12).

(a) Show that $\underline{x}^T \underline{x} \hat{\boldsymbol{\beta}} = \underline{x}^T \underline{y}$.

(b) By expanding the product $(\underline{y} - \underline{x}\boldsymbol{\beta})^T (\underline{y} - \underline{x}\boldsymbol{\beta})$ and completing the square, show that

$$(\underline{y} - \underline{x}\boldsymbol{\beta})^T (\underline{y} - \underline{x}\boldsymbol{\beta}) = (\boldsymbol{\beta} - \hat{\boldsymbol{\beta}})^T \underline{x}^T \underline{x} (\boldsymbol{\beta} - \hat{\boldsymbol{\beta}}) + \underline{y}^T \underline{y} - \hat{\boldsymbol{\beta}}^T \underline{x}^T \underline{x} \hat{\boldsymbol{\beta}}.$$

(c) Show that $\underline{y}^T \underline{y} - \hat{\boldsymbol{\beta}}^T \underline{x}^T \underline{x} \hat{\boldsymbol{\beta}} = (\underline{y} - \underline{x}\hat{\boldsymbol{\beta}})^T (\underline{y} - \underline{x}\hat{\boldsymbol{\beta}})$ and hence establish identity (9.11).

We use hat notation in definition (9.12) because $\hat{\boldsymbol{\beta}}$ is the maximum likelihood (or least squares) estimate of $\boldsymbol{\beta}$. To see this, note that the scalar product $(\boldsymbol{\beta} - \hat{\boldsymbol{\beta}})^T \underline{x}^T \underline{x} (\boldsymbol{\beta} - \hat{\boldsymbol{\beta}})$ – which must be non-negative – achieves its minimum value when $\boldsymbol{\beta} = \hat{\boldsymbol{\beta}}$. This corresponds to the maximum value of the likelihood $\mathcal{L}(\boldsymbol{\beta}, \sigma)$ for given σ. The second term on the right-hand side of (9.11) may be written in expanded form as

$$(\underline{y} - \underline{x}\hat{\boldsymbol{\beta}})^T (\underline{y} - \underline{x}\hat{\boldsymbol{\beta}}) = \sum_{i=1}^{n} \left(y_i - \hat{\beta}_0 - \sum_{j=1}^{p} \hat{\beta}_j x_{ij} \right)^2,$$

from which we see that it is equal to the residual sum of squares obtained using the maximum likelihood (least squares) coefficient vector. An unbiased frequentist estimate of σ^2 is

$$s^2 = \frac{(\underline{y} - \underline{x}\hat{\boldsymbol{\beta}})^T (\underline{y} - \underline{x}\hat{\boldsymbol{\beta}})}{\nu}, \tag{9.13}$$

where $\nu = n - p - 1$. When performing frequentist inference, this estimate is often used in preference to the maximum likelihood estimate $\widehat{\sigma^2} = \mathrm{RSS}/n$, since the latter is biased (on average, it will be an under-estimate). We may now write the likelihood as

$$\mathcal{L}(\boldsymbol{\beta}, \sigma) = \frac{1}{(2\pi)^{n/2} \sigma^n} \exp\left(-\frac{1}{2\sigma^2} (\boldsymbol{\beta} - \hat{\boldsymbol{\beta}})^T \underline{x}^T \underline{x} (\boldsymbol{\beta} - \hat{\boldsymbol{\beta}}) \right) \exp\left(-\frac{\nu s^2}{2\sigma^2} \right). \tag{9.14}$$

By comparison with the standard form of the multivariate normal density (Box 9.1), we can rewrite the likelihood as

$$\mathcal{L}(\boldsymbol{\beta}, \sigma) = \frac{1}{(2\pi)^{\nu/2} \sigma^{\nu} \sqrt{\det(\underline{x}^T \underline{x})}} \exp\left(-\frac{\nu s^2}{2\sigma^2} \right) \mathcal{N}(\boldsymbol{\beta} | \hat{\boldsymbol{\beta}}, \sigma^2 (\underline{x}^T \underline{x})^{-1}). \tag{9.15}$$

Now that we have transformed the likelihood into the product of a multivariate normal density in $\boldsymbol{\beta}$ and a simple function of σ, it will be much easier to identify the posterior as a known density in the parameters. This justifies the effort expended to obtain expression (9.15).

■ ***Exercise*** 9.2 Derive expression (9.15) for the likelihood, starting from equation (9.14). You may make use of the fact that if $\underline{\boldsymbol{A}}$ is an $n \times n$ matrix with inverse $\underline{\boldsymbol{A}}^{-1}$ and $c \in \mathbb{R}$ then $\det(\underline{\boldsymbol{A}}^{-1}) = 1/\det(\underline{\boldsymbol{A}})$ and $\det(c\underline{\boldsymbol{A}}) = c^n \det(\underline{\boldsymbol{A}})$.

To obtain a posterior, we must choose our priors. In the absence of any substantive prior knowledge we can opt for a *vague* (or *diffuse* or *non-informative*) prior which has minimal influence on the posterior. In this way we 'let the data do the talking' via the likelihood function. A common choice is the 'flat' or uniform prior

$$f_{\boldsymbol{B}}(\boldsymbol{\beta}) = c_{\boldsymbol{B}}, \quad \boldsymbol{\beta} \in \mathbb{R}^{p+1},$$

where $c_{\boldsymbol{B}}$ is a constant. A potential problem with this choice is that the uniform distribution cannot be normalised on an infinite domain, meaning that the distribution is *improper*. In practice, improper priors can be used provided we can normalise the posterior. Of course we could make our flat prior proper by specifying an extremely large domain, but since the resulting posterior would be no different numerically from that obtained using an improper prior, there is no benefit in doing this.

A convenient way to define a non-informative prior for Σ that respects the fact that it is a positive quantity is to assume that $\log \Sigma$ is uniform, or equivalently that $\Sigma = e^U$ where U is uniform. According to this assumption, the prior density of Σ is

$$f_\Sigma(\sigma) = \frac{c_\Sigma}{\sigma}, \quad \sigma > 0.$$

According to these assumptions the posterior is

$$f_{\boldsymbol{B},\Sigma|\underline{Y},\underline{\boldsymbol{X}}}(\boldsymbol{\beta}, \sigma|\underline{y}, \underline{\boldsymbol{x}}) = c\sigma^{-\nu-1} \exp\left(-\frac{\nu s^2}{2\sigma^2}\right) \mathcal{N}(\boldsymbol{\beta}|\hat{\boldsymbol{\beta}}, \sigma^2(\underline{\boldsymbol{x}}^T\underline{\boldsymbol{x}})^{-1}), \tag{9.16}$$

where c is a normalising constant. In principle we could find c by imposing normalisation

$$\int f_{\boldsymbol{B},\Sigma|\underline{Y},\underline{\boldsymbol{X}}}(\boldsymbol{\beta}, \sigma|\underline{y}, \underline{\boldsymbol{x}}) d\boldsymbol{\beta} d\sigma = 1.$$

However, computing the multidimensional integral over $\boldsymbol{\beta}$ and σ is impractical. The difficulty of posterior normalisation, and the related problem of sampling from the posterior, are recurring themes in Bayesian inference, and we will introduce general computational methods for tackling them in Chapter 10. In special cases, including the current one and our lizard sex ratio example (see Section 6.5.3), we can write the posterior in terms of known distributions which are normalised by definition. In the current case, this can be achieved by using the chain rule of probability to factorise the posterior as follows:

$$f_{\boldsymbol{B},\Sigma|\underline{Y},\underline{\boldsymbol{X}}}(\boldsymbol{\beta}, \sigma|\underline{y}, \underline{\boldsymbol{x}}) = f_{\boldsymbol{B}|\Sigma,\underline{Y},\underline{\boldsymbol{X}}}(\boldsymbol{\beta}|\sigma, \underline{y}, \underline{\boldsymbol{x}}) f_{\Sigma|\underline{Y},\underline{\boldsymbol{X}}}(\sigma|\underline{y}, \underline{\boldsymbol{x}})$$

corresponding to the DAG

$$(\Sigma) \rightarrow (\boldsymbol{B}). \tag{9.17}$$

Every DAG describes a procedure for sampling from the multivariate distribution it represents: first generate samples of the variables with no parents, use these values to determine the conditional distributions of their children, then sample from the latter, and so on. In the DAG above this procedure is very simple: we first generate a realisation of Σ, which determines the conditional distribution of $\boldsymbol{B}$, from which we generate a realisation of $\boldsymbol{B}$. By comparison with equation (9.16), we see that

$$f_{\boldsymbol{B}|\Sigma,\underline{Y},\underline{\boldsymbol{X}}}(\boldsymbol{\beta}|\sigma,\underline{y},\underline{\boldsymbol{x}}) = \mathcal{N}(\boldsymbol{\beta}|\hat{\boldsymbol{\beta}},\sigma^2(\underline{\boldsymbol{x}}^T\underline{\boldsymbol{x}})^{-1}), \tag{9.18}$$

$$f_{\Sigma|\underline{Y},\underline{\boldsymbol{X}}}(\sigma|\underline{y},\underline{\boldsymbol{x}}) = c\sigma^{-(\nu+1)}\exp\left(-\frac{\nu s^2}{2\sigma^2}\right). \tag{9.19}$$

The conditional distribution of $\boldsymbol{B}$ is known, and there are efficient algorithms for sampling from it. We now show that the distribution of Σ, with density given by equation (9.19), is related in a simple way to the *inverse gamma distribution* described in Box 9.3.

★ Box 9.3 The gamma and inverse gamma distributions

Let X be a random variable with probability density function

$$f_X(x;a,b) = \frac{b^a}{\Gamma(a)}x^{a-1}e^{-xb} := \text{Ga}(x|a,b),$$

where $a,b > 0$ and $\Gamma(a) = \int_0^\infty t^{a-1}e^{-t}dt$ is the gamma function. We say that X is 'gamma distributed with shape a and rate b', written $X \sim \text{Gamma}(a,b)$. We have

$$\mathbb{E}(X) = \frac{a}{b}, \quad \text{Var}(X) = \frac{a}{b^2}$$

and

$$\text{mode}(X) = \frac{a-1}{b} \text{ for } a \geq 1,$$

while $\text{mode}(X) = 0$ for $0 < a < 1$.

Suppose $Z = 1/X$, where $X \sim \text{Gamma}(a,b)$. We say Z is 'inverse gamma distributed with shape a and scale b', or $Z \sim \text{Inv-Gamma}(a,b)$, with probability density function

$$f_Z(z;a,b) = \frac{b^a}{\Gamma(a)}z^{-(a+1)}e^{-b/z} := \text{IG}(z|a,b).$$

We have

$$\mathbb{E}(Z) = \frac{b}{a-1}, \quad \text{mode}(Z) = \frac{b}{a+1}, \quad \text{Var}(Z) = \frac{b^2}{(a-1)^2(a-2)},$$

where $\mathbb{E}(Z)$ and $\text{Var}(Z)$ exist provided $a > 1$ and $a > 2$, respectively.

To establish the connection between the posterior density of Σ and the inverse gamma distribution, we define the new random variable $Z = \Sigma^2$ (Z is the variance of the conditional distribution of Y). By changing variables from Σ to Z in equation (9.19), we obtain

$$f_{Z|\underline{Y},\underline{X}}(z|\underline{y},\underline{x}) = \frac{c}{2}z^{-(\frac{\nu}{2}+1)}\exp\left(-\frac{\nu s^2}{2z}\right) = \text{IG}\left(z\middle|\frac{\nu}{2},\frac{\nu s^2}{2}\right), \tag{9.20}$$

where IG is the inverse gamma density defined in Box 9.3. That is, conditional on the data, Σ^2 is inverse gamma distributed:

$$\Sigma^2\middle|\left\{\underline{Y}=\underline{y},\underline{X}=\underline{x}\right\} \sim \text{Inv-Gamma}\left(\frac{\nu}{2},\frac{\nu s^2}{2}\right).$$

By changing variables back to $\Sigma = \sqrt{Z}$, we obtain the posterior density of Σ in closed form,

$$f_{\Sigma|\underline{Y},\underline{X}}(\sigma|\underline{y},\underline{x}) = 2\sigma\text{IG}\left(\sigma^2\middle|\frac{\nu}{2},\frac{\nu s^2}{2}\right), \tag{9.21}$$

from which we could, if we wished, identify the normalising constant c in equation (9.19). To draw samples $(\boldsymbol{B},\Sigma)$ from the posterior, we do the following:

(a) generate $Z \sim \text{Inv-Gamma}\left(\frac{\nu}{2},\frac{\nu s^2}{2}\right)$ and let $\Sigma = \sqrt{Z}$;

(b) generate $\boldsymbol{B} \sim \mathcal{N}\left(\hat{\boldsymbol{\beta}},\Sigma^2(\underline{\boldsymbol{x}}^T\underline{\boldsymbol{x}})^{-1}\right)$.

These steps can be carried out using standard sampling algorithms.

■ ***Exercise*** 9.3 Make use of the change of random variable technique (Appendix A) to derive the density given in equation (9.20).

■ ***Exercise*** 9.4 By expanding the right-hand side of equation (9.21), find an expression for the normalising constant c in equation (9.19).

9.2 Application to Body Fat Data

Let us fit our Bayesian linear model to the body fat dataset we worked with in Chapter 7. Consider first the case of a single predictor, the standardised abdomen circumference Z_{Abd}. In this case, we have three model parameters (latent variables), B_0, B_1 and Σ. A single sample (β_0,β_1,σ) from the posterior corresponds to the model

$$Y = \beta_0 + \beta_1 Z_{\text{Abd}} + \mathcal{E},$$

where $\mathcal{E} \sim \mathcal{N}(0,\sigma^2)$. By generating many samples from the posterior we can examine the range of plausible models which are consistent with our data. We view each such model as a particular linear relationship (a regression function) together with a noise magnitude. Figure 9.1 shows 50 linear relationships sampled from the posteriors obtained from two datasets – one small, one large – sampled uniformly at random from the full dataset. Also shown are the corresponding posterior distributions of Σ. From Figure 9.1 we see that the variability in the linear relationship is substantially greater for the smaller dataset (a) than the larger (b). This reflects the fact that the more data we have the more concentrated (or 'sharply peaked') in latent space the posterior becomes, reflecting increasing certainty about parameter values. We saw the same effect when computing posterior distributions of the fraction of male lizards in Chapter 6. In Figure 9.1c we see that our posterior for Σ also becomes more concentrated

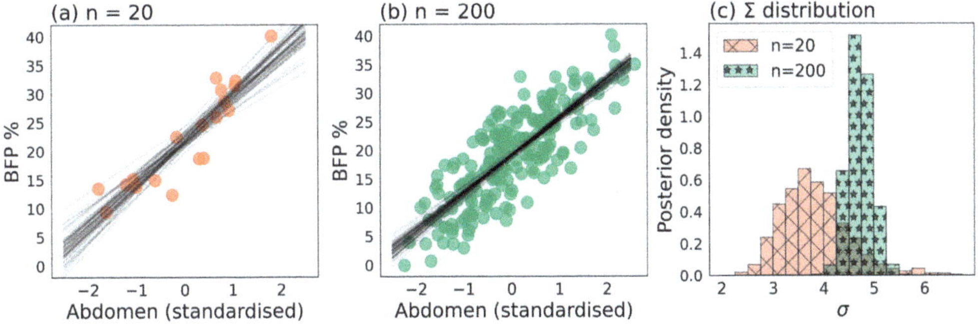

Figure 9.1 (a) and (b) each show 50 posterior samples of the regression function $y = B_0 + B_1 x$ based on datasets of size (a) $n = 20$ and (b) $n = 200$. (c) Histograms of 10^3 samples of Σ from the same posteriors used in (a) and (b).

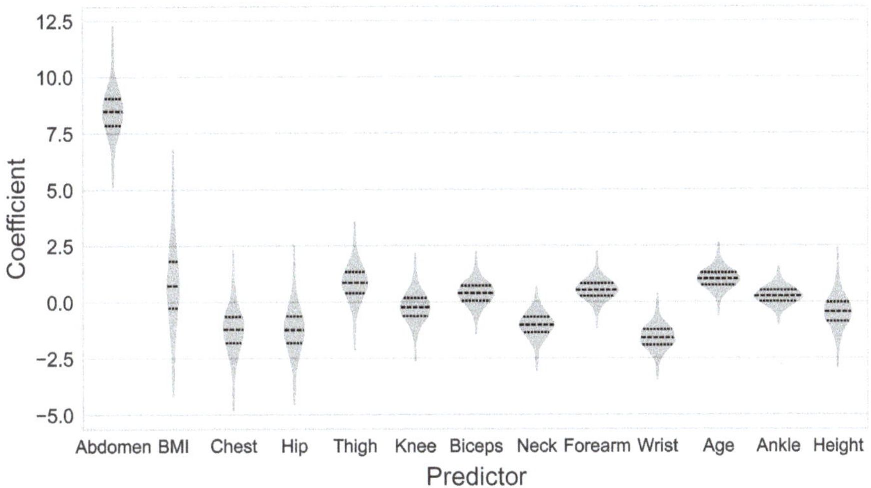

Figure 9.2 Marginal posterior distributions of the model coefficients.

when more data is used to compute it. Note that the posterior concentrates around larger Σ values for the larger dataset. While part of this difference may be explained by the particular selection of datapoints used, we also expect to see smaller noise estimates for smaller datasets because the model is partially fitting to the noise.

Now consider the case where all standardised predictors are used, so a single sample $(\boldsymbol{\beta}, \sigma)$ from the posterior corresponds to the model

$$Y = \boldsymbol{\beta}^T \mathbf{Z} + \mathcal{E}, \tag{9.22}$$

where $\mathbf{Z} = (1, Z_1, \ldots, Z_p)$ and $\mathcal{E} \sim \mathcal{N}(0, \sigma^2)$. Figure 9.2 shows the marginal posterior distributions of the model coefficients estimated by sampling the posterior obtained using the full dataset ($n = 248$). These take a similar form to the bootstrap sampling distributions obtained for the frequentist version of this model in Chapter 7. For the majority of coefficients, there is substantial posterior probability assigned to both positive *and* negative values. In

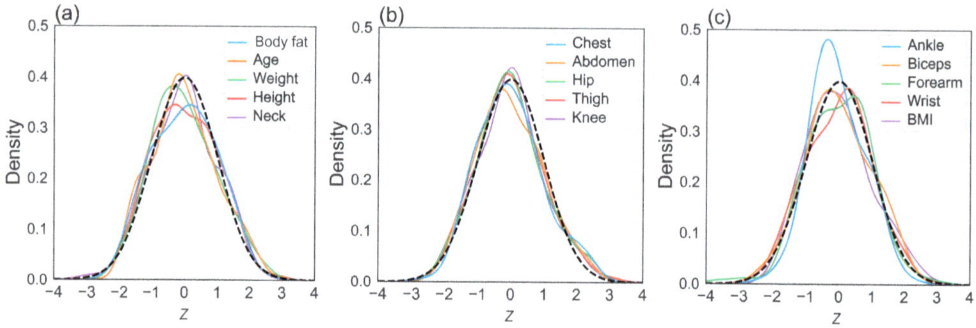

Figure 9.3 (a–c) Estimated probability densities for each standardised variable in the body fat dataset. The dashed black curves show the standard normal density.

these cases, substantially more data would be required to be confident of the presence of a systematic relationship. The high degree of uncertainty in the sign of the model coefficients suggests that restricting the complexity of the model – *regularising* it – might improve its predictive power. We will show in Section 9.3 how regularisation can be implemented via a suitable choice of prior on the coefficients.

We can gain a deeper understanding of the posterior distribution of the standardised predictor model (9.22) by establishing a connection between posterior correlations and *conditional* predictor correlations. Conditioning on the variance of the noise term $\Sigma = \sigma$, from equation (9.18) we have

$$\underline{\boldsymbol{\Sigma}}_{\boldsymbol{B}} = \sigma^2(\underline{z}^T\underline{z})^{-1},$$

where $\underline{z}$ denotes the standardised design matrix. The *empirical covariance matrix* (see Box 9.5) of the standardised predictors is

$$\hat{\underline{\boldsymbol{\Sigma}}}_Z = \frac{1}{n}\underline{z}^T\underline{z}$$

with inverse $\hat{\underline{\boldsymbol{\Omega}}}_Z := \hat{\underline{\boldsymbol{\Sigma}}}_Z^{-1}$. The inverse of a covariance matrix is called a *precision matrix*. Combining these definitions, we have

$$\underline{\boldsymbol{\Sigma}}_{\boldsymbol{B}} = \frac{\sigma^2}{n}\hat{\underline{\boldsymbol{\Sigma}}}_Z^{-1} = \frac{\sigma^2}{n}\hat{\underline{\boldsymbol{\Omega}}}_Z.$$

We now construct a simple model of the predictor distribution. Figure 9.3 shows estimated probability density functions for standardised versions of every variable in our dataset.[1] Their close match to the normal density, combined with the elliptical pairwise scatter plots in Chapter 7, suggests that the joint predictor distribution is approximately multivariate normal. According to this assumption, the *conditional correlation* between any two predictors given

[1] Estimates obtained using *kernel density estimation*. Given a dataset $\{x_i\}_{i=1}^n$ of observations of X, the kernel density estimate of f_X is $\hat{f}_X(x) = \frac{1}{nh}\sum_{i=1}^n \phi((x - x_i)/h)$, where ϕ is the standard normal density and h is a 'bandwidth' (chosen via 'Scott's rule' $h = n^{-\frac{1}{5}}\hat{\sigma}$ where $\hat{\sigma}$ is the sample standard deviation).

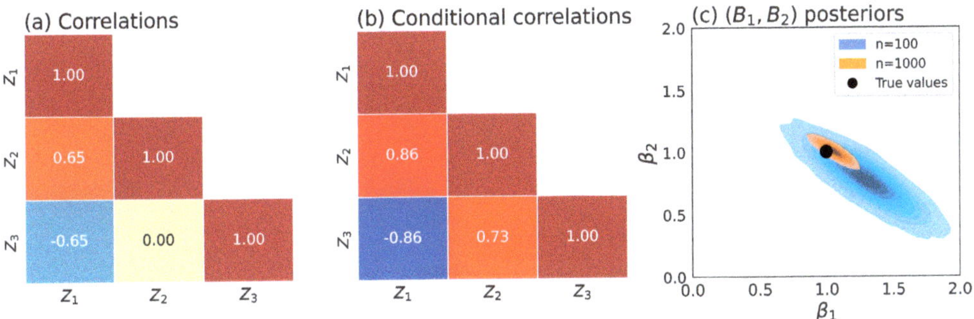

Figure 9.4 (a) Correlation/covariance matrix for $\mathbf{Z} \sim \mathcal{N}(\mathbf{0}, \underline{\mathbf{\Sigma}})$. (b) Conditional correlations $\rho(Z_i, Z_j|\mathbf{Z} \setminus \{Z_i, Z_j\})$. (c) Marginal posterior of (B_1, B_2) for the model $Y = B_1Z_1 + B_2Z_2 + B_3Z_3 + \mathcal{E}$ based on $n = 100$ and $n = 1,000$ samples from the model $Y = Z_1 + Z_2 + Z_3 + \mathcal{E}$.

the values of all the others is

$$\rho(X_i, X_j|\boldsymbol{X} \setminus \{X_i, X_j\}) = \rho(Z_i, Z_j|\mathbf{Z} \setminus \{Z_i, Z_j\}) \approx -\frac{[\hat{\underline{\mathbf{\Omega}}}_{\mathbf{Z}}]_{ij}}{\sqrt{[\hat{\underline{\mathbf{\Omega}}}_{\mathbf{Z}}]_{ii}}\sqrt{[\hat{\underline{\mathbf{\Omega}}}_{\mathbf{Z}}]_{jj}}},$$

where $\boldsymbol{X} \setminus \{X_i, X_j\}$ denotes the set of predictors excluding X_i and X_j (see Box 9.4). Here we made use of the fact that linear rescaling of variables does not change the correlations between them. Now, since $\underline{\mathbf{\Sigma}}_{\boldsymbol{B}} = \frac{\sigma^2}{n}\hat{\underline{\mathbf{\Omega}}}_{\mathbf{Z}}$, we have that if $\boldsymbol{B}$ is the coefficient vector of a standardised predictor model, then

$$\rho(B_i, B_j) = \frac{[\underline{\mathbf{\Sigma}}_{\boldsymbol{B}}]_{ij}}{\sqrt{[\underline{\mathbf{\Sigma}}_{\boldsymbol{B}}]_{ii}[\underline{\mathbf{\Sigma}}_{\boldsymbol{B}}]_{jj}}} = \frac{[\hat{\underline{\mathbf{\Omega}}}_{\mathbf{Z}}]_{ij}}{\sqrt{[\hat{\underline{\mathbf{\Omega}}}_{\mathbf{Z}}]_{ii}[\hat{\underline{\mathbf{\Omega}}}_{\mathbf{Z}}]_{jj}}} \approx -\rho(X_i, X_j|\boldsymbol{X} \setminus \{X_i, X_j\}). \tag{9.23}$$

Let us consider the implications of this relationship. Suppose two predictors have a strong positive conditional correlation. According to equation (9.23), their coefficients will have a strong *negative* correlation in the posterior, meaning we can increase one coefficient and decrease the other whilst retaining a plausible model. As a concrete example, consider the data generating process $Y = Z_1 + Z_2 + Z_3 + \mathcal{E}$, where $\mathcal{E} \sim \mathcal{N}(0, 1)$ and $\mathbf{Z} \sim \mathcal{N}(\mathbf{0}, \underline{\mathbf{\Sigma}})$, with $\underline{\mathbf{\Sigma}}$ defined in Figure 9.4a. The conditional correlations in this model are shown in Figure 9.4b. Now suppose we fit a Bayesian model of the form $Y = B_1Z_1 + B_2Z_2 + B_3Z_3 + \mathcal{E}$ to a dataset from our generating model. As a consequence of the high conditional correlation between Z_1 and Z_2, the posterior correlation between B_1 and B_2 will be strongly negative, producing the elongated posterior in Figure 9.4c. For small sample sizes, this reduces *identifiability* of our model – a wide range of parameter combinations explain the data similarly well. By collecting more data, we can shrink the overall scale of the posterior, reducing problems of identifiability. We used standardised predictors in our analysis to make the posterior easier to interpret. Similar issues occur if we don't standardise, but interpretation of posterior correlations is less straightforward.

Conditional correlations within the body fat predictors are shown in Figure 9.5. Comparison with the corresponding Pearson correlations shown in Figure 7.2 reveals substantial

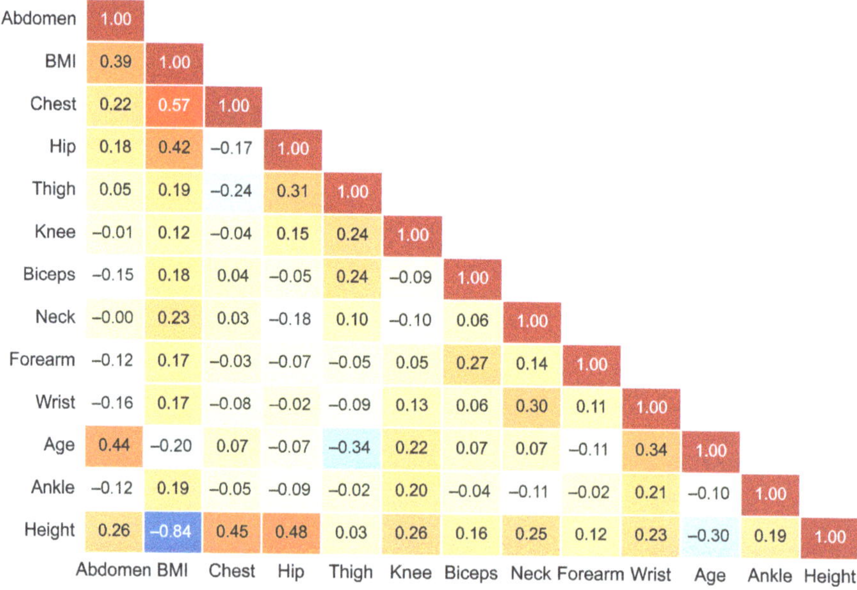

Figure 9.5 Conditional correlations between predictors in the body fat dataset, estimated from the precision matrix of the predictors.

differences. For example, height and BMI have a near-zero Pearson correlation but a strongly negative conditional correlation. This makes sense: if all other body measurements are held fixed, most notably weight, then increasing height lowers BMI. As just discussed, high posterior correlations between parameters indicate strong dependencies among them within the set of *plausible* models (i.e. those with high posterior probability). Such dependencies may make it hard to separate out the effects of different predictors on the response; they may also be a sign that some predictors could be discarded without reducing the performance of the model. These issues were discussed in the frequentist context, where the sampling distribution plays the role the posterior plays here, in Chapter 7.

★ Box 9.4 Conditional correlations in the multivariate normal distribution

Suppose we partition the elements of the random vector $\boldsymbol{X} \sim \mathcal{N}(\boldsymbol{\mu}, \underline{\boldsymbol{\Sigma}})$ into two sets, A and B, corresponding to two vectors $\boldsymbol{X}_A$ and $\boldsymbol{X}_B$. Writing the mean vector, covariance matrix and precision matrix in partitioned (block) form,

$$\boldsymbol{\mu} = \begin{pmatrix} \boldsymbol{\mu}_A \\ \boldsymbol{\mu}_B \end{pmatrix}, \quad \underline{\boldsymbol{\Sigma}} = \begin{pmatrix} \underline{\boldsymbol{\Sigma}}_{AA} & \underline{\boldsymbol{\Sigma}}_{AB} \\ \underline{\boldsymbol{\Sigma}}_{BA} & \underline{\boldsymbol{\Sigma}}_{BB} \end{pmatrix}, \quad \underline{\boldsymbol{\Omega}} = \underline{\boldsymbol{\Sigma}}^{-1} = \begin{pmatrix} \underline{\boldsymbol{\Omega}}_{AA} & \underline{\boldsymbol{\Omega}}_{AB} \\ \underline{\boldsymbol{\Omega}}_{BA} & \underline{\boldsymbol{\Omega}}_{BB} \end{pmatrix},$$

we have $\boldsymbol{X}_A | \{\boldsymbol{X}_B = \boldsymbol{x}_B\} \sim \mathcal{N}(\boldsymbol{\mu}_{A|B}, \underline{\boldsymbol{\Sigma}}_{A|B})$, where

$$\boldsymbol{\mu}_{A|B} = \underline{\boldsymbol{\Sigma}}_{A|B}(\underline{\boldsymbol{\Omega}}_{AA}\boldsymbol{\mu}_A - \underline{\boldsymbol{\Omega}}_{AB}(\boldsymbol{x}_B - \boldsymbol{\mu}_B)), \quad \underline{\boldsymbol{\Sigma}}_{A|B} = \underline{\boldsymbol{\Sigma}}_{AA} - \underline{\boldsymbol{\Sigma}}_{AB}\underline{\boldsymbol{\Sigma}}_{BB}^{-1}\underline{\boldsymbol{\Sigma}}_{BA} = \underline{\boldsymbol{\Omega}}_{AA}^{-1}.$$

See Murphy (2022) or Mardia et al. (2024) for a derivation of this result. Suppose our partition consists of the sets $A = \{X_i, X_j\}$ and $B = \boldsymbol{X} \setminus \{X_i, X_j\}$. Then, using the result above, we have

$$\underline{\Sigma}_{A|B} = \begin{pmatrix} \Omega_{ii} & \Omega_{ij} \\ \Omega_{ij} & \Omega_{jj} \end{pmatrix}^{-1} = \frac{1}{\Omega_{ii}\Omega_{jj} - \Omega_{ij}^2} \begin{pmatrix} \Omega_{jj} & -\Omega_{ij} \\ -\Omega_{ij} & \Omega_{ii} \end{pmatrix}.$$

The conditional correlation between X_i and X_j given the value of $\boldsymbol{X} \setminus \{X_i, X_j\}$ is therefore

$$\rho(X_i, X_j | \boldsymbol{X} \setminus \{X_i, X_j\}) = \frac{[\underline{\Sigma}_{A|B}]_{ij}}{\sqrt{[\underline{\Sigma}_{A|B}]_{ii}[\underline{\Sigma}_{A|B}]_{jj}}} = -\frac{\Omega_{ij}}{\sqrt{\Omega_{ii}\Omega_{jj}}}.$$

★ Box 9.5 Estimating covariance matrices

Let $\underline{\boldsymbol{x}} = (\boldsymbol{x}_1, \ldots, \boldsymbol{x}_n)^T$ be i.i.d. observations of a random vector $\boldsymbol{X} = (X_1, \ldots, X_p)$. An unbiased estimator of the covariance matrix $\underline{\Sigma}$ with elements $\Sigma_{ij} = \text{Cov}(X_i, X_j)$ is the *sample covariance matrix* $\underline{S}$ with elements

$$S_{ij} = \frac{1}{n-1} \sum_{k=1}^{n} (x_{ki} - \bar{x}_i)(x_{kj} - \bar{x}_j),$$

where x_{ki} is the ith element of the kth observation, $\boldsymbol{x}_k$, of $\boldsymbol{X}$ and $\bar{x}_i = n^{-1} \sum_{k=1}^{n} x_{ki}$. Under the assumption that $\boldsymbol{X} \sim \mathcal{N}(\boldsymbol{\mu}, \underline{\Sigma})$ for some $\boldsymbol{\mu}$ and $\underline{\Sigma}$, the maximum likelihood estimate of $\underline{\Sigma}$ is $\hat{\underline{\Sigma}} = \left(\frac{n-1}{n}\right) \underline{S}$. To distinguish $\hat{\underline{\Sigma}}$ from $\underline{S}$, we will refer to $\hat{\underline{\Sigma}}$ as the *empirical covariance matrix*. Typically n will be large so the difference between the two estimates will be small. The empirical matrix has the advantage of decluttering calculations. If our observations have mean $\mathbf{0}$ (for example if they have been standardised), then $\hat{\underline{\Sigma}} = n^{-1}\underline{\boldsymbol{x}}^T\underline{\boldsymbol{x}}$.

9.3 Gaussian Priors and Ridge Regression

When using all available body fat predictors we found that the marginal posterior distributions of many of their coefficients were peaked close to zero and had comparatively high variance. This means that the sensitivity of the response to these predictors is both *small* and *uncertain*, and therefore that coefficient estimates will be sensitive to noise in the data – an example of overfitting. We observed this in Chapter 7, when we analysed the multiple linear regression model using a frequentist approach. In that context, we found that excluding predictors improved cross-validation performance.

An alternative Bayesian way to tackle overfitting is to impose a *shrinkage prior* on the coefficients. The aim is to shrink coefficient estimates associated with the weaker (i.e. less predictively useful) predictors, thereby reducing the propensity to overfit. We can view such priors as describing our scepticism about the importance of individual predictors. That is, before collecting any data, our belief is that there is most likely no relationship between any given predictor and the response.

To keep our analysis simple, we will consider a model in which the variance σ^2 of the

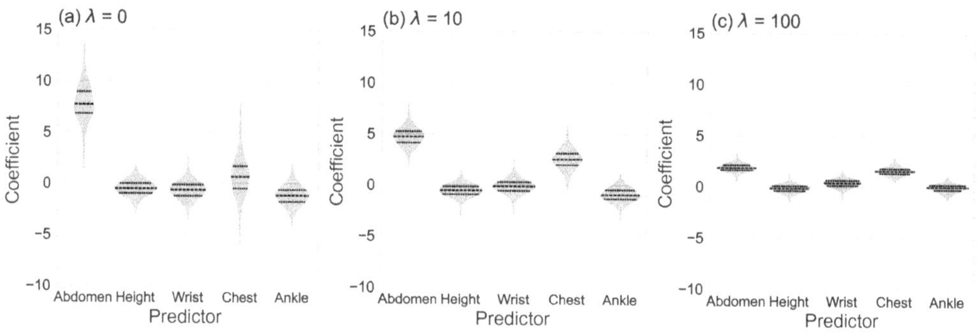

Figure 9.6 (a–c) Marginal posteriors of model coefficients in a four-predictor model with three different values of the ridge shrinkage hyperparameter λ.

noise is known. In practice, we can use the unbiased estimate given by equation (9.13). This model corresponds to the DAG

B

$x_i \rightarrow Y_i \leftarrow \sigma$

$i = 1, \ldots, n$. (9.24)

Our first example of a shrinkage prior is to assume that the coefficients $B_1, \ldots, B_p$ are i.i.d. normal with zero mean and variance σ^2/λ, and that B_0 is normal with variance $\sigma^2/(\lambda\epsilon)$, so

$$f_{\boldsymbol{B}}(\boldsymbol{\beta}) = \mathcal{N}\left(\boldsymbol{\beta}\middle|\mathbf{0}, \frac{\sigma^2}{\lambda}\underline{\boldsymbol{I}}_{\epsilon}^{-1}\right),$$

where $\underline{\boldsymbol{I}}_{\epsilon} = \mathrm{diag}(\epsilon, 1, \ldots, 1)$ and λ is the *shrinkage hyperparameter*.[2] Increasing λ concentrates the prior closer to zero. Importantly, we assume here that the predictors have been standardised so that they vary over similar scales. If we did not do this, features with a smaller range of variation would, in effect, be subject to a more severe shrinkage penalty. The penalty $\epsilon\lambda$ on B_0 is included so that the prior can be expressed as a normalised density, but we take $\epsilon \to 0$ in the posterior corresponding to a flat prior on B_0. The posterior is given by

$$f_{\boldsymbol{B}|\underline{Y},\underline{\boldsymbol{X}}}(\boldsymbol{\beta}|\underline{y},\underline{\boldsymbol{x}}) = c\exp\left(-\frac{(\underline{y}-\underline{\boldsymbol{x}}\boldsymbol{\beta})^T(\underline{y}-\underline{\boldsymbol{x}}\boldsymbol{\beta}) + \lambda\boldsymbol{\beta}^T\underline{\boldsymbol{I}}_{\epsilon}\boldsymbol{\beta}}{2\sigma^2}\right), \tag{9.25}$$

where c is a normalising constant. It is possible, following similar steps to those in Section 9.1, to express the posterior as a normal density from which we can sample $\boldsymbol{B}$ (see Exercise 9.5). We illustrate the effect of shrinkage on the posterior by example. Figure 9.6 shows the marginal posteriors for a four-predictor model with different shrinkage hyperparameter values. The prior shrinks both coefficient magnitudes and variances, *constraining* the model

[2] The notation $\underline{\boldsymbol{M}} = \mathrm{diag}(m_1, \ldots, m_n)$ means that $\underline{\boldsymbol{M}}$ is a diagonal $n \times n$ matrix (off-diagonal elements are zero) with diagonal elements given by $M_{ii} = m_i$ for $i \in \{1, \ldots, n\}$.

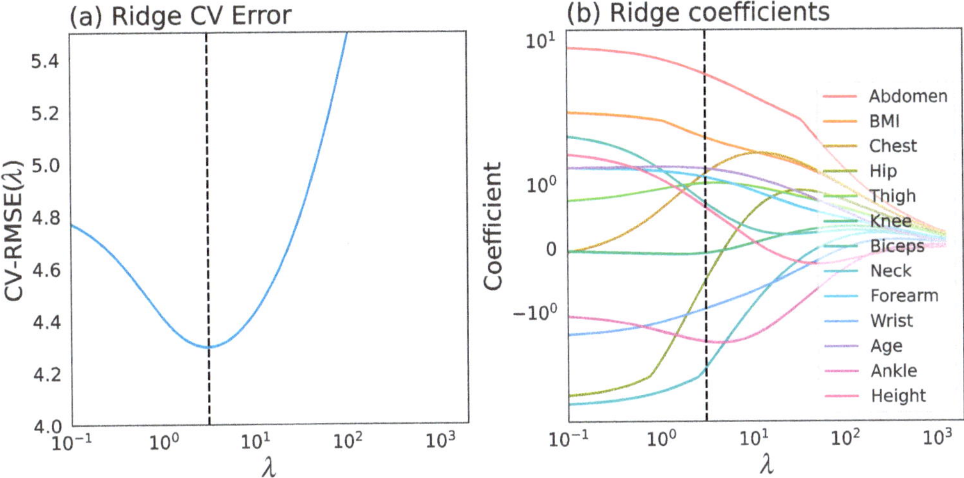

Figure 9.7 (a) Cross-validated RMSE in the all-predictor model of BFP as a function of ridge shrinkage hyperparameter λ for a (standardised) dataset of size $n = 50$ randomly selected from the full dataset. (b) Dependence of predictor coefficients on λ.

so that it is it less flexible. It would clearly be a mistake to push this constraint too far, but the hope is that just the right amount of shrinkage will optimise model performance.

■ ***Exercise*** 9.5 Show that the posterior (9.25) is multivariate normal with covariance matrix $\underline{\underline{\Sigma}} = \sigma^2(\underline{\underline{x}}^T\underline{\underline{x}} + \lambda\underline{\underline{I}}_\epsilon)^{-1}$ and mean vector $\boldsymbol{\mu} = \sigma^{-2}\underline{\underline{\Sigma}}\,\underline{\underline{x}}^T\boldsymbol{y}$.

Let us investigate the effect of shrinkage on the predictive power of the all-predictor model, as measured by cross-validation. The maximum a posteriori (MAP) estimate of the coefficient vector is as follows:

$$\hat{\boldsymbol{\beta}} = \arg\min_{\boldsymbol{\beta}} \left(\sum_{i=1}^{n} \left(y_i - \beta_0 - \sum_{k=1}^{p} \beta_k x_{ik} \right)^2 + \lambda \sum_{k=1}^{p} \beta_k^2 \right). \tag{9.26}$$

The objective to be minimised in equation (9.26) is the negative exponent of the posterior (9.25), written here in component form. It is equal to the RSS plus a 'shrinkage penalty' originating from the prior. MAP coefficient estimation (with shrinkage prior) can therefore be regarded as a penalised form of ordinary least squares regression. When the shrinkage penalty takes the form shown in equation (9.26), the technique is called *ridge regression.* We are *regularising* the loss function using the sum of the squared coefficient magnitudes (their 'L2 norm'). Figure 9.7a shows the cross-validation error in predictions made using the MAP estimate, as a function of the shrinkage hyperparameter. We used a small dataset here to exaggerate the propensity of the model to overfit. The error in Figure 9.7a follows the U-shaped curve characteristic of a trade-off between bias and variance. For low penalty, the model is more flexible and therefore capable of capturing a wider range of possible predictor–response associations but also more sensitive to noise in the dataset. This is the low-bias, high-noise scenario. When the penalty is high, the model is less flexible but also less sensitive to noise – the high-bias, low-variance scenario. Between these two extremes is a sweet spot where predictive power is optimised, shown by the dashed line in Figure 9.7a. Figure 9.7b shows how the regression coefficients depend on the size of the shrinkage hyperparameter. As regularisation strength increases, the coefficients shrink to zero.

We have presented ridge regression as a Bayesian method, but it can equally be viewed as a method for producing biased frequentist estimates, given by equation (9.26), of the coefficients of the underlying conditional probabilistic model. In this case, the bias is justified based on the advantage it brings in reducing overfitting. Selecting the prior by cross-validation, as we have done, is in fact not a fully Bayesian method, because the prior is selected *after* observing the data. A Bayesian approach is to specify a 'hyper-prior' on the shrinkage hyperparameter, viewed as a random variable Λ. The shrinkage prior then becomes a conditional density $f_{\boldsymbol{B}|\Lambda}(\boldsymbol{\beta}|\lambda)$ dependent on the realised value of Λ. The DAG corresponding to this model is

$$\sigma \to Y_i; \quad \Lambda \to \boldsymbol{B} \to Y_i \leftarrow \boldsymbol{x}_i, \quad i = 1, \ldots, n. \tag{9.27}$$

This is an example of a hierarchical Bayesian model. We will see such models again in Chapter 10.

9.4 Laplace Priors and the Lasso

In ridge regression, we penalise the squared magnitudes of the regression coefficients. We saw in Section 9.3 that this penalty may be understood as arising from a normal shrinkage prior on the coefficients. An alternative is to assume a prior distribution under which the coefficients $B_1, \ldots, B_p$ are i.i.d. and follow a *Laplace distribution*,

$$f_{B_k}(\beta_k) = \frac{1}{2b} \exp\left(-\frac{|\beta_k|}{b}\right),$$

implying $\mathbb{E}(B_k) = 0$ and $\mathrm{Var}(B_k) = 2b^2$. Smaller values of b correspond to prior distributions more tightly concentrated around zero. We can express b in terms of a shrinkage hyperparameter λ, as $b = 2\sigma^2/\lambda$. Assuming a flat (improper) prior on B_0, we have

$$f_{\boldsymbol{B}}(\boldsymbol{\beta}) \propto \exp\left(-\frac{\lambda}{2\sigma^2} \sum_{k=1}^{p} |\beta_k|\right),$$

yielding a posterior

$$f_{\boldsymbol{B}|\underline{Y},\underline{\boldsymbol{X}}}(\boldsymbol{\beta}|\underline{y},\underline{\boldsymbol{x}}) = c \exp\left(-\frac{(\underline{y} - \underline{\boldsymbol{x}}\boldsymbol{\beta})^T(\underline{y} - \underline{\boldsymbol{x}}\boldsymbol{\beta}) + \lambda \sum_{k=1}^{p} |\beta_k|}{2\sigma^2}\right),$$

where c is a normalising constant. The MAP estimate of the coefficient vector is

$$\hat{\boldsymbol{\beta}} = \arg\min_{\boldsymbol{\beta}} \left(\sum_{i=1}^{n} \left(y_i - \beta_0 - \sum_{k=1}^{p} \beta_k x_{ik}\right)^2 + \lambda \sum_{k=1}^{p} |\beta_k|\right). \tag{9.28}$$

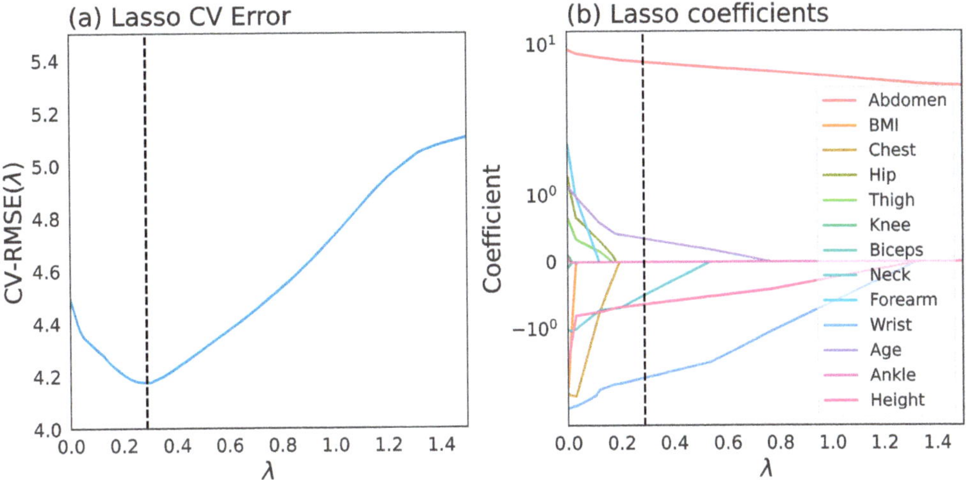

Figure 9.8 (a) Cross-validated RMSE in the all-predictor model of BFP as a function of lasso shrinkage hyperparameter λ for a dataset of size $n = 100$ randomly selected from the full dataset. (b) Dependence of predictor coefficients on λ.

As with ridge regression the objective is a penalised residual sum of squares, but now we penalise the sum of the *magnitudes* (rather than squared magnitudes) of the coefficients. Estimating coefficients using this objective is known as *lasso regression.* Figure 9.8a shows how varying the lasso shrinkage hyperparameter affects cross-validation error in the all-predictor BFP model. The error follows the familiar U-shaped curve seen in the ridge case. As with ridge regression, λ acts as a tunable dial controlling model complexity, helping us find a sweet spot between overfitting and underfitting. Lasso and ridge differ in the patterns of coefficients they produce – as can be seen by comparing Figures 9.7b and 9.8b. In both cases, coefficients shrink towards zero as regularisation strength increases. With ridge, they remain non-zero; with lasso, an increasing number of them become *exactly* zero. The lasso penalty has the effect of selecting a *subset* of the predictors, producing a *sparse* model. Sparse models are useful when we want explanations humans can understand easily. Exercise 9.7 invites you to explain why the lasso penalty can yield coefficients which are exactly zero.

9.5 Chapter Summary

In this chapter, we have defined the simple and multiple Bayesian linear regression models, and derived the posterior of the multiple predictor model in explicit form, assuming uniform priors on $\boldsymbol{B}$ and $\log \Sigma$. We used Gaussian and Laplace distributions as shrinkage priors on model coefficients. Obtaining MAP coefficient estimates in these two cases is known respectively as ridge and lasso regression. These are examples of *regularisation* methods designed to reduce overfitting by penalising complexity.

- Simple and multiple **Bayesian linear regression** are defined by promoting model parameters to be latent (unobservable) random variables.

- Due to the simplicity of these models, their posteriors can be found in closed form. When using flat priors, the posterior closely resembles the frequentist sampling distribution.
- We can penalise model complexity by using **shrinkage priors**.
- MAP parameter estimates obtained using a zero-mean Gaussian prior are equivalent to L2 regularised least squares estimates. This technique is known as **ridge regression**.
- Combining MAP estimation with a Laplace prior is equivalent to L1 regularised least squares, known as **lasso regression**.
- Whereas the ridge penalty shrinks all coefficient estimates towards zero, the lasso penalty tends to produce a subset of zero coefficient estimates, leading to a **sparse model**.

9.6 Further Exercises

■ ***Exercise*** 9.6 Let $D = \{(x_i, y_i)\}_{i=1}^n$ where the pairs (X_i, Y_i) are i.i.d. copies of the random variable pair (X, Y). Suppose we assume the conditional probabilistic model

$$Y = \beta X + \mathcal{E},$$

where $\mathcal{E} \sim \mathcal{N}(0, 1)$.

(a) Derive a formula for the maximum likelihood estimate of β in terms of D.
(b) Suppose that β is the realisation of a random variable $B \sim \mathcal{N}(0, c)$ which is independent of X. Draw the DAG corresponding to this model of the data-generating process, treating the X variables as known constants.
(c) Find an expression for the MAP estimate of β.
(d) Explain the connection between this model and ridge regression. What is the role of the parameter c?

■ ***Exercise*** 9.7 Let $D = \{(x_i, y_i)\}_{i=1}^n$ be a set of observations of the random variable pair (X, Y). Suppose we want to predict Y given X using the prediction function $y = \beta x$. One way to learn β is to minimise the lasso regularised MSE,

$$\ell(\beta) = \frac{1}{n} \sum_{i=1}^{n} (y_i - \beta x_i)^2 + \lambda |\beta|.$$

We are given the sample means $n^{-1} \sum_{i=1}^n x_i^2 = 1, n^{-1} \sum_{i=1}^n x_i y_i = 0.5$.

(a) Compute $\ell'(\beta)$.
(b) Suppose $\lambda = 0.25$. Plot $\ell'(\beta)$ against β, and find the value of β for which $\ell'(\beta) = 0$.
(c) Find the smallest λ^* such that for all $\lambda > \lambda^*$, the value of β which minimises the lasso regularised loss is $\hat{\beta} = 0$.

10

Bayesian Methods

In Chapter 6, we introduced the principles of Bayesian inference using the example of lizard sex ratio estimation. In Chapter 9, we generalised multiple linear regression to the Bayesian setting. We now expand our Bayesian toolkit, introducing general principles of model construction, fitting, testing and comparison.

10.1 The Components of a Bayesian Model

We begin by summarising the components of a Bayesian statistical model, contrasting with the frequentist approach. In both cases, we have an observed dataset D viewed as a single realisation of a random dataset $\mathcal{D}$. This is a collection of random variables with one of two structures:

$$\mathcal{D} = \begin{cases} \{X_i\}_{i=1}^n \text{ or } \underline{X} & \text{for density estimation,} \\ \{(X_i, Y_i)\}_{i=1}^n \text{ or } (\underline{X}, \underline{Y}) & \text{for regression or classification.} \end{cases}$$

In the frequentist setting, the *model* is a either a density, $f_X(\boldsymbol{x}; \boldsymbol{\theta})$, or a conditional density, $f_{Y|X}(y|\boldsymbol{x}; \boldsymbol{\theta})$, where $\boldsymbol{\theta}$ is a parameter vector. For simplicity we have assumed that X is continuous; if it takes a discrete set of values, then the model will be a mass function rather than a density. Parameter estimates $\hat{\theta}$ are computed using an *estimator function*, $\boldsymbol{t}$, where $\hat{\theta} = \boldsymbol{t}(D)$. The distribution of the random vector $\boldsymbol{t}(\mathcal{D})$ is the *sampling distribution* of the estimator. It is used to describe uncertainty in $\hat{\boldsymbol{\theta}}$. In the Bayesian setting, $\boldsymbol{\theta}$ is promoted to be a random vector, and the frequentist model is replaced with a conditional density, either $f_{X|\Theta}(\boldsymbol{x}|\boldsymbol{\theta})$ or $f_{Y|X,\Theta}(y|\boldsymbol{x}, \boldsymbol{\theta})$. We also define a prior $f_\Theta(\boldsymbol{\theta})$ which reflects our beliefs about the values of model parameters *before* observing D. The likelihood function is defined in one of two ways, depending on whether we are carrying out density estimation or regression:

$$\mathcal{L}(\boldsymbol{\theta}) = \begin{cases} f_{\underline{X}|\Theta}(\underline{x}|\boldsymbol{\theta}) = \prod_{i=1}^n f_{X|\Theta}(\boldsymbol{x}_i|\boldsymbol{\theta}) & \text{unconditional likelihood,} \\ f_{\underline{Y}|\underline{X},\Theta}(\underline{y}|\underline{x}, \boldsymbol{\theta}) = \prod_{i=1}^n f_{Y|X,\Theta}(y_i|\boldsymbol{x}_i, \boldsymbol{\theta}) & \text{conditional likelihood.} \end{cases}$$

The above expressions for the likelihood as products rely on assumptions about independence. In the unconditional likelihood, used for density estimation (unsupervised learning), we assume that the observations $X_1, \dots, X_n$ are independent. In the conditional likelihood, used for regression or classification (supervised learning), we assume that the response observations $Y_1, \dots, Y_n$ are independent, conditional on the predictors $X_1, \dots, X_n$. Our knowledge *after* observing D is contained in the *posterior*, obtained by applying Bayes' rule,

$$\left.\begin{array}{l} f_{\Theta|\underline{X}}(\theta|\underline{x}) \\ f_{\Theta|\underline{Y},\underline{X}}(\theta|\underline{y},\underline{x}) \end{array}\right\} \propto \mathcal{L}(\theta) f_\Theta(\theta) = \begin{cases} f_{\underline{X}|\Theta}(\underline{x}|\theta) f_\Theta(\theta), \\ f_{\underline{Y}|\underline{X},\Theta}(\underline{y}|\underline{x},\theta) f_\Theta(\theta). \end{cases}$$

The posterior can be used to produce parameter estimates, measure uncertainty in parameters, and generate random samples of those parameters. Note that the proportional relationship

$$\text{posterior} \propto \text{likelihood} \times \text{prior}$$

gives the posterior only up to a normalising constant, which depends on D (see Exercise 10.1). We will see later that we do not necessarily need to find this constant, and that in some cases it may not be practically possible to do so.

■ ***Exercise*** 10.1 By applying Bayes' rule, show that for density estimation problems the normalising constant, c, in the general expression for the posterior

$$f_{\Theta|\underline{X}}(\theta|\underline{x}) = \frac{\mathcal{L}(\theta) f_\Theta(\theta)}{c}$$

is given by the marginal distribution of $\underline{X}$ evaluated at $\underline{X} = \underline{x}$. That is,

$$c = f_{\underline{X}}(\underline{x}) = \int f_{\underline{X}|\Theta}(\underline{x}|\theta) f_\Theta(\theta) d\theta.$$

Find a similar expression for the normaliser in a regression or classification problem.

Finally, we highlight an important distinction in terminology between frequentist and Bayesian inference. For frequentists, the *model* is the probability density we assumed for the data. For Bayesians, the model includes both the conditional probability density of the data given the parameter(s) – the analogue of the frequentist model – *and* the prior.

10.2 Modelling Rainfall Statistics

You are planning to take your family camping in South West England during the last two weeks of August. Will the weather be OK? Figure 10.1 illustrates the pattern of daily rainfall depths over the course of a single year in South West England.[1] Days when rainfall exceeded 5 mm are marked with a red border; we'll call these 'rainy' days. Within out dataset, covering 1931–2024, the proportion of rainy days is 20%. Let X be the number of rainy days during your holiday. If we had a model, $\hat{f}_X(x)$, of the probability mass function of X, we could use it to address questions such as

- What is the most likely number of rainy days?
- What is the chance there will be no rainy days at all?
- What is the chance there will be more than k rainy days?

Figure 10.2a shows the empirical distribution of X based on data from the last two weeks of August between 1931 and 2024. Let us start with a simple Bayesian model of X which

[1] Data extracted from the Met Office UK regional precipitation series, Alexander and Jones (2000).

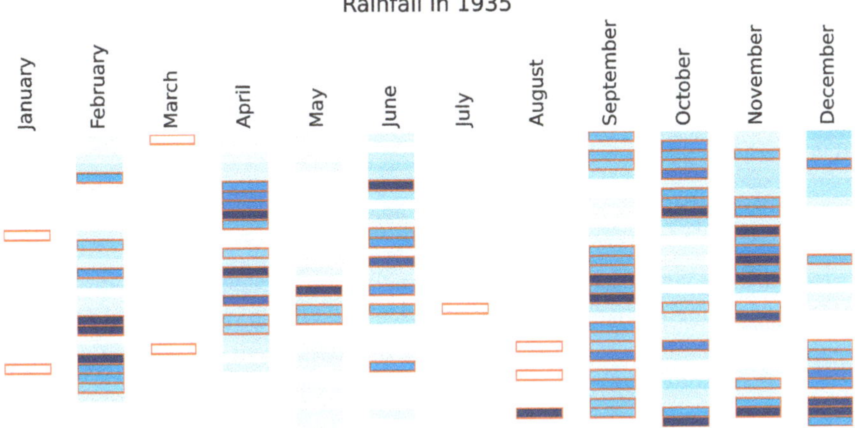

Figure 10.1 Daily rainfall pattern in Southwest England though 1935. Each rectangle corresponds to a single day with depth of colour indicating depth of rain. The days outlined in red exceed 5 mm rain.

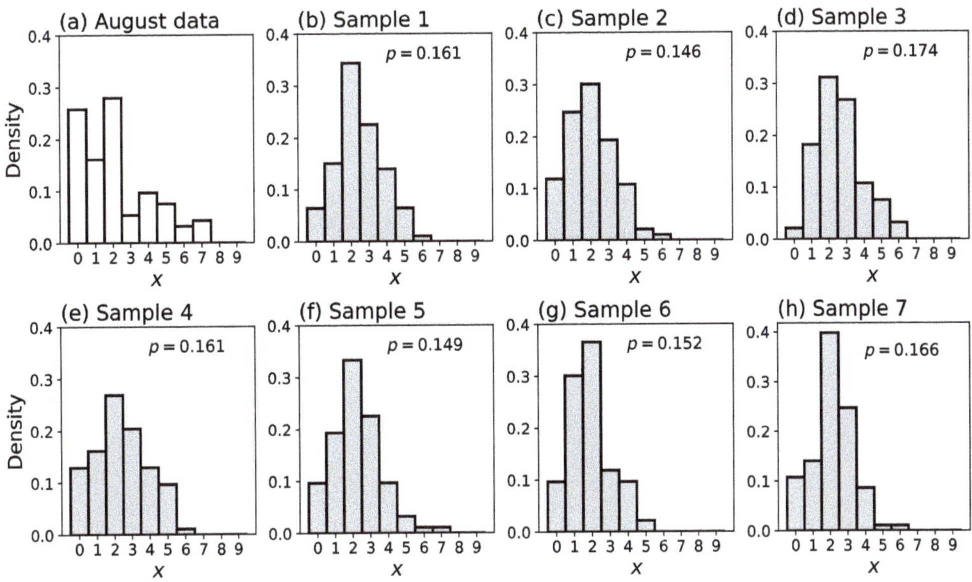

Figure 10.2 Probability distribution of the number of rainy days (> 5 mm) in the last two weeks of August. (a) Historical data (1931–2024). (b)–(h) Samples from the posterior predictive distribution using the binomial model with uniform prior.

will serve as a baseline against which we can compare more sophisticated alternatives. For each day $k \in \{1, \ldots, 14\}$, we define the *indicator* random variable

$$H_k = \begin{cases} 1 & \text{if day } k \text{ is rainy,} \\ 0 & \text{otherwise.} \end{cases}$$

We then have

$$X = \sum_{k=1}^{14} H_k.$$

We will assume that for all k, H_k has marginal distribution Bernoulli(P) conditional on latent variable $P \in [0, 1]$ (the probability of a rainy day). We will also assume that $H_i \perp\!\!\!\perp H_j$ for all $i \neq j$, so that X is the number of successes from 14 independent Bernoulli trials. Therefore, $X|\{P = p\} \sim \text{Binomial}(14, p)$, and

$$f_{X|P}(x|p) = \binom{14}{x} p^x (1-p)^{14-x}.$$

An advantage of this model is that provided we make an appropriate choice of prior, the posterior can be computed analytically. Let us assume that the prior on P is a Beta(α, β) distribution (see Box 6.6 for reference) having density

$$f_P(p) = \frac{p^{\alpha-1}(1-p)^{\beta-1}}{B(\alpha, \beta)}.$$

Our dataset consists of a series of observations of X, written $\underline{x} = (x_1, x_2, \ldots, x_n)^T$. We view $\underline{x}$ as the realisation of a random vector $\underline{X} = (X_1, \ldots, X_n)^T$, the components of which are i.i.d. copies of X. The DAG describing this model is

$$X_i \leftarrow P, \quad i = 1, \ldots, n \qquad (10.1)$$

Letting $s_n := \sum_{i=1}^n x_i$, the likelihood function may be written as follows:

$$\begin{aligned} \mathcal{L}(p) &= f_{\underline{X}|P}(\underline{x}|p) \\ &= \prod_{i=1}^n f_{X|P}(x_i|p) \\ &= \prod_{i=1}^n \binom{14}{x_i} p^{x_i}(1-p)^{14-x_i} \\ &= \left(\prod_{i=1}^n \binom{14}{x_i}\right) p^{s_n}(1-p)^{14n-s_n}. \end{aligned}$$

Notice that the product of combinatorial factors in the final line does not depend on p. The posterior distribution of P therefore satisfies

$$f_{P|\underline{X}}(p|\underline{x}) \propto f_P(p)\mathcal{L}(p) \propto p^{s_n+\alpha-1}(1-p)^{14n-s_n+\beta-1}.$$

Recognising the posterior as a member of the beta distribution family, we have

$$P|\{\underline{X} = \underline{x}\} \sim \text{Beta}(\alpha + s_n, \beta + 14n - s_n).$$

The fact that the posterior belongs to the same distribution family as the prior means the prior is conjugate to the likelihood. The beta and binomial distributions form a *conjugate pair.*

Our data covers the years 1931 to 2024, giving $n = 94$ years of data. The total number of rainy days during the last two weeks of August during this period was $s_n = 205$. Assuming a uniform prior on P (so $\alpha = \beta = 1$), we get the posterior

$$P|\{\underline{X} = \underline{x}\} \sim \text{Beta}(206, 1112).$$

As a simple sense check, the MAP estimate of P is $\hat{p} = s_n/(14n) = 0.156$. This is the historical fraction of rainy days in the last two weeks of August, unsurprisingly somewhat less than the probability of a rainy day without reference to the time of year (20%).

Why did we use a uniform prior? It is tempting to say that this prior is *uninformative*, and hence that it must be the methodologically sound choice in a situation where we have no relevant background knowledge. When we consider prior choice more carefully in Section 10.6, we will see that 'the uninformative prior' is an elusive concept. Nevertheless, in practice, priors which are intended to express open-mindedness are widely used.

10.3 Posterior Predictive Checks

When evaluating a fitted model, it is often a good idea to use the model to generate *synthetic* datasets and to compare them with the real data. To do this for our rainfall model, we need to consider the nature of the uncertainties captured by the posterior, $f_{P|\underline{X}}$, and by the conditional mass function, $f_{X|P}$. Broadly, uncertainty is either *aleatory* or *epistemic.* Epistemic uncertainty is a lack of knowledge about things we could in principle know, and which therefore must already have been decided. Aleatory or 'stochastic' uncertainty is a lack of knowledge of the outcome of future random events.

According to the assumptions of our model, a single realised value of P is responsible for *all* our observed data. The more data we collect, the more our posterior distribution will become concentrated around this value. Since (according to our assumptions) this realised value of P was decided *once, in the past*, our uncertainty about it is epistemic. Our uncertainty about future values of X is aleatory. To generate synthetic datasets, we repeat the process which we assumed was responsible for creating our observations. To generate the ith synthetic dataset we execute two steps.

1 **Epistemic step**. Generate a sample, p_i, from the posterior $f_{P|\underline{X}}(p|\underline{x})$.
2 **Aleatory step**. Generate n samples $x_{i1}, \ldots, x_{in}$ from the datapoint-generating model $f_{X|P}(x|p_i)$.

For now we will assume that we have the computational tools we need to carry out these sampling steps. Let random dataset $\underline{\tilde{X}}$ be the output of the two-step process. We have

$$f_{\underline{\tilde{X}}|\underline{X}}(\underline{\tilde{x}}|\underline{x}) = \int f_{\underline{X}|P}(\underline{\tilde{x}}|p) f_{P|\underline{X}}(p|\underline{x}) dp.$$

We will refer to this distribution as the *posterior dataset predictive distribution.* Seven samples from it are shown in Figure 10.2b–h. These reveal some problems with our model. It appears that particularly wet or particularly dry holiday periods are less common in the synthetic data than they are in the empirical data. Greater variability in observational data than in the model is known as *overdispersion*.

A quantitative way to measure discrepancies between model and data is to define a *statistic*, T, which can be computed from both our real and synthetic datasets, and which captures some aspect of the model which we deem to be important. For example, suppose that T_k is the number of holiday periods where there are at most k rainy days,

$$T_k(\underline{X}) = \sum_{i=1}^{n} \mathbf{1}_{\{X_i \leq k\}}.$$

Here $\mathbf{1}_A$ is the *indicator function*, equal to 1 if statement A is true and 0 otherwise. Using this statistic we can define a Bayesian 'p-value', equal to the probability that the statistic calculated for a sample from the posterior dataset predictive distribution is greater than or equal to the statistic calculated using the original data,

$$p_B(k) = \mathbf{Pr}\left(T_k(\underline{\tilde{X}}) \geq T_k(\underline{x})\right).$$

Figure 10.3a shows estimated p-values for our binomial model, obtained using 10^4 samples from the posterior dataset predictive. The smaller the value of our estimate $\hat{p}_B(k)$, the fewer synthetic datasets contain k or fewer rainy days. For example, we have $\hat{p}_B(0) = 0$, meaning that *none* of the datasets produced by the model contain as many holiday periods with no rainy days as were seen in the empirical data. That is, the model is producing unrealistically small numbers of very dry holidays. Similarly, when $k = 3$ and $k = 4$ almost all the datasets produced by the model contain more holiday periods with k or fewer rainy days than are present in the empirical data. That is, the model seems also to be failing to produce realistic numbers of particularly rainy holidays. The estimated p-values in Figure 10.3a (the black dots) are consistent with our initial observation of overdispersion in Figure 10.2.

In the preceding discussion, our suspicions were raised by p-values close to 0 or 1, but what is a 'reasonable' p-value? To answer this question, consider the following thought experiment. Suppose that our dataset was in fact a sample, $\underline{\tilde{X}}_0$, from the posterior dataset predictive. The p-value calculated using this sample may then be viewed as a random variable whose value is determined by $\underline{\tilde{X}}_0$

$$p_B(k; \underline{\tilde{X}}_0) = \mathbf{Pr}\left(T_k(\underline{\tilde{X}}) < T_k(\underline{\tilde{X}}_0) | \underline{\tilde{X}}_0\right).$$

If we repeatedly sample this random variable, we will obtain the distribution of p-values that we would expect to see *if the model were a good description of reality*. These distributions are illustrated in Figure 10.3a by means of box-and-whisker plots, with whiskers ranging from the 5th to the 95th percentiles. For $k \in \{0, 1, 3, 4\}$, the p-values we have observed would be very surprising if our posterior dataset predictive distribution, or at least a mechanism statistically very similar to it, was indeed responsible for generating our observed data. These observations suggest that it would be worth our while trying to a construct an improved model of X, and we will do this in Section 10.4.

Before moving on, we clarify some important data sampling terminology. Our synthetic data was created by generating *one* realisation of the parameter, followed by *many* realisations of the data. An alternative is to use a new posterior sample for each datapoint. Doing so produces samples from the 'posterior predictive distribution', defined for a general model in Box 10.1.

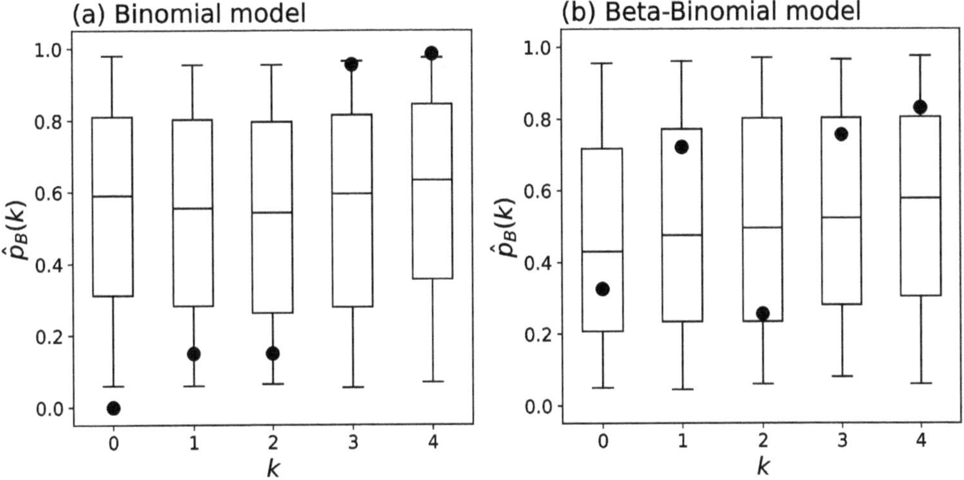

Figure 10.3 Dots show estimated Bayesian p-values $\mathbf{Pr}\left(T_k(\underline{\tilde{X}}) \geq T_k(\underline{x})\right)$ for (a) the binomial model and (b) the beta-binomial model. For each k, $p_B(k)$ was estimated by generating 10^4 samples from the posterior dataset predictive. Box-and-whisker plots show the distribution of p-values which would be obtained by replacing the empirical data with a sample from the posterior dataset predictive. The boxes show the middle 50% of these simulated p-values: they stretch from the first quartile to the third quartile. The black lines inside boxes mark medians. The whiskers show 5th and 95th percentiles.

★ Box 10.1 Posterior predictive distribution

Let $\underline{X}$ be a dataset of n observations of a random vector $\boldsymbol{X}$, and let $f_{\boldsymbol{X}|\boldsymbol{\Theta}}(\boldsymbol{x}|\boldsymbol{\theta})$ be a probability model of $\boldsymbol{X}$ given a parameter vector $\boldsymbol{\Theta}$. Assuming our dataset consists of i.i.d. realisations of $\boldsymbol{X}$ then our model of the dataset given the value of $\boldsymbol{\Theta}$ is as follows:

$$f_{\underline{X}|\boldsymbol{\Theta}}(\underline{x}|\boldsymbol{\theta}) = \prod_{i=1}^{n} f_{\boldsymbol{X}|\boldsymbol{\Theta}}(\boldsymbol{x}_i|\boldsymbol{\theta}) = \mathcal{L}(\boldsymbol{\theta}).$$

Let $f_{\boldsymbol{\Theta}|\underline{X}}(\boldsymbol{\theta}|\underline{x})$ be the posterior density of $\boldsymbol{\Theta}$. Let $\tilde{\boldsymbol{X}}$ be the random vector produced by drawing a *single* sample of the parameter vector from the posterior, and then drawing a *single* sample from the model of $\boldsymbol{X}$ using this parameter vector. The generating process of $\tilde{\boldsymbol{X}}$ is represented by the DAG

$$\tilde{X} \leftarrow \Theta \leftarrow \underline{x}. \tag{10.2}$$

The density of $\tilde{X}$ is the *posterior predictive* density,

$$f_{\tilde{X}|\underline{X}}(\tilde{x}|\underline{x}) = \int f_{X|\Theta}(\tilde{x}|\theta) f_{\Theta|\underline{X}}(\theta|\underline{x}) d\theta.$$

Now let $\underline{\tilde{X}}$ be the random dataset produced by drawing a *single* sample of the parameter vector from the posterior, and then drawing *multiple* samples from the model of X using this parameter vector. The generating process of $\underline{\tilde{X}}$ is represented by the DAG

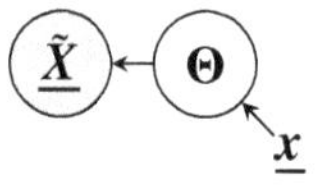

or, equivalently, 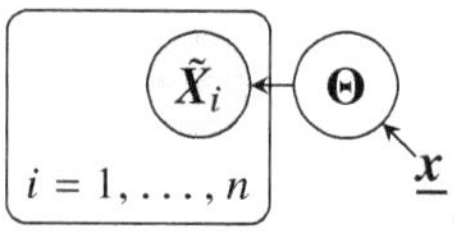 (10.3)

The density of $\underline{\tilde{X}}$ is the posterior **dataset** predictive,

$$f_{\underline{\tilde{X}}|\underline{X}}(\underline{\tilde{x}}|\underline{x}) = \int f_{\underline{X}|\Theta}(\underline{\tilde{x}}|\theta) f_{\Theta|\underline{X}}(\theta|\underline{x}) d\theta.$$

10.4 A New Model

Our binomial model of X contained the assumption that a *single* realised value of P was responsible for all the observed data. That is, we assumed that the chance of a rainy day during the last two weeks of August is always the same, irrespective of the year or the day in question. We discovered that the real data was *overdispersed* compared to this model. One possible explanation for this failure of the model is that the underlying propensity for rainy days in the last two weeks of August changes from one year to the next.

A simple way to adapt our original model to allow for such changes is to assume that each realisation of X is drawn from a binomial model with a different probability, P, of rain. To keep things simple, we will assume that the distribution from which P is drawn is the same every year. (We will not attempt to account for systematic long-term climate change.) Let's suppose that $P \sim \text{Beta}(\alpha, \beta)$ where α and β are unknown. Since we are being Bayesian, we express our (epistemic) uncertainty about their values by viewing them as realisations of random variables A and B which have some prior distribution, to be decided. This new model of the dataset-generating process is represented by the DAG

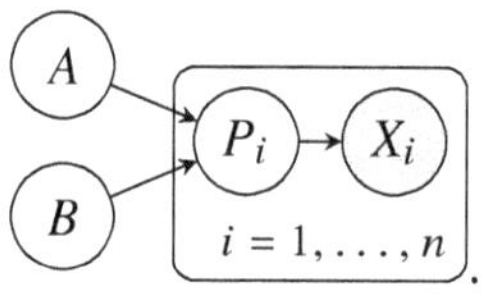

(10.4)

In this model, the variables A, B and P are all latent variables in the sense of being unobservable. However, A and B are of a different character to P. Whereas A and B are realised once, each realisation of X requires a new realisation of P. When it comes to estimating the distribution of *future* realisations of X, there is nothing to be gained by trying to estimate posterior distributions for *past* values of P. What we actually need is the posterior distribution of A and B given observations $\underline{x}$,

$$f_{A,B|\underline{X}}(\alpha, \beta|\underline{x}) \propto f_{\underline{X}|A,B}(\underline{x}|\alpha, \beta) f_{A,B}(\alpha, \beta),$$

where $f_{A,B}$ is a prior on A and B. Consider the likelihood

$$f_{\underline{X}|A,B}(\underline{x}|\alpha,\beta) = \prod_{i=1}^{n} f_{X|A,B}(x_i|\alpha,\beta).$$

Here, $f_{X|A,B}$ is the *marginal* distribution of X conditional on A and B. That is, the distribution of X given A and B, *without reference* to P. We compute it by integrating the joint distribution of X and P given A and B over all possible values of P, giving

$$f_{X|A,B}(x|\alpha,\beta) = \int_0^1 f_{X,P|A,B}(x,p|\alpha,\beta)dp.$$

Integrating over p in this way is known as 'marginalising out' the variable P. Making use of DAG (10.4), we have

$$\int_0^1 f_{X,P|A,B}(x,p|\alpha,\beta)dp = \int_0^1 f_{X|P}(x|p)f_{P|A,B}(p|\alpha,\beta)dp.$$

The expression on the right-hand side is the average of a binomial mass function over a beta-distributed success parameter. It is known as the 'beta-binomial' distribution, details of which are given in Box 10.2.

■ ***Exercise*** 10.2 Write down the factorisation of $f_{X,P,A,B}(x,p,\alpha,\beta)$ implied by DAG 10.4. Hence, show that $f_{X,P|A,B}(x,p|\alpha,\beta) = f_{X|P}(x|p)f_{P|A,B}(p|\alpha,\beta)$.

★ Box 10.2 Beta-binomial distribution

Let X be a binomial random variable where the success probability is unknown and drawn from a beta distribution. That is

$$X|\{P=p\} \sim \text{Binomial}(n,p),$$
$$P \sim \text{Beta}(\alpha,\beta).$$

The joint distribution of X and P is represented by the DAG

$$\alpha \rightarrow (P) \rightarrow (X), \quad \beta \rightarrow (P) \qquad (10.5)$$

The marginal distribution of X is

$$\begin{aligned} f_X(x;\alpha,\beta) &= \int_0^1 \text{Binomial}(x|n,p)\text{Beta}(p|\alpha,\beta)dp \\ &= \frac{1}{B(\alpha,\beta)}\binom{n}{x}\int_0^1 p^{x+\alpha-1}(1-p)^{n-x+\beta-1}dp \\ &= \binom{n}{x}\frac{B(x+\alpha,n-x+\beta)}{B(\alpha,\beta)}, \end{aligned}$$

where $B(u,v) = \Gamma(u)\Gamma(v)/\Gamma(u+v)$ is the beta function (see Box 6.6). The marginal

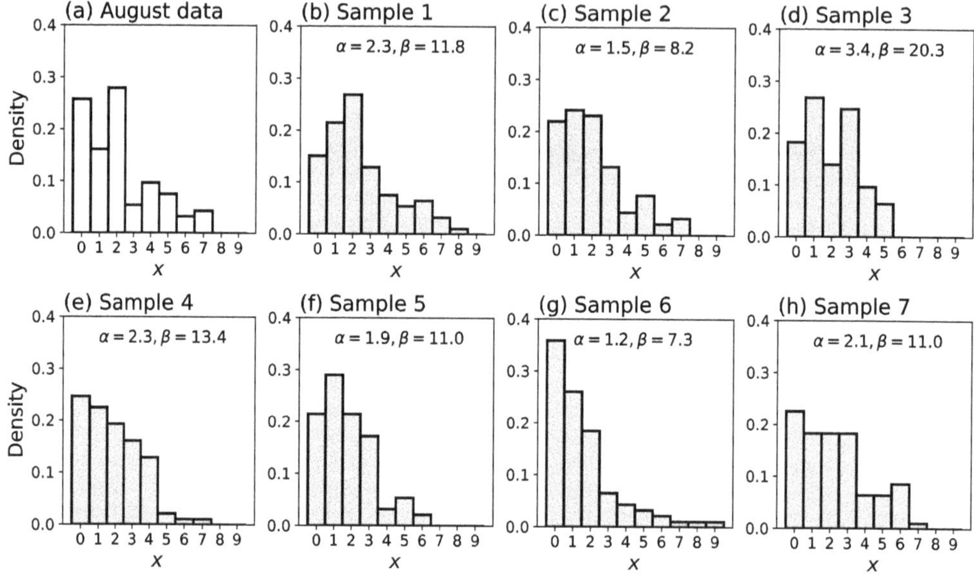

Figure 10.4 Probability distribution of the number of rainy days (> 5 mm) in the last two weeks of August. (a) Historical data (1931–2024). (b)–(h) Samples from the posterior predictive distribution using the beta-binomial model with half-normal priors $A \sim B \sim \text{HalfNormal}(50)$.

mass function of X is known as the beta-binomial mass function, $\text{BetaBin}(x|n, \alpha, \beta)$. If X is beta-binomial distributed we write $X \sim \text{BetaBin}(n, \alpha, \beta)$.

Since the marginal distribution of X given A and B is beta-binomial,

$$X|\{A = \alpha, B = \beta\} \sim \text{BetaBin}(14, \alpha, \beta),$$

the posterior distribution of A and B is

$$f_{A,B|\underline{X}}(\alpha, \beta|\underline{x}) = \frac{1}{c} f_{A,B}(\alpha, \beta) \prod_{i=1}^{n} \text{BetaBin}(x_i|14, \alpha, \beta),$$

where c is a constant required to normalise the posterior. In our first (purely binomial) model of X, it was possible to choose a prior which was conjugate to the model, meaning that the posterior belonged to a distribution family (the beta distributions) with well-understood statistical properties and easily accessible sampling algorithms.[2] With our more sophisticated beta-binomial model, this is no longer the case. This presents us with a challenge: how do we generate random samples from this new probability distribution, or estimate its statistical properties?

In Section 10.5, we introduce a general-purpose method for sampling from arbitrary probability distributions, giving us complete flexibility when choosing priors. Principles of prior choice will be covered in Section 10.6. Before making this detour, we provide a brief

[2] The SciPy library in Python, for example, is able to generate random variates from most known distributions.

preview of the results. Figure 10.4b–h shows samples from the posterior dataset predictive distribution of our new model, using a 'weakly informative' prior (see Section 10.6). It appears from these samples that we have cured the overdispersion problem that we saw in the baseline binomial model. Figure 10.3b shows Bayesian p-values for the new model. The observed p-values are not surprising given the hypothesis that the new model was responsible for generating the data. So, the new model looks like it may be an improvement on the baseline. In Section 10.8, we will show how the models can be compared quantitatively, using cross-validation.

10.5 Markov Chain Monte Carlo

Let us take a step back from our rainfall example and consider the general form of a Bayesian posterior. For density estimation this is

$$f_{\Theta|\underline{X}}(\theta|\underline{x}) = \frac{f_{\underline{X}|\Theta}(\underline{x}|\theta)f_{\Theta}(\theta)}{f_{\underline{X}}(\underline{x})}.$$

The denominator of this expression, known as the *marginal likelihood*, is given by

$$f_{\underline{X}}(\underline{x}) = \int f_{\underline{X}|\Theta}(\underline{x}|\theta)f_{\Theta}(\theta)d\theta.$$

Mathematically, the marginal likelihood is just a normalising constant for the posterior, but it may also be interpreted as a measure of the *evidence* for a model, and used for model comparison. While there are cases where this is an appropriate thing to do, such comparisons can be very sensitive to prior choice and we avoid them in this book.[3] Box 10.3 explains the concept of evidence in more detail, for the interested reader.

★ Box 10.3 Evidence

For simplicity we consider the case where X can take a discrete set of values. The model defined by the conditional mass function, $f_{\underline{X}|\Theta}$, and the prior, f_{Θ}, may be viewed as a *hypothesis*, H, about the data-generating process. In the language of probability, H is the *event* that the hypothesis is true. The quantity $f_{\underline{X}}(\underline{x})$ is then the (marginal) probability of observing the data $\underline{x}$, given hypothesis H,

$$f_{\underline{X}}(\underline{x}) = \mathbf{Pr}(\underline{X} = \underline{x}|H).$$

Now suppose we have two alternative models representing two alternative hypotheses $H_1 = (f^{(1)}_{\underline{X}|\Theta_1}, f^{(1)}_{\Theta_1})$ and $H_2 = (f^{(2)}_{\underline{X}|\Theta_2}, f^{(2)}_{\Theta_2})$. The marginal likelihoods of these two models give the conditional probabilities of the observed data, given the two different hypotheses. That is $f^{(k)}_{\underline{X}}(\underline{x}) = \mathbf{Pr}(\underline{X} = \underline{x}|H_k)$, where $k \in \{1, 2\}$. Bayes' theorem then allows us to compare the probabilities of the hypotheses given the data,

$$\frac{\mathbf{Pr}(H_1|\underline{X} = \underline{x})}{\mathbf{Pr}(H_2|\underline{X} = \underline{x})} = \frac{\mathbf{Pr}(H_1)\mathbf{Pr}(\underline{X} = \underline{x}|H_1)}{\mathbf{Pr}(H_2)\mathbf{Pr}(\underline{X} = \underline{x}|H_2)},$$

[3] See Gelman et al. (2024) for a detailed discussion.

where $\mathbf{Pr}(H_k)$ is the prior probability of hypothesis H_k. If we assume that both hypotheses are equally likely a priori ($\mathbf{Pr}(H_1) = \mathbf{Pr}(H_2)$), then the ratio of the probability of H_1 to the probability of H_2 given that $\underline{X} = \underline{x}$ is just the ratio of the marginal likelihoods of the two models. For this reason, the marginal likelihood is referred to as the *evidence* for a model.

If the posterior does not belong to a well-known distribution family then exploring its statistical properties can be challenging. There are at least two things we can do.

1 We can compute the evidence integral numerically, and then compute expectations by performing further numerical integrations.
2 We can use a general-purpose algorithm to generate random samples from the posterior and posterior predictive. These samples can then be used to estimate the statistical properties of the model.

While numerical integration is effective in low dimensions (a small number of parameters), it quickly becomes unfeasibly slow as the dimensionality increases. This is because the number of function evaluations required to evaluate a d-dimensional integral increases as the dth power of the number of evaluations needed to compute a one-dimensional integral. This is an example of the *curse of dimensionality*. Moreover, even if we can get numerical integration to work, we still need the ability to generate samples. The second approach – based entirely on random sampling – therefore looks more promising. We will introduce a general-purpose random sampling method known as 'Markov Chain Monte Carlo' (MCMC). Although other sampling methods exist, many of them are susceptible to the curse of dimensionality, or require tailoring to individual applications.[4] MCMC, which was discovered by physicists who needed a way to simulate random systems containing thousands of variables, requires nothing more than the unnormalised form of the density we are sampling from.

Taking a step back from Bayesian inference, we consider the problem of generating samples of a random vector $\boldsymbol{X} \in \mathbb{R}^p$ with density function given by

$$f_{\boldsymbol{X}}(\boldsymbol{x}) = \frac{f^*_{\boldsymbol{X}}(\boldsymbol{x})}{c}. \tag{10.6}$$

Here, $f^*_{\boldsymbol{X}}$ is an unnormalised density and c is a normalising constant which may or may not be known. We call $f_{\boldsymbol{X}}$ our *target density*. MCMC sampling works by performing a 'biased random walk' through p-dimensional sample space. In general, a random walk is a path created by taking random steps between locations. A particular observed sequence of locations may be written as follows:

$$\{\boldsymbol{x}_0, \boldsymbol{x}_1, \ldots\}.$$

We usually interpret the indices of these locations as discrete times. An observed sequence may be viewed as a single realisation of a sequence of random vectors,

$$\mathcal{S} = \{\boldsymbol{X}_0, \boldsymbol{X}_1, \boldsymbol{X}_2, \ldots\}.$$

In general, sequences of random vectors are known as *discrete time stochastic processes.*

[4] See, for example, Press et al. (2007) or MacKay (2003).

Suppose we have observed the first n steps of $\mathcal{S}$. The conditional probability density of the next location, $\boldsymbol{X}_{n+1}$, given the previous ones is

$$f_{\boldsymbol{X}_{n+1}|\boldsymbol{X}_0,\ldots,\boldsymbol{X}_n}(\boldsymbol{x}_{n+1}|\boldsymbol{x}_0,\ldots,\boldsymbol{x}_n).$$

If this conditional probability depends *only* on the immediately preceding location, $\boldsymbol{x}_n$, so that for all possible values of $\boldsymbol{x}_0,\ldots,\boldsymbol{x}_n$ we have

$$f_{\boldsymbol{X}_{n+1}|\boldsymbol{X}_0,\ldots,\boldsymbol{X}_n}(\boldsymbol{x}_{n+1}|\boldsymbol{x}_0,\ldots,\boldsymbol{x}_n) = f_{\boldsymbol{X}_{n+1}|\boldsymbol{X}_n}(\boldsymbol{x}_{n+1}|\boldsymbol{x}_n),$$

then we say that $\mathcal{S}$ is a *Markov process* or *Markov chain* with transition probability density $f_{\boldsymbol{X}_{n+1}|\boldsymbol{X}_n}(\boldsymbol{x}_{n+1}|\boldsymbol{x}_n)$. This is the probability that the walker will move to position $\boldsymbol{x}_{n+1}$ at the next step, given that she is currently at position $\boldsymbol{x}_n$. The probability density of the first n locations of a Markov process may be written in factorised form

$$f_{\boldsymbol{X}_0,\ldots,\boldsymbol{X}_n}(\boldsymbol{x}_0,\ldots,\boldsymbol{x}_n) = f_{\boldsymbol{X}_0}(\boldsymbol{x}_0)\prod_{k=0}^{n-1} f_{\boldsymbol{X}_{k+1}|\boldsymbol{X}_k}(\boldsymbol{x}_{k+1}|\boldsymbol{x}_k),$$

corresponding to the DAG

$$X_0 \rightarrow X_1 \rightarrow \cdots \rightarrow X_{n-1} \rightarrow X_n. \qquad (10.7)$$

■ ***Exercise*** 10.3 Suppose that $\mathcal{S} = \{\boldsymbol{X}_0, \boldsymbol{X}_1, \boldsymbol{X}_2, \ldots\}$ is a Markov process. Beginning from the factorisation

$$f_{\boldsymbol{X}_0,\ldots,\boldsymbol{X}_n}(\boldsymbol{x}_0,\ldots,\boldsymbol{x}_n) = f_{\boldsymbol{X}_n|\boldsymbol{X}_0,\ldots,\boldsymbol{X}_{n-1}}(\boldsymbol{x}_n|\boldsymbol{x}_0,\ldots,\boldsymbol{x}_n)f_{\boldsymbol{X}_0,\ldots,\boldsymbol{X}_{n-1}}(\boldsymbol{x}_0,\ldots,\boldsymbol{x}_{n-1}),$$

use the Markov property to show that

$$f_{\boldsymbol{X}_0,\ldots,\boldsymbol{X}_n}(\boldsymbol{x}_0,\ldots,\boldsymbol{x}_n) = f_{\boldsymbol{X}_0}(\boldsymbol{x}_0)\prod_{k=0}^{n-1} f_{\boldsymbol{X}_{k+1}|\boldsymbol{X}_k}(\boldsymbol{x}_{k+1}|\boldsymbol{x}_k).$$

If our transition probabilities do not depend on n (i.e. they are independent of time), we say that the process is *stationary*. In that case, we can simplify our notation to

$$f_{\boldsymbol{X}_{n+1}|\boldsymbol{X}_n}(\boldsymbol{x}|\boldsymbol{y}) = p(\boldsymbol{x}|\boldsymbol{y}),$$

where $\boldsymbol{x}$ and $\boldsymbol{y}$ are any two points in $\mathbb{R}^p$. Here $p(\boldsymbol{x}|\boldsymbol{y})$ is the transition probability density from $\boldsymbol{y}$ to $\boldsymbol{x}$ *in one step*.

Suppose that the form of our transition probability density is such that the walker is attracted to a certain region of $\mathbb{R}^p$. If the walker starts at position $\boldsymbol{x}_0$, then the conditional probability density of her location at step n is $f_{\boldsymbol{X}_n|\boldsymbol{X}_0}(\boldsymbol{x}|\boldsymbol{x}_0)$. As $n \to \infty$, this conditional density will become

- Progressively less sensitive to her start point.
- Concentrated in the region to which she is attracted.

In the limit $n \longrightarrow \infty$, the probability density of the walker's location will no longer depend on $\boldsymbol{X}_0$. Formally we define the long-run probability density of the walker's location to be

$$\pi(\boldsymbol{x}) = \lim_{n\to\infty} f_{\boldsymbol{X}_n|\boldsymbol{X}_0}(\boldsymbol{x}|\boldsymbol{x}_0).$$

Here $f_{\boldsymbol{X}_n|\boldsymbol{X}_0}(\boldsymbol{x}|\boldsymbol{x}_0)$ may be understood as an n step transition probability density, which can be written in terms of the $n-1$ step transition probability, $f_{\boldsymbol{X}_{n-1}|\boldsymbol{X}_0}(\boldsymbol{y}|\boldsymbol{x}_0)$, and the one-step transition probability, $p(\boldsymbol{x}|\boldsymbol{y})$, as follows:

$$f_{\boldsymbol{X}_n|\boldsymbol{X}_0}(\boldsymbol{x}|\boldsymbol{x}_0) = \int p(\boldsymbol{x}|\boldsymbol{y}) f_{\boldsymbol{X}_{n-1}|\boldsymbol{X}_0}(\boldsymbol{y}|\boldsymbol{x}_0) d\boldsymbol{y}.$$

Taking the limit $n \longrightarrow \infty$, we obtain

$$\pi(\boldsymbol{x}) = \int p(\boldsymbol{x}|\boldsymbol{y})\pi(\boldsymbol{y}) d\boldsymbol{y}. \tag{10.8}$$

This is an important relationship. It can be understood as follows. Imagine that our walker has been moving around unobserved for a very long time, long enough for the probability density describing her location to have converged to a stable form, $\pi(\boldsymbol{x})$. If she happens currently to be at position $\boldsymbol{y}$ and takes one more step then the probability density of her new location will be $p(\boldsymbol{x}|\boldsymbol{y})$. Averaging over all possible values of her current location $\boldsymbol{y}$, weighted by their probability densities, $\pi(\boldsymbol{y})$, we obtain the probability density of her new location, equal to the right-hand side of (10.8). Since we have assumed that the probability density of her location is stable, it must remain unchanged after taking this one extra step, and so must remain equal to $\pi(\boldsymbol{x})$. Equation (10.8) expresses this invariance. A distribution which satisfies (10.8) is said to be an *equilibrium* distribution of the Markov process with transition density $p(\boldsymbol{x}|\boldsymbol{y})$.

The idea of MCMC is to design a transition density so that the equilibrium distribution of the process it describes is equal to the probability density we are trying to sample from. The set of locations visited by a random walk generated using this transition density, provided it contains enough steps, will then represent a large sample from our target density. However, we need to be careful because the locations visited by the walker will not be independent of one another – MCMC is a *dependent* sampling method. We will return to this point later. To design $p(\boldsymbol{x}|\boldsymbol{y})$, we use the idea of *detailed balance.* A probability density $\tilde{\pi}$ is said to satisfy detailed balance with respect to the transition probability $p(\boldsymbol{x}|\boldsymbol{y})$ if

$$p(\boldsymbol{x}|\boldsymbol{y})\tilde{\pi}(\boldsymbol{y}) = p(\boldsymbol{y}|\boldsymbol{x})\tilde{\pi}(\boldsymbol{x})$$

for all $\boldsymbol{x}, \boldsymbol{y} \in \mathbb{R}^p$. Suppose $\tilde{\pi}$ satisfies detailed balance. Then

$$\int p(\boldsymbol{x}|\boldsymbol{y})\tilde{\pi}(\boldsymbol{y}) d\boldsymbol{y} = \tilde{\pi}(\boldsymbol{x}) \underbrace{\int p(\boldsymbol{y}|\boldsymbol{x}) d\boldsymbol{y}}_{=1} = \tilde{\pi}(\boldsymbol{x}).$$

Hence, $\tilde{\pi}$ is an equilibrium of the chain. Our goal is to choose $p(\boldsymbol{x}|\boldsymbol{y})$ so that the target density $f_{\boldsymbol{X}}$ (equation (10.6)) is an equilibrium. Setting $\tilde{\pi} = f_{\boldsymbol{X}}$ in our detailed balance equation, we obtain

$$p(\boldsymbol{x}|\boldsymbol{y}) f^*_{\boldsymbol{X}}(\boldsymbol{y}) = p(\boldsymbol{y}|\boldsymbol{x}) f^*_{\boldsymbol{X}}(\boldsymbol{x}), \tag{10.9}$$

where we have cancelled the normalising constant from both sides. We now define the *Metropolis Algorithm*, for generating a stationary Markov chain $\boldsymbol{X}_0, \boldsymbol{X}_1, \boldsymbol{X}_2, \ldots$ We state the algorithm in Box 10.4 before demonstrating that its transition density satisfies (10.9).

★ Box 10.4 Metropolis algorithm

Choose $\boldsymbol{X}_0$. Suppose we have generated $\boldsymbol{X}_1, \boldsymbol{X}_2, \ldots, \boldsymbol{X}_n$. Let $r(\boldsymbol{y}|\boldsymbol{x})$ be a 'proposal' distribution which is *symmetric* so that $r(\boldsymbol{x}|\boldsymbol{y}) = r(\boldsymbol{y}|\boldsymbol{x})$. If $\boldsymbol{X}_n = \boldsymbol{x}$, to generate $\boldsymbol{X}_{n+1}$ we

1 Generate a 'candidate' random vector $\boldsymbol{Y}$ from the proposal density $r(\boldsymbol{y}|\boldsymbol{x})$.
2 Evaluate the *acceptance probability*

$$a(\boldsymbol{x}, \boldsymbol{y}) = \min\left\{\frac{f_{\boldsymbol{X}}^*(\boldsymbol{y})}{f_{\boldsymbol{X}}^*(\boldsymbol{x})}, 1\right\},$$

where $\boldsymbol{y}$ is the realised value of $\boldsymbol{Y}$ in step 1.
3 Set

$$\boldsymbol{X}_{n+1} = \begin{cases} \boldsymbol{y} & \text{with probability } a(\boldsymbol{x}, \boldsymbol{y}), \\ \boldsymbol{x} & \text{with probability } 1 - a(\boldsymbol{x}, \boldsymbol{y}). \end{cases}$$

The Metropolis algorithm with a general acceptance probability $a(\boldsymbol{x}, \boldsymbol{y})$ corresponds to a transition density satisfying

$$p(\boldsymbol{y}|\boldsymbol{x}) = a(\boldsymbol{x}, \boldsymbol{y}) r(\boldsymbol{y}|\boldsymbol{x})$$

for any $\boldsymbol{x} \neq \boldsymbol{y}$. Substituting into equation (10.9), and making use of the symmetry of the proposal distribution, we obtain the condition we want the acceptance probability to satisfy:

$$a(\boldsymbol{y}, \boldsymbol{x}) f_{\boldsymbol{X}}^*(\boldsymbol{y}) = a(\boldsymbol{x}, \boldsymbol{y}) f_{\boldsymbol{X}}^*(\boldsymbol{x}).$$

You can now check that the acceptance probability defined in Box 10.4 satisfies the condition.

■ ***Exercise*** 10.4 By considering the separate cases $f_{\boldsymbol{X}}^*(\boldsymbol{x}) > f_{\boldsymbol{X}}^*(\boldsymbol{y})$ and $f_{\boldsymbol{X}}^*(\boldsymbol{y}) \geq f_{\boldsymbol{X}}^*(\boldsymbol{x})$, show that the acceptance probability $a(\boldsymbol{x}, \boldsymbol{y}) = \min\{f_{\boldsymbol{X}}^*(\boldsymbol{y})/f_{\boldsymbol{X}}^*(\boldsymbol{x}), 1\}$ is a solution to $a(\boldsymbol{y}, \boldsymbol{x}) f_{\boldsymbol{X}}^*(\boldsymbol{y}) = a(\boldsymbol{x}, \boldsymbol{y}) f_{\boldsymbol{X}}^*(\boldsymbol{x})$.

Figure 10.5 shows three random walks, each generated using 500 steps of the Metropolis algorithm, with $f_{\boldsymbol{X}}^*$ given by the sum of two normal densities. The proposal distributions in these examples are normal,

$$\boldsymbol{Y}|\{\boldsymbol{X}_n = \boldsymbol{x}\} \sim N(\boldsymbol{x}, b^2 \underline{\boldsymbol{I}}),$$

where b determines the average step size. The walk in each sub-figure was generated with a different b value. The step size determines how efficiently the chain explores the target distribution, or 'mixes'. The plots are annotated with the *acceptance rate*, $\bar{a}$, equal to the fraction of candidate locations which were accepted. In Figure 10.5a, the step size is too small, so even though almost all candidate locations are accepted, the chain doesn't explore much of the target density. In Figure 10.5c, the steps size is too large, causing most candidate locations to be in regions of very low probability density. These candidates typically have

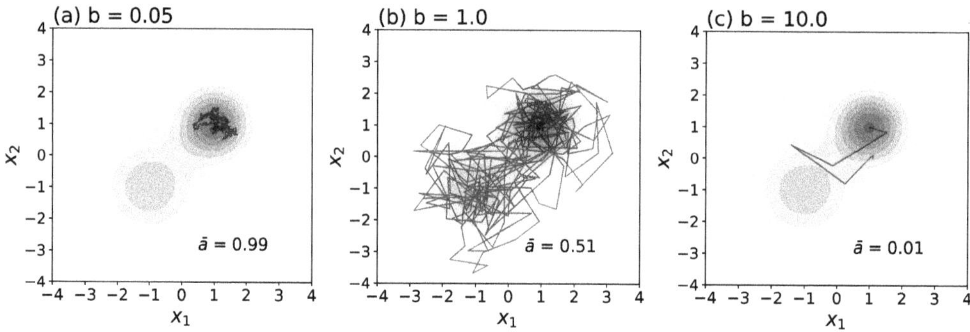

Figure 10.5 (a–c) Random walks generated using 500 Metropolis steps with normal proposal distributions $r(\boldsymbol{y}|\boldsymbol{x}) = \mathcal{N}(\boldsymbol{y}|\boldsymbol{x}, b^2\underline{\boldsymbol{I}})$. The unnormalised density, $f^*_{\boldsymbol{X}}(\boldsymbol{x})$, is the sum of two normal densities with parameters $\boldsymbol{\mu}_1 = (-1, -1)^T$, $\sigma_1 = 1$, $\boldsymbol{\mu}_2 = (1, 1)^T$, $\sigma_2 = 0.75$. Chains are annotated with the acceptance rate $\bar{a}$.

a low acceptance probability so the acceptance rate is also low. In both cases, we would need to run the chain for a very long time to generate a substantial sample from $f_{\boldsymbol{X}}$. When selecting b we need to find a sweet spot between excessively small values which mean the walker moves too slowly around sample space, and excessively large values which mean that she takes very few steps. A useful rule of thumb is to set b so that about half of the proposed steps are accepted. From Figure 10.5b, we see that when $b = 1.0$, the acceptance rate is $\bar{a} = 0.51$, and the chain is well mixed.

■ ***Exercise*** 10.5 Demonstrate that if $f^*_{\boldsymbol{X}}(\boldsymbol{x})$ is a continuous function then the Metropolis acceptance probability approaches one as the step size approaches zero.

It is useful to think intuitively about how the Metropolis algorithm works. Suppose hypothetically that we accepted every proposed new location.[5] The resulting random walk would, over time, explore further and further from its starting position, its probability density becoming increasingly spread out over sample space. By introducing the Metropolis acceptance probability, we make it more likely that the chain will step towards locations where the target probability density is higher. In other words, the Metropolis walk is biased so that it tends to move 'uphill' in sample space, and therefore spends more time in regions where the target density is higher. Moreover, the strength of this bias is just right: the expected time spent in a region is proportional to the total probability in that region.

The Metropolis algorithm is a simple general method for generating MCMC samples. In practice, we often use more sophisticated alternatives. When the conditional target density of each variable given the others is known, we can cycle through each variable in turn, sampling new locations from the corresponding conditional densities. This approach, known as 'Gibbs sampling', produces a move in sample space at *every* step. A state-of-the-art family of methods known as 'Hamiltonian Monte Carlo' (HMC) add momentum to the random walk, allowing it to explore the target density more efficiently in high dimensions.

[5] Equivalent to Metropolis sampling from an improper density with $f^*(\boldsymbol{x}) = \text{const.}$

10.5.1 Mixing Time Lower Bound

Now that we have an MCMC algorithm, we need to know how many steps must be taken until the probability density of the location of the random walker is 'close' to its equilibrium form, which by design is the target density. The number of steps needed to achieve this is the *mixing time* of the chain.

We will start by estimating a **lower bound** on the mixing time.[6] Suppose we want to sample from the density function, f_X, of a single random variable. Let L be the size of an interval which contains most of the probability density of f_X. We think of this interval as the plausible *state space* of X. We call L the *length scale* of the distribution. For a normal distribution with variance σ^2, a reasonable choice would be $L = 6\sigma$. That is, three standard deviations either side of the mean (accounting for 99.7% of the probability density). Equilibrium has been reached once the probability density of the walker's location does not depend on the initial condition. If the chain hasn't been running long enough to travel a distance L, then its current position must be dependent on where it started. The time taken to cover a distance L is therefore a *lower bound* on the mixing time.

To estimate the time needed to travel a distance L, we need to know how far, on average, a chain will travel in T steps. Calculating the expected distance moved by a Metropolis random walk of a given number of steps is not possible in general. However, we can estimate it by approximating the acceptance probability as a constant equal to its average over many steps $a(x, y) \approx \bar{a}$. In that case, a Metropolis walk of T steps is equivalent to a (symmetric) random walk of $N = \bar{a}T$ steps with transition density

$$r(y|x) = \mathcal{N}(y|x, b^2).$$

To distinguish this random walk from our Metropolis walk, we write its locations $S_0, S_1, \ldots, S_N$. Then, letting $\Delta_i = S_i - S_{i-1}$ we have

$$S_N = S_0 + \sum_{i=1}^{N} \Delta_i,$$
$$\Delta_i \sim \mathcal{N}(0, b^2),$$

with $\Delta_i \perp\!\!\!\perp \Delta_j$ for $i \neq j$. The expectation of the squared displacement of the chain from its starting point is

$$\mathbb{E}((S_N - S_0)^2) = \mathbb{E}\left(\left(\sum_{i=1}^{N} \Delta_i\right)^2\right) = Nb^2,$$

so $\mathbb{E}(|S_N - S_0|) \approx \sqrt{\mathbb{E}((S_N - S_0)^2)} = b\sqrt{N}$. For a Metropolis walk of T steps, we therefore have $\mathbb{E}(|X_T - X_0|) = \sqrt{\bar{a}Tb^2}$. Setting this distance equal to the length scale, L, we obtain the following 'rule of thumb' lower bound on the mixing time

$$T_{\text{mix}} \approx \frac{1}{\bar{a}}\left(\frac{L}{b}\right)^2. \tag{10.10}$$

In $d > 1$ dimensions, suppose $f_X(\boldsymbol{x})$ is concentrated inside a hypercube of side L. Then the

[6] We take a similar approach to MacKay (2003).

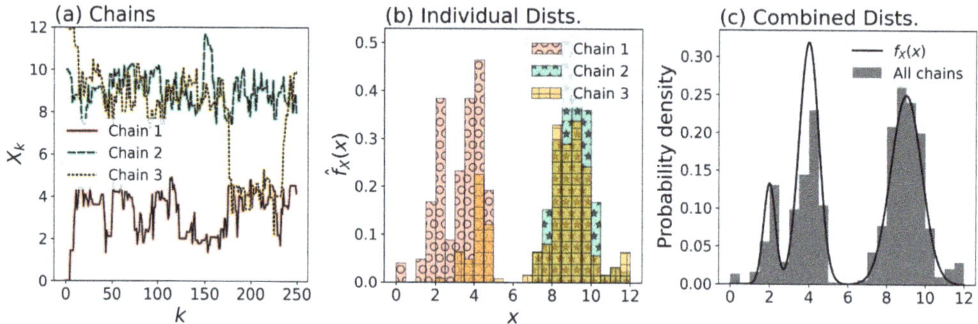

Figure 10.6 (a) Trace plots of three chains sampling from $f_X(x)$ (shown in panel (c)) with $b = 1$ ($\bar{a} \approx 0.6$). (b) Individual histograms of the three chains. (c) Density $f_X(x)$ and histogram of combined chains.

distance across the cube is $\sqrt{d}L$. We can approximate our Metropolis walk with a symmetric random walk described by position vector $\boldsymbol{S}_i$ and step vectors $\boldsymbol{\Delta}_i = (\boldsymbol{S}_i - \boldsymbol{S}_{i-1}) \sim \mathcal{N}(\boldsymbol{0}, b^2\underline{\boldsymbol{I}})$. In this case we find that $\mathbb{E}(\|\boldsymbol{S}_N - \boldsymbol{S}_0\|^2) = Ndb^2$ (see Exercise 10.6). For a Metropolis walk of T steps we therefore have

$$\mathbb{E}(\|\boldsymbol{X}_T - \boldsymbol{X}_0\|) \approx \sqrt{\bar{a}Tdb^2}.$$

Setting this equal to the distance across the hypercube, we obtain the same rule of thumb (10.10) as we did in one dimension.

■ ***Exercise*** 10.6 Let $\boldsymbol{\Delta}_1, \boldsymbol{\Delta}_2, \ldots, \boldsymbol{\Delta}_N$ be d-dimensional i.i.d. random walk step vectors with common distribution $\boldsymbol{\Delta}_i \sim \mathcal{N}(\boldsymbol{0}, b^2\underline{\boldsymbol{I}})$. Let $\boldsymbol{S} = \sum_{i=1}^{N} \boldsymbol{\Delta}_i$ be the position after N steps, starting from the origin.

(a) When $d = 1$, the steps are scalars $\Delta_i \sim \mathcal{N}(0, b^2)$. Show that $\mathbb{E}(S) = 0$ and hence that $\text{Var}(S) = \mathbb{E}(S^2)$. Hence show that $\mathbb{E}(S^2) = Nb^2$.
(b) When $d > 1$ we have $\mathbb{E}(\boldsymbol{\Delta}_i^T \boldsymbol{\Delta}_j) = db^2\delta_{ij}$. Show that $\mathbb{E}(\|\boldsymbol{S}\|^2) = \mathbb{E}(\boldsymbol{S}^T\boldsymbol{S}) = Ndb^2$.

Let us apply rule (10.10) to the chains illustrated in Figure 10.5. A generous estimate for the length scale is $L \approx 10$. When $b = 0.05$ we find that $\bar{a} = 0.99$ so $T_{\text{mix}} \approx 4 \times 10^4$. We would therefore need to run the chain for about a hundred times as many steps as were used in Figure 10.5a in order to reach equilibrium. When $b = 10$ we have $\bar{a} = 0.01$ so $T_{\text{mix}} \approx 100$. This is the expected time for *one* candidate point to be accepted, which makes sense because one step can take the walk *anywhere* in the plausible state space. When $b = 1$ we have $\bar{a} = 0.51$ so $T_{\text{mix}} \approx 200$. According to our approximations, the chains in Figures 10.5b and c can both be considered to have reached equilibrium.

10.5.2 The Need for Multiple Chains

Our rule of thumb (10.10) is a lower bound. When the target density forms a single 'blob' in sample space, then the mixing time will typically be close to this estimate. If not, then the

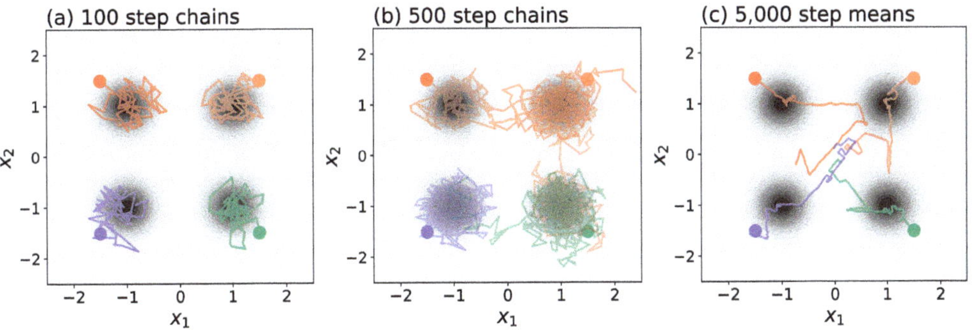

Figure 10.7 (a) Chains sampling a four-blob density. Large dots show initial locations. (b) The same chains but with more steps. (c) Trajectories of mean locations over first n steps for $n \in \{1, \ldots, 5{,}000\}$.

real mixing time can be much larger. Figure 10.6a shows *trace plots* of three chains sampling from the density in Figure 10.6c, which contains three peaks, one of which is well separated from the other two. To pass through the low-density region in between requires a 'lucky' sequence of moves with low acceptance probability, causing chains to get stuck on one side of the gap for extended periods of time. This trapping effect means that the influence of the initial location of the chain can last much longer than the lower bound we previously estimated for the mixing time.

Trace plots of multiple chains are a useful way to detect slow mixing. If we see different traces getting stuck in different locations, this is a sign that further checks are needed to be confident that we are running our chains for long enough. If we are sampling a multivariate density, then we need to examine trace plots for every variable. We can also compare the histograms of the points sampled by each chain. Although we do not expect different chains to produce identical histograms – they are only meant to be finite *samples* from the target density – if there are prominent peaks in some histograms but not others, this is a sign that the chains have not individually had long enough to reach equilibrium. We can see this phenomenon in Figure 10.6b. Conversely, if the histograms of all our chains look very similar, this is a sign that we are comfortably beyond the mixing time.

10.5.3 Practical MCMC

Mixing time estimates and visual inspection of trace plots are useful ways to check for convergence for simple densities in low dimensions. A more general approach is to use a *convergence metric*. A widely adopted metric is $\hat{R}$ ('R-hat'), proposed by Gelman and Rubin (1992). We will provide a plausibility argument for the metric as a convergence measure. Readers should consult Gelman and Rubin (1992) for a rigorous derivation.

Suppose we simulate m chains, each of n steps, initialised at locations which are *overdispersed* with respect to the target. Consider a single parameter, Θ, typically a coordinate of a chain, and let σ^2 be its variance under the target distribution (usually a Bayesian posterior). We define the *within-chain variance*, W, to be the average of the sample variances of Θ *within* each of the individual chains. The within-chain variance will tend to underestimate σ^2 for finite n because the individual chains will not have had time to range over all of the target distribution. For example, Figure 10.7a shows four chains exploring a density consisting of four Gaussian blobs. Early on, each chain is trapped in a single blob, so that $W \ll \sigma^2$. Eventually the chains will escape and explore other blobs (Figure 10.7b), so W will increase.

We now define a second variance measure. First, let $\hat{\sigma}^2_{\bar{\theta}}$ be the sample variance of the set of mean values of Θ for each chain. This variance measures the variability *between* chains. For large n, we will have

$$\hat{\sigma}^2_{\bar{\theta}} \approx \frac{\sigma^2_A}{n},$$

where $\sigma^2_A \neq \sigma^2$ because successive chain locations are not independent (see Box 10.5). Typically $\sigma^2_A > \sigma^2$. Figure 10.7c shows the mean locations of four chains. Over time, these means converge to the true mean, but in the earlier stages their differences (and hence their sample variance) will be large because the chains were started in overdispersed locations and have not had time to explore the target. The *between-chain variance* is then defined $B = n\hat{\sigma}^2_{\bar{\theta}}$. For the reasons just discussed, B substantially overestimates σ^2 while the chains have not fully explored the target.

Let us take stock. The within-chain and between-chain variances, respectively, underestimate and overestimate the variance. If we were to consider only within-chain variance, there is a danger our chains could be failing to explore the target – they could be stuck in localised blobs – and we would have no way of knowing. To avoid this problem, we should also consider the discrepancy between W and B. The idea of the R-hat metric is to first define a variance estimate which incorporates this discrepancy, $\hat{V} = W + n^{-1}(B - W)$, and then to compare this with the pure within-chain estimate as follows:

$$\hat{R} = \sqrt{\frac{\hat{V}}{W}} = \sqrt{\frac{W + n^{-1}(B - W)}{W}}.$$

The better mixed the chains, the closer $\hat{R}$ will be to one. A recommended recipe for generating well-converged samples is to simulate a set of chains, discard the first halves and declare what remains to be 'well converged' provided that $\hat{R} < 1.01$.

As well as checking for convergence, we also need to consider whether we have enough samples to estimate the quantities we are interested in. The fact that MCMC produces dependent samples means we typically need more of them to achieve a given level of accuracy. The *effective sample size*, n_{eff}, of a chain is the number of *independent* samples it would take to match the accuracy of estimates computed using a chain of length n. To calculate n_{eff}, consider an MCMC sample of a random variable Θ. The sample is a realisation of the stationary stochastic process $\Theta_0, \Theta_1, \ldots$. In a stationary chain, the dependence between states that are τ steps apart is measured by the *autocorrelation function*

$$\rho(\tau) = \frac{\text{Cov}(\Theta_i, \Theta_{i+\tau})}{\text{Var}(\Theta_i)},$$

where $\text{Var}(\Theta_i) = \text{Var}(\Theta)$ for all i. MCMC states which are closer together in time are more strongly dependent, so $\rho(\tau)$ is a decreasing function, with $\rho(0) = 1$. The correlation time, τ_c, is the time separation beyond which two states are approximately independent. Given a chain $\underline{\Theta} = (\Theta_1, \ldots, \Theta_n)$, define the mean estimator $t(\underline{\Theta}) = n^{-1} \sum_{i=1}^{n} \Theta_i$. When $n \gg \tau_c$ the variance of this estimator – its squared standard error – is given by

$$\mathrm{Var}(t(\underline{\Theta})) \approx \frac{\mathrm{Var}(\Theta)}{n} \left(1 + 2 \sum_{\tau=1}^{n-1} \rho(\tau) \right) = \mathrm{se}^2_{\mathrm{dep}}(n). \tag{10.11}$$

This result may be derived using the results in Box 10.5. If the states of the chain were independent, then we would have $\rho(\tau) = 0$ for $\tau \geq 1$, reducing (10.11) to $\mathrm{Var}(t(\underline{\Theta})) = n^{-1}\mathrm{Var}(\Theta) = \mathrm{se}^2_{\mathrm{ind}}(n)$. We find n_{eff} by matching the independent and dependent sample errors $\mathrm{se}_{\mathrm{ind}}(n_{\mathrm{eff}}) = \mathrm{se}_{\mathrm{dep}}(n)$, so

$$n_{\mathrm{eff}} = \frac{n}{1 + 2 \sum_{\tau=1}^{n-1} \rho(\tau)}.$$

A ‘plug-in’ estimate of n_{eff} may be obtained by replacing the autocorrelation function values with sample estimates, with the caveat that the sum over τ is truncated at the point where these estimates become too noisy. Our estimate of n_{eff} can be used to determine how many MCMC steps are required in order to generate a given number of independent samples, or to estimate standard errors in sample statistics.

★ Box 10.5 Autocorrelation and standard error

Given a sequence of n states of a stationary process $\underline{\Theta} = (\Theta_1, \Theta_2, \ldots, \Theta_n)$, an estimator of the mean of Θ is $t(\underline{\Theta}) = n^{-1} \sum_{i=1}^{n} \Theta_i$. Using Bienaymé’s identity (Box 7.1), the variance of this estimator may be written in terms of the autocorrelation function as

$$\mathrm{Var}(t(\underline{\Theta})) = \frac{1}{n^2} \sum_{i=1}^{n} \sum_{j=1}^{n} \mathrm{Cov}(\Theta_i, \Theta_j) = \frac{\mathrm{Var}(\Theta)}{n} \left(1 + 2 \sum_{\tau=1}^{n-1} \left(1 - \frac{\tau}{n}\right) \rho(\tau) \right). \tag{10.12}$$

For uncorrelated samples $\rho(\tau > 0) = 0$, we have $\mathrm{Var}(t(\underline{\Theta})) = n^{-1}\mathrm{Var}(\Theta)$.

■ ***Exercise*** 10.7 Establish identity (10.12) in Box 10.5.

10.6 How to Choose Priors

In the Bayesian setting, specifying a model requires specifying *both* a rule that maps possible parameter values to conditional probability distributions over observables given the parameter values (the frequentist ‘model’) *and* a prior distribution over the parameters. Whereas the mapping from parameters to conditional distributions is usually chosen from a discrete set of alternative families, the prior is selected from a continuous space of functions – the space of probability densities over the parameters. It will be useful to formulate some general principles for making this selection.

A distinction is sometimes made between *informative* and *uninformative* priors. The idea here is that we use an informative prior when our goal is to incorporate background knowledge into our inferences, and an uninformative prior if we want to avoid ‘polluting’ our inferences with assumptions. Unfortunately, the concept of an uninformative prior, although attractive, has no entirely satisfactory definition. The root of the problem is that if we make a change of variables in the parameters of a model, then this changes the prior. For example, suppose we take 'uninformative' to mean 'flat'. In our rainfall model, we would express a lack of prior knowledge about P by using the prior $f_P(p) = 1$. Now suppose we decide to parametrise the chance of rain using the 'log-odds' $\psi = \log(p/(1-p))$. Then our uniform prior on P corresponds to the prior

$$f_\Psi(\psi) = \frac{e^\psi}{(1+e^\psi)^2}$$

on Ψ. Since we know nothing about P, then we must also know nothing about Ψ, but the prior on Ψ is not flat, and therefore does not conform to our definition of uninformative! It turns out that there is a procedure, invented by Harold Jeffreys,[7] for constructing priors which leave posterior inferences invariant to changes of variable in the parameters. While Jeffreys' priors are uninformative in this particular sense, they are difficult to define for multi-parameter models and are often improper.

Given the difficulty in unambiguously defining uninformative priors, we take the more practical approach of categorising priors according to the *degree* to which they are informative. We have the following options:

(a) A **flat** or otherwise **minimally informative** prior. This is a (possibly improper) prior which, to the extent that it is possible, avoids the incorporation of prior knowledge. If the prior is flat, then the posterior is just a normalised version of the likelihood. We may also use a Jeffreys prior, for example.
(b) A **weakly informative** prior. This is a proper prior designed to provide *less* information than the prior knowledge actually available. Given a reasonably large amount of data, the specific choice of weakly informative prior will be unimportant to inferences.
(c) An **informative** prior. This is a prior that reflects our knowledge of the range of plausible parameter values learned from previous experience. An informative prior is useful when data is sparse, and we want to exploit knowledge of similar data-generating mechanisms to produce a better model of the system of interest.

If we want to construct a weakly informative prior for a parameter θ that could in principle lie anywhere in an infinite or semi-infinite interval I_θ, then the constraint that the prior must be proper means that we cannot use a flat distribution. In this case, we would use a distribution with location and scale parameters chosen so that the range of plausible values according to the prior is substantially wider than the interval in which we expect the parameter to lie. It is recommended to avoid distributions that assign zero probability to any region $I_0 \subset I_\theta$, even if we think the parameter is extremely unlikely to belong to this region. Doing so forces the posterior to be zero in I_0, so even if the likelihood becomes more and more sharply peaked in I_0 as more and more data accumulates, it will be impossible to 'wash-out' our mistake.

10.7 MCMC for the New Model

We are now ready to complete the specification of our beta-binomial model for the number of rainy days (X) in the first two weeks of August. We will explore the model's properties

[7] A British geophysicist who made important contributions to the theory of probability and inference.

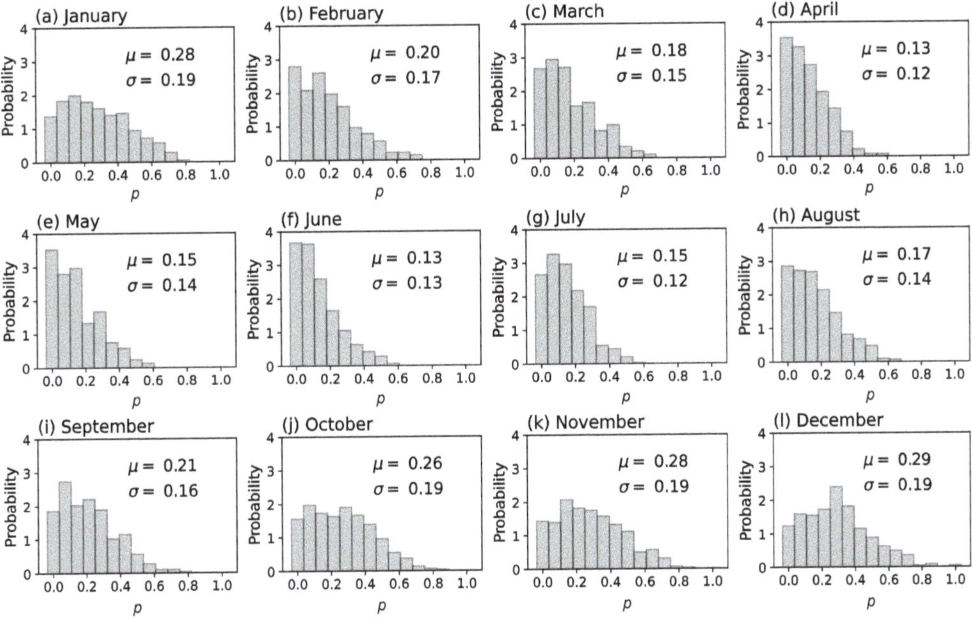

Figure 10.8 (a–l) Histograms of the proportion of rainy days per fortnight, grouped by month, over the period 1931–2024.

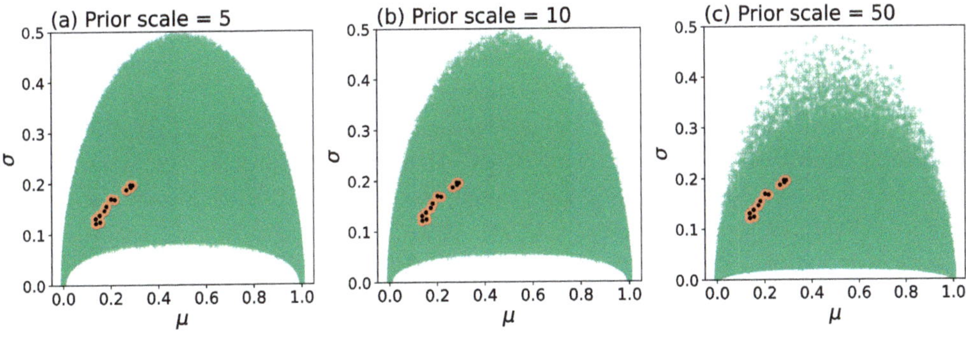

Figure 10.9 Green scatters show means and standard deviations of 10^7 beta distributions with parameters $A, B \sim \text{HalfNormal}(s)$ and $A \perp\!\!\!\perp B$, where (a) $s = 5$, (b) $s = 10$ and (c) $s = 50$. Orange points with black centres show means and standard deviations of empirical fortnightly P distributions for each month.

using Markov Chain Monte Carlo and compare its performance with our baseline binomial model.

Our first step is to specify priors on the parameters, A and B, of our model. We will aim for weakly informative priors. The assumption of our beta-binomial model is that the probability, P, of a rainy day is a beta-distributed random variable which is realised once during the fortnight. Although this variable is latent, we can estimate what values it has taken historically – under the assumptions of our model – using maximum likelihood estimates

Table 10.1 *Summary MCMC statistics for the second halves of $m = 4$ chains of $n = 2000$ steps (first half discarded) sampled from the posterior of the beta-binomial model with half-normal priors $A \sim \text{HalfNormal}(50)$ and $B \sim \text{HalfNormal}(50)$.*

Parameter	Mean	Std. dev.	se(Mean)	se(Std. Dev.)	n_{eff}	$\hat{R}$
α	1.69	0.521	0.019	0.014	878	1.0053
β	9.20	2.89	0.103	0.077	868	1.0057

for P during different fortnights. Figure 10.8 shows histograms of these estimates for each month over the period 1931–2024. Examining all months provides us with a measure of the variability in the distribution of P. The means of these empirical distributions range from 0.14 to 0.29; the standard deviations range from 0.12 to 0.20. We wish to choose priors on A and B which produce beta distributions with statistics lying in substantially wider ranges. Since the parameters of the beta distribution are positive, a reasonable choice for their prior is

$$A, B \sim \text{HalfNormal}(s), A \perp\!\!\!\perp B,$$

where the scale parameter s controls the range over which the parameters vary. Figure 10.9 shows scatter plots of the means and standard deviations of a large number of beta distributions with parameters sampled from this prior using different scales $s \in \{5, 10, 50\}$. Also shown are the statistics of the 12 empirical P-distributions. Increasing the scale allows the prior to generate beta distributions with *smaller* standard deviations and places greater prior weight on these distributions. The upper limit on the standard deviation that can be achieved is not sensitive to s. Since we want a weakly informative prior, we will opt for $s = 50$, which accommodates standard deviations that are substantially lower than those observed empirically (as low as $\sigma \approx 0.03$ when $\mu = 0.2$).

Having specified the prior, our model definition is complete and we can sample from the posterior using MCMC. We do this using PyMC, an open-source probabilistic programming language which implements state-of-the-art samplers.[8] Figure 10.10 shows the second half of $m = 4$ traces of $n = 2{,}000$ steps (the first half of which have been discarded), along with the marginal distributions of each trace. The statistics of these chains are summarised in Table 10.1. Since $\hat{R} < 1.01$ for both parameters, we can be satisfied that our chains are sampling the target. The fact that $n_{\text{eff}} \approx n/2$ in this case means that autocorrelation is low in the chains. The posterior mean parameter values given in the table produce a beta distribution with $\mu = 0.155$ and $\sigma = 0.105$. The posterior means of α and β are both significantly less than the scale parameter s of our half-normal prior, indicating that the impact of our choice of $s = 50$ on inferences was minimal.

In Section 10.8, we will perform a quantitative comparison between the baseline (binomial) and new (beta-binomial) model. Before we do that, let us briefly return to the questions that motivated our modelling exercise in the first place.

- What is the most likely number of rainy days?
- What is the chance there will be no rainy days at all?
- What is the chance there will be more than k rainy days?

[8] We use the No U-Turn Sampler (NUTS), which is an extension of HMC.

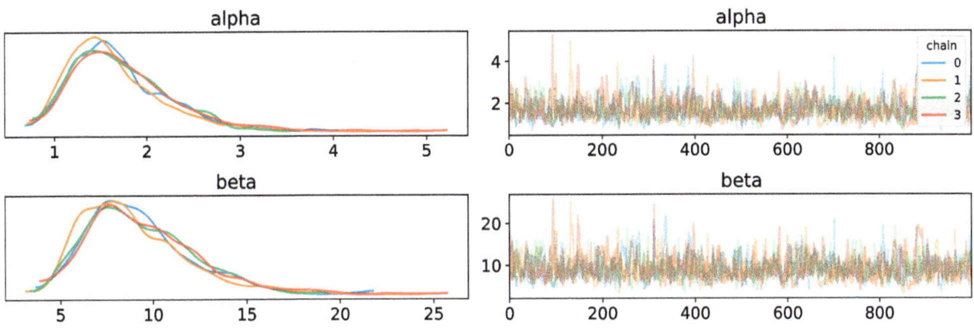

Figure 10.10 The second halves of the traces of $m = 4$ chains of length $n = 2{,}000$ (right) and their marginal distributions (left).

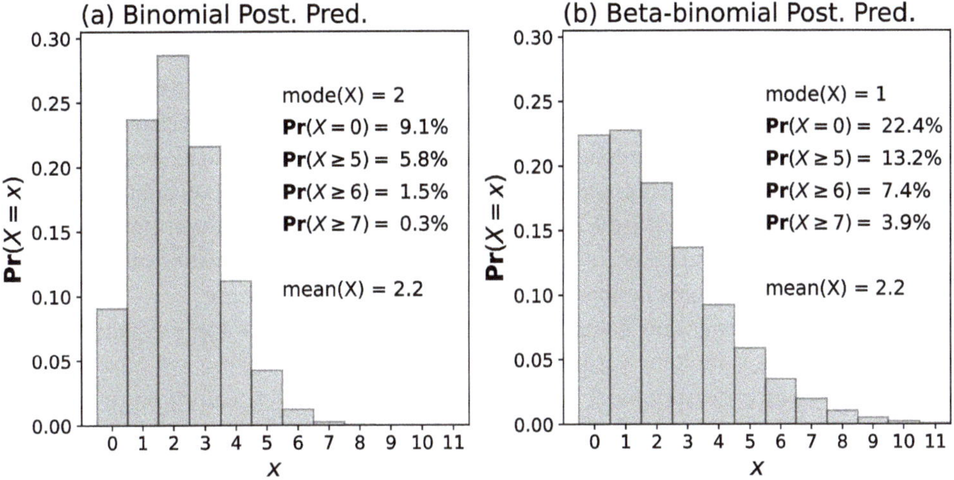

Figure 10.11 Posterior predictive distributions of X under (a) binomial and (b) beta-binomial models.

From the Bayesian perspective, there are two kinds of uncertainty we should account for when using a specific model to answer these questions. There is *epistemic* uncertainty in the parameter values of our model, and there is (in addition) *aleatory* uncertainty in the particular value of X this coming August. (That is, even given a particular set of parameter values, there is still aleatory uncertainty about X.) These two sources of uncertainty are combined in the *posterior predictive* distribution (see Box 10.1). This is the distribution of the result we obtain by first sampling the parameter(s) from the posterior, and then sampling from the conditional mass function of X given the sampled parameter(s). Figure 10.11 shows the posterior predictive distributions for our two models, annotated with the answers to our questions of interest. There is a substantial difference between the two models. The binomial model assumes that the probability of a rainy day is *always the same*. Although there is epistemic uncertainty about the value of P, it is quite small (the standard deviation of the posterior is ≈ 0.01). The beta-binomial model assumes that P is *different every year*, with variations modelled using beta distribution $P|\{A = \alpha, B = \beta\} \sim \text{Beta}(\alpha, \beta)$. Here the epistemic uncertainty is in the values of A and B, while the uncertainty in this year's P has an aleatory component. The magnitude of the total uncertainty in this year's P may be measured by averaging the standard deviation of the beta distribution from which P is drawn over the posterior distribution of the beta parameters. This average standard deviation is ≈ 0.11, 10 times larger than the epistemic uncertainty in the binomial model. As a consequence the variation in the number of rainy days is much larger in the beta-binomial model, resulting in substantial differences in the answers to our questions. Given these differences, we need to decide which model provides a better description of reality.

10.8 Model Comparison

We have two competing models, and we would like to decide which one is 'better'. We have addressed the model selection problem several times already in both the supervised setting (the Queen's new weapon in Chapter 2; body fat prediction in Chapter 7) and the unsupervised setting (the hidden treasure in Chapter 3). In those cases, we chose between models by using cross-validation, to assess how well each was likely to perform on *unseen data*. In order to use cross-validation we need to define a performance measure for our models. In the supervised setting, prediction accuracy is the obvious choice, but our current modelling problem is unsupervised. It is similar in form to the hidden treasure model from Chapter 3, where we measured performance using the summed log densities of observations not used for training. We now define a Bayesian version of this measure.

We will define the measure using densities, but our definitions apply equally to mass functions. Let X be a random variable with true density $\mathfrak{f}_X$. Suppose we have a dataset $\underline{x} = (x_1, \ldots, x_n)$ of observations of X. We view this dataset as a realisation of a random vector $\underline{X} = (X_1, \ldots, X_n)$ containing i.i.d. copies of X. Consider a model of X, with posterior predictive density $f_{X|\underline{X}}(-|\underline{x})$. One way to measure how well this model describes $\mathfrak{f}_X$ is to observe a second *test* dataset, $\underline{\tilde{x}}$, of i.i.d. observations of the true X, having density $\mathfrak{f}_X$. We view this second dataset as a realisation of the random vector $\underline{\tilde{X}}$. The logarithm of the probability density of this new dataset, according to the posterior predictive we obtained from the original dataset, is

$$\log f_{\underline{\tilde{X}}|\underline{X}}(\underline{\tilde{x}}|\underline{x}) = \sum_{i=1}^{n} \log f_{X|\underline{X}}(\tilde{x}_i|\underline{x}).$$

Intuitively, we expect that, on average, a better model of $\mathfrak{f}_X$ will assign a higher probability density to datasets sampled from $\mathfrak{f}_X$. This is the intuition behind the maximum likelihood (ML) principle. According to this intuition, the greater the log posterior predictive density assigned to new data, the more plausible that density is as a model of $\mathfrak{f}_X$. Of course, this measure is subject to noise in the data-generating process. Taking its expectation over all possible realisations of the test dataset, we obtain the *expected log pointwise predictive density*[9]

$$\text{elpd} = \int \mathfrak{f}_{\underline{X}}(\underline{u}) \log f_{\underline{\tilde{X}}|\underline{X}}(\underline{u}|\underline{x}) d\underline{u} = n \int \mathfrak{f}_X(x) \log f_{X|\underline{X}}(x|\underline{x}) dx.$$

[9] Some authors use the abbreviation elppd. We follow the conventions in Vehtari et al. (2017).

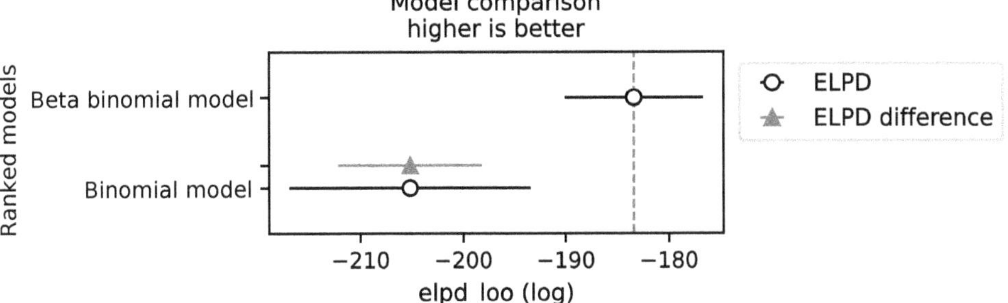

Figure 10.12 Circles and error bars show elpd estimates and standard errors for our two models estimated using leave-one-out cross-validation. The grey triangle with error bars shows estimated elpd difference (relative to the dashed line) and standard error estimates. Plot and estimates output from PyMC.

The negative of the expectation on the right-hand side defines a *functional* (a function of a one or more functions), which we call the *cross-entropy* of our posterior predictive $f_{X|\underline{X}}(-|\underline{x})$ relative to the true density $\mathfrak{f}_X$. It is denoted

$$\mathbb{H}\left[f_{X|\underline{X}}(-|\underline{x}), \mathfrak{f}_X\right] = -\int \mathfrak{f}_X(x) \log f_{X|\underline{X}}(x|\underline{x})dx.$$

The cross-entropy of one distribution relative to another is minimised when they are equal. Therefore, the elpd attains its maximum possible value when the posterior predictive matches the true density. We will have more to say about the cross-entropy in Chapter 12.

Unfortunately, we cannot compute the elpd exactly. In the best-case scenario, we have access to a second, unseen dataset, but more often we won't have this, so we approximate the elpd using cross-validation. Let $\underline{x}_{-i}$ be the training dataset with the ith observation removed. If we calculate the posterior predictive using this dataset, then we can treat x_i as an unseen datapoint, representing a sample from the true density $\mathfrak{f}_X$. The log of the posterior predictive probability evaluated at this datapoint is $\widehat{\text{elpd}}_i = \log f_{X|\underline{X}_{-i}}(x_i|\underline{x}_{-i})$. Letting each datapoint take its turn as 'unseen', we obtain the following elpd estimate and standard error,[10]

$$\widehat{\text{elpd}} = \sum_{i=1}^{n} \widehat{\text{elpd}}_i, \quad \text{se}(\widehat{\text{elpd}}) = \sqrt{\frac{n}{n-1}\sum_{i=1}^{n}\left(\widehat{\text{elpd}}_i - \frac{1}{n}\widehat{\text{elpd}}\right)^2}.$$

Figure 10.12 shows elpd estimates and standard errors for our two rainfall models. As we might have expected, the elpd of the beta-binomial model is higher than the binomial model, but is the difference 'real' or just a fluke of the data? Reassuringly, the standard errors suggest that the sampling distributions of our estimators are well separated. If we had to make a choice between the models, we should certainly choose the beta-binomial, but can we quantify our level of confidence in doing so?

Rather than focus on the elpds of the individual models, let us consider the statistical properties of their difference. Given two models, A and B, we define the difference in their

[10] See Vehtari et al. (2017).

elpd values $\Delta = \text{elpd}_A - \text{elpd}_B$. Letting $\Delta_i = \widehat{\text{elpd}}_{A,i} - \widehat{\text{elpd}}_{B,i}$, we define the following estimator for Δ and its standard error,

$$\widehat{\Delta} = \frac{1}{n}\sum_{i=1}^{n}\Delta_i = \widehat{\text{elpd}}_A - \widehat{\text{elpd}}_B, \quad \text{se}(\widehat{\Delta}) = \sqrt{\frac{n}{n-1}\sum_{i=1}^{n}\left(\Delta_i - \widehat{\Delta}\right)^2}.$$

We note that the standard error in $\widehat{\Delta}$ will typically satisfy $\text{se}^2(\widehat{\Delta}) < \text{se}^2(\widehat{\text{elpd}}_A) + \text{se}^2(\widehat{\text{elpd}}_B)$ because the values of $\widehat{\text{elpd}}_{A,i}$ and $\widehat{\text{elpd}}_{B,i}$ will typically be positively correlated (see Exercise 10.8).

■ ***Exercise*** 10.8 Let X_A and X_B be random variables with variances σ_A^2, σ_B^2 and correlation ρ. Show that $\text{Var}(X_A - X_B) = \sigma_A^2 + \sigma_B^2 - 2\rho\sigma_A\sigma_B$.

Letting model A be the beta-binomial and B the binomial, we obtain the estimates $\widehat{\Delta} = 21.8$ and $\text{se}^2(\widehat{\Delta}) = 7.0$ (using PyMC). Under the assumption that our estimator is normally distributed,[11] that is, $\widehat{\Delta} \sim \mathcal{N}(\Delta, \text{se}^2(\widehat{\Delta}))$, the 95% (frequentist) confidence interval for the elpd is $[14.8, 28.8]$. This gives us confidence that there is a genuine difference in the performance of the two models – that is, $\text{elpd}_A > \text{elpd}_B$. Another useful concept here is the frequentist p-value. Suppose we hypothesise that there is *no difference* in the ability of the models to describe the data-generating process, so that $\Delta = 0$. According to this null hypothesis, denoted H_0, we have $\widehat{\Delta} \sim \mathcal{N}(0, \text{se}^2(\widehat{\Delta}))$. The frequentist p-value is the probability of observing a value of $\widehat{\Delta}$ which is *at least as extreme* as the value we have observed, assuming H_0 is true (see Box 6.5). According to H_0, the standardised variable $z = \widehat{\Delta}/\text{se}(\widehat{\Delta})$ is a realisation of an $\mathcal{N}(0, 1)$ variable, Z. Therefore,

$$p = \mathbf{Pr}\left(Z > \frac{\widehat{\Delta}}{\text{se}(\widehat{\Delta})}\right) = \mathbf{Pr}(Z > 3.1) \approx 0.001.$$

That is, it would have been *extremely* surprising (one in a thousand chance) to see a difference as large as ours if there were no difference in the performance of the models. We therefore firmly declare in favour of model A, the beta-binomial model.

10.9 The Queen's Archers

We now leave rainy England and return to our medieval kingdom. With the Queen's new weapon still in development, the primary means of launching projectiles remains the longbow. Archers are evaluated by having them shoot arrows at a circular target from 25 yards away. The score of a single shot is equal to the distance from the centre of the target to the point where the arrow hits. A lower score is, of course, better.

A single shot at the target yields a position vector $\boldsymbol{r} = (x, y)$, giving the horizontal and vertical coordinates of the hit point relative to the centre of the target. This hit point may be viewed as the realisation of a random vector $\boldsymbol{R} = (X, Y)$. For an individual archer, this random vector is approximately normal,

$$\boldsymbol{R} \sim \mathcal{N}(\mathbf{0}, \sigma^2\underline{\boldsymbol{I}}),$$

[11] See Vehtari et al. (2017) for a discussion.

where σ determines how accurate, on average, the archer's shots are. For a given shot, the distance from the centre of the target – the score – is as follows:

$$R = \sqrt{X^2 + Y^2}.$$

After a suitable change of variables (see Exercise 10.9) we find that R has marginal density

$$f_R(r;\sigma) = \frac{r}{\sigma^2} e^{-\frac{r^2}{2\sigma^2}}. \tag{10.13}$$

This is known as the *Rayleigh distribution* after the physicist Lord Rayleigh. If R is Rayleigh distributed, we write $R \sim \text{Rayleigh}(\sigma)$. The expected value of R is

$$\mu = \mathbb{E}(R) = \sqrt{\frac{\pi}{2}}\sigma.$$

We refer to μ as the *skill level* of the archer. We will assume that the scores obtained by an archer with current skill level μ are i.i.d. draws from a Rayleigh distribution with parameter $\sqrt{2/\pi}\mu$.

■ ***Exercise*** 10.9 Suppose that $\boldsymbol{R} = (X, Y) \sim \mathcal{N}(\boldsymbol{0}, \sigma^2 \underline{\boldsymbol{I}})$, so

$$f_{X,Y}(x, y) = \frac{e^{-\frac{x^2+y^2}{2\sigma^2}}}{2\pi\sigma^2}.$$

Making the change of variables $X = R\cos(\Psi), Y = R\sin(\Psi)$, where $\Psi \in [0, 2\pi]$ is an angular coordinate, show that

$$f_{R,\Psi}(r, \psi) = \frac{re^{-\frac{r^2}{2\sigma^2}}}{2\pi\sigma^2}.$$

Hence, show that the marginal density of R is the Rayleigh distribution.

■ ***Exercise*** 10.10 Show that if $R \sim \text{Rayleigh}(\sigma)$, then $\mathbb{E}(R) = \sqrt{\pi/2}\sigma$.

10.10 Archery Training and Informative Priors

Almost all young people train at archery and take part in local competitions. The Queen's scouts visit these competitions to sample and record individual scores; the aim, of course, is to identify promising potential recruits for the royal army. The data available for each archer consists of a set of times and scores,

$$D = \{(t_i, r_i)\}_{i=1}^n.$$

By convention, time is measured in months since the archer's fourteenth birthday (when they begin using a full-size bow). Money for paying scouts has been tight recently, so the number of recorded datapoints is quite small for each potential recruit. Much more detailed records exist for archers from earlier decades. Given a potential recruit, there are two important questions:

- What skill level will they achieve eventually?

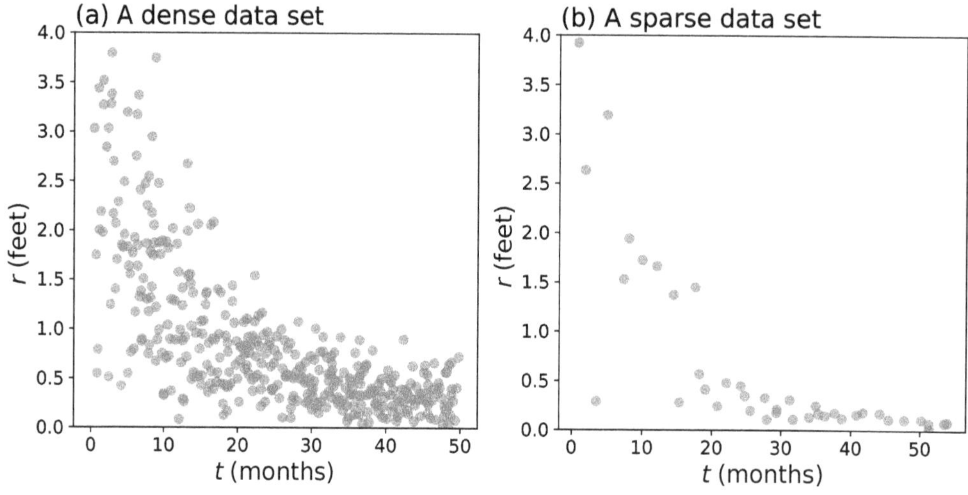

Figure 10.13 (a, b) Competition records for two archers.

- How long will it take for them to reach this level?

Currently, regimental commanders guess the answers by 'eyeballing' the scouts' data. They are interested so see if you can do better.

Figure 10.13 shows examples of the competition data collected for two young archers. From these examples, it appears that when an archer starts using a full-size bow, their accuracy is poor, producing high competition scores. These scores decline – perhaps exponentially – over time, settling down to a stable equilibrium as the archer reaches their late teens. Given these observations, a plausible model for the development of skill level over time is

$$\mu(t; \mu_0, \mu_1, \lambda) = \mu_1 + (\mu_0 - \mu_1)e^{-\lambda t}. \tag{10.14}$$

Here μ_0 is the initial skill level, μ_1 is the lowest (i.e. best) skill level that the archer is capable of and λ is a learning rate that determines how long the archer will take to reach μ_1. Given the values of these parameters, we model the probability distribution of R at time t as follows:

$$R \sim \text{Rayleigh}\left(\sqrt{2/\pi}\,\mu(t; \mu_0, \mu_1, \lambda)\right).$$

In the Bayesian setting, the parameters μ_0, μ_1 and λ are viewed as realisations of random variables M_0, M_1 and Λ, and our model may be represented by the DAG

M_0 M_1 Λ

$t_i \longrightarrow R_i$

$i = 1, \ldots, n$. (10.15)

We view our dataset as a single realisation of the random vectors $\underline{T} = (T_1, T_2, \ldots, T_n)$ and $\underline{R} = (R_1, R_2, \ldots, R_n)$, where the scores are conditionally independent given the times,

$R_i \perp\!\!\!\perp R_j | \underline{T}$. We are not interested in the distribution of the times because we are constructing a model of R conditional on the time, T. In other words, this is a *regression* problem.

For given M_0, M_1, Λ and T, the probability density of R, according to our model, is

$$f_{R|M_0,M_1,\Lambda,T}(r|\mu_0,\mu_1,\lambda,t) = f_R\left(r; \sqrt{2/\pi}\mu(t;\mu_0,\mu_1,\lambda)\right),$$

where the right-hand side is the Rayleigh density defined in equation (10.13). The likelihood is the conditional probability density of the scores $\underline{r}$ given the times $\underline{t}$,

$$\mathcal{L}(\mu_0,\mu_1,\lambda) = f_{\underline{R}|M_0,M_1,\Lambda,\underline{T}}(\underline{r}|\mu_0,\mu_1,\lambda,\underline{t}) = \prod_{i=1}^{n} f_{R|M_0,M_1,\Lambda,T}(r_i|\mu_0,\mu_1,\lambda,t_i),$$

yielding the posterior

$$f_{M_0,M_1,\Lambda|\underline{R},\underline{T}}(\mu_0,\mu_1,\lambda|\underline{r},\underline{t}) \propto \mathcal{L}(\mu_0,\mu_1,\lambda) f_{M_0,M_1,\Lambda}(\mu_0,\mu_1,\lambda).$$

To keep life simple, we will assume the prior factorises as

$$f_{M_0,M_1,\Lambda}(\mu_0,\mu_1,\lambda) = f_{M_0}(\mu_0) f_{M_1}(\mu_1) f_\Lambda(\lambda).$$

This is not especially realistic. For example, before examining the data, we would surely be prepared to bet that M_0 and M_1 are positively correlated. (After all, archers with great potential were probably also better-than-average beginners.) In a real-world research project, we would explore strategies for articulating this hunch mathematically – if only to demonstrate that it didn't, in the end, make much difference to the posterior. Here we will just note that the factorised prior still lets us inject *some* relevant background knowledge. Since the parameters of the model are positive, a reasonable choice of family for the priors is the gamma family (see Box 9.3). With this choice, we can summarise the complete model as follows:

$$\begin{aligned} M_0 &\sim \text{Gamma}(a_0, b_0), \\ M_1 &\sim \text{Gamma}(a_1, b_1), \\ \Lambda &\sim \text{Gamma}(a_\lambda, b_\lambda), \\ R|\{M_0 = \mu_0, M_1 = \mu_1, \Lambda = \lambda, T = t\} &\sim \text{Rayleigh}\left(\sqrt{2/\pi}\mu(t,\mu_0,\mu_1,\lambda)\right). \end{aligned}$$

We note that the Rayleigh distribution transforms to an exponential distribution under the change of variable $Z = R^2$. This transformation is useful when using simulation tools which do not implement the Rayleigh distribution by default, or calculating posteriors analytically.

■ ***Exercise*** 10.11 Show that if $R \sim \text{Rayleigh}(\sigma)$ and $Z = R^2$, then $Z \sim \text{Exponential}\left(\frac{1}{2\sigma^2}\right)$.

The regimental commanders are keen for us to make the most of the sparse data available on current trainee archers. Perhaps the denser data available on *past* trainee archers will be helpful? Fortunately, the tools of Bayesian inference are particularly useful in this situation. Each new archer who begins training with a full-size bow will have their own initial skill level, learning rate and final potential. We want our prior to approximate the distribution of these characteristics in the population. One method is to obtain multiple historical datasets

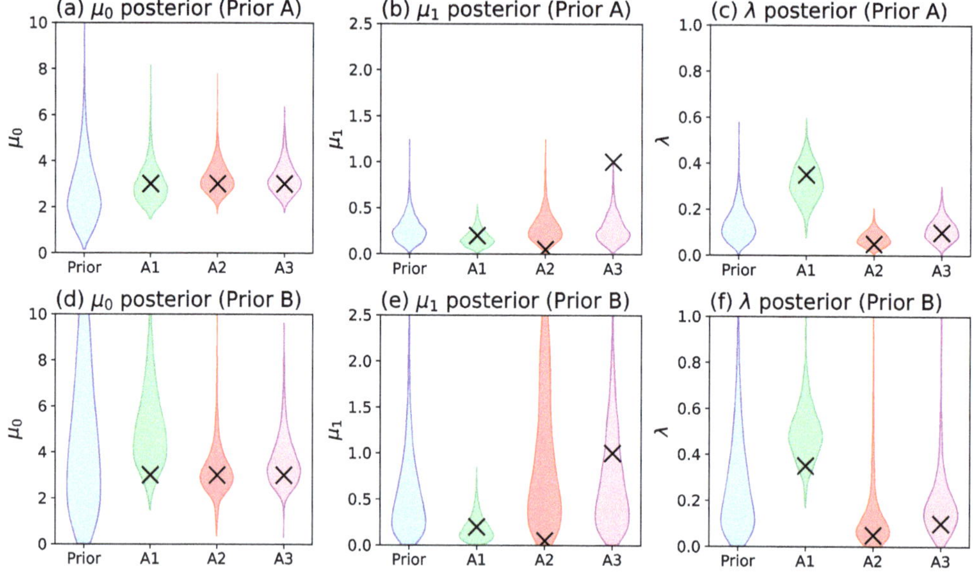

Figure 10.14 (a)–(c) Informative Prior A and corresponding posteriors for three archers. (d)–(f) Weakly informative Prior B and corresponding posteriors for the same archers. Black crosses show true parameter values. Posteriors generated using MCMC, implemented in PyMC.

for previous archers – denser than today's records – and fit our model to each one (using a weak prior). This will yield a dataset of estimated parameter vectors,

$$D_{\text{prior}} = \{(\mu_{0i}, \mu_{1i}, \lambda_i)\}_{i=1}^{N},$$

where $(\mu_{0i}, \mu_{1i}, \lambda_i)$ are the parameters estimated for the ith archer and N is the number of archers analysed. We then fit the parameters of our prior models to this data. We will assume that this process yielded the following results:

$$M_0 \sim \text{Gamma}(2, 1), \quad M_1 \sim \text{Gamma}(0.2, 0.1), \quad \Lambda \sim \text{Gamma}(0.1, 0.05). \tag{10.16}$$

This is our informative prior – *Prior A*.

■ ***Exercise*** 10.12 Use the properties of the gamma distribution (Box 9.3) to calculate the prior means and standard deviations of M_0, M_1 and Λ according to the informative prior (10.16).

If data is not available to construct a prior, we can use expert opinion. The idea is to ask experts to estimate various statistical properties of the population and then to find a prior which is consistent with this information. For example, suppose the scouts report the following:

- The most typical initial and final skill levels and learning rate are $\mu_0 = 2$ft, $\mu_1 = 0.2$ft and $\lambda = 0.1$/month ($\approx$ 10% improvement towards the final skill level per month).
- The least able quarter of archers have skill parameters in excess of $\mu_0 = 4$ft and $\mu_1 = 0.4$ft. An archer is in the top quarter of fastest learners if $\lambda > 0.2$/month.

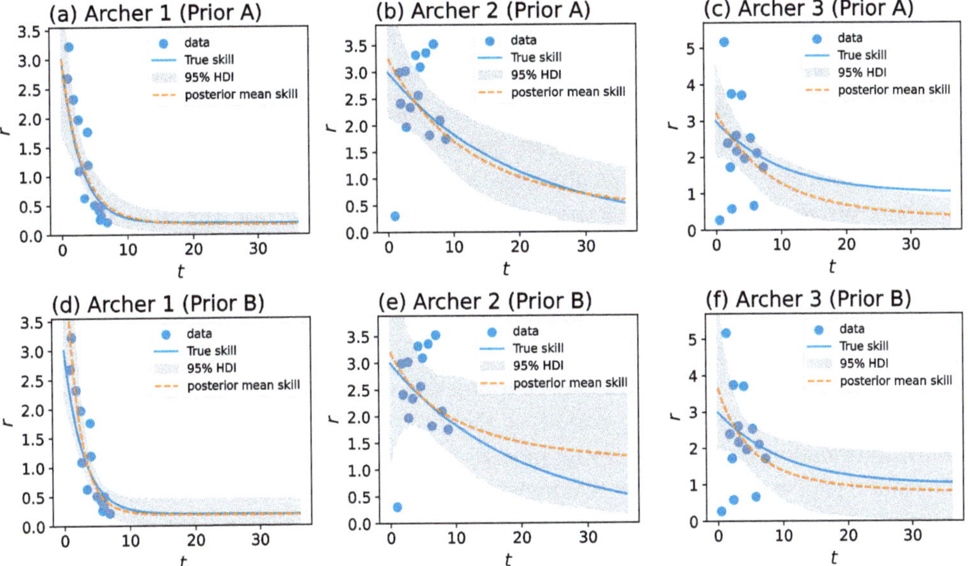

Figure 10.15 (a–f) Posterior predictions of the skill levels of three archers based on 14 datapoints collected over approximately seven months. Grey bands show 95% 'highest density intervals' for the posterior distribution of skill curves. An HDI is a credible interval for which every value inside the interval has a higher density than every value outside it.

This information gives us the mode and upper quartile of the prior for each parameter. The gamma parameters of each prior may then be adjusted to match these statistics (see Exercise 10.13 for some other examples). We note that the modes and upper quartiles of the data-driven Prior A (10.16) are $\mu_0 : \{2, 3.47\}$, $\mu_1 : \{0.2, 0.347\}$, and $\lambda : \{0.1, 0.173\}$, reassuringly close to the scout's estimates. Due to this close match, we won't separately analyse the model using this 'expert prior'. However, we will define a weakly informative prior – *Prior B* – by quadrupling the standard deviations of Prior A. This will allow us to explore the impact of prior knowledge on model predictions. The individual prior distributions for each model parameter are shown in Figure 10.14a–c (Prior A) and Figure 10.14d–f (Prior B).

■ ***Exercise*** 10.13 Let $\Theta \sim \text{Gamma}(a, b)$ be a Bayesian model parameter.

(a) Suppose the mean μ and variance σ^2 of Θ are known. Determine a and b in terms of μ and σ.

(b) Suppose the mode $m > 0$ and variance σ^2 of Θ are known. Determine a and b in terms of m and σ.

To illustrate the predictive abilities of our model, we consider three archers:

- A1: a fairly fast learner, but otherwise typical.
- A2: a fairly slow learner, but otherwise typical.
- A3: an 'outlier' with unusually high μ_1.

Suppose we have seven months of data for each archer, and estimate posteriors using MCMC for both priors. The data for each archer is plotted in Figure 10.15, along with the evolution of their true skill level. The posterior mean skill curves based on the two priors, and 95% credible intervals for each curve, are also plotted for each archer. The marginal posteriors estimated using each of the two priors are shown in Figure 10.14. The posteriors obtained from the informative prior are typically narrower, meaning that parameter estimates will be more severely *shrunk* towards their prior means. Consider archer A1. Her high learning rate means that she completed her journey from high to low skill (recall that low is better) during the data collection period. As a consequence, observations of this archer contain plenty of information about all three of her parameters. Therefore, her posteriors are less sensitive to her priors, and her predicted skill level over time is accurately estimated using either prior. Now consider archer A2. Her low learning rate means that there is little change in her skill level during the observation period, so the observational data does not contain a great deal of information about her learning rate. In this situation, the informative prior fills the information gap, yielding a model that makes more accurate predictions of future skill levels (Figure 10.15e). Finally, consider archer A3, with an unusually high final skill parameter. In this case, the more informative prior makes less accurate predictions because it introduces a bias towards more typical parameter values.

In summary, when data is sparse, well-chosen informative priors *typically* lead to more accurate predictions, but the shrinkage effect can produce less accurate results if the real data-generating mechanism is *atypical* according to the prior.

10.11 Accuracy Estimation and Hierarchical Models

A group of archers arrive by ship from a foreign kingdom. Their shooting technique is different to that of the archers in the Queen's regiments, and their bows are made of different wood. The regimental commanders wish to assess the skill levels of the foreign archers, so each archer is asked to shoot a number of arrows at a target. Let R_{ij} be the score of the jth shot by the ith archer. We will assume that the archers are mature in the sense that they have reached their maximum potential. Writing M_i for the latent skill level of archer i, we have

$$R_{ij}|\{M_i = \mu_i\} \sim \text{Rayleigh}\left(\sqrt{2/\pi}\mu_i\right).$$

Our goal is to find posterior distributions for M_i for each archer. Let us start by considering the archers individually. Let $\underline{r}_i = (r_{i1}, \ldots, r_{in_i}\}$ be the scores of archer i, understood as a realisation of the random vector $\underline{R}_i = (R_{i1}, \ldots, R_{in_i})$, where n_i is the number of shots taken. The likelihood function for the archer's skill is

$$\mathcal{L}_i(\mu_i) = f_{\underline{R}|M^2}(\underline{r}_i|\mu_i^2) = \prod_{j=1}^{n_i} \frac{\pi r_{ij}^2}{2\mu_i^2} \exp\left(-\frac{\pi r_{ij}^2}{4\mu_i^2}\right) \propto \frac{1}{(\mu_i^2)^{n_i}} \exp\left(-\frac{\pi}{4\mu_i^2} \sum_{j=1}^{n_i} r_{ij}^2\right),$$

where μ_i-independent factors have been dropped from the final expression on the right-hand side. Suppose our prior is flat. The posterior is then an inverse-gamma distribution (see Box 9.3),

$$f_{M_i^2|\underline{R}_i}(\mu_i^2|\underline{r}_i) = \text{IG}\left(\mu_i^2 \middle| n_i - 1, n_i \pi \overline{r^2}_i/4\right),$$

where $\overline{r^2}_i = n_i^{-1} \sum_j r_{ij}^2$ and the inverse-gamma density is defined as follows:

$$\mathrm{IG}(x|a,b) = \frac{b^a}{\Gamma(a)} x^{-(a+1)} e^{-b/x}.$$

The MAP estimate of μ_i^2 is then

$$\widehat{\mu^2}_{i,\mathrm{FLAT}} = \frac{\pi}{4}\overline{r^2}_i.$$

From the Bayesian perspective, we ought to be able to improve on this using prior knowledge. Unfortunately, this is the first time we have met archers of this type, so our knowledge is rather limited. We could perhaps justify using a weakly informative prior based on our own native archers, but this would have limited impact on our estimates. A more promising approach is to use the collective statistical properties of the foreign archers who arrived on the ship to learn about the distribution of skill levels in the population they came from. We first illustrate how this can be done in a practical but not fully Bayesian way, before introducing a fully Bayesian approach which can be applied to a wide range of problems.

10.11.1 Empirical Bayes

We wish to estimate the prior and the posteriors using the same dataset. The first method we present is simple enough that all computations can be carried out analytically. We'll work with a transformed skill parameter $\lambda = \pi/(4\mu^2)$ and the squared-score variable $Z = R^2$, where

$$R \sim \text{Rayleigh}\left(\sqrt{2/\pi}\mu\right).$$

We will revert to our original version of the model when we present the fully Bayesian approach in Section 10.11.2. The density of Z (see Exercise 10.11) is

$$f_{Z|\Lambda}(z|\lambda) = \lambda e^{-\lambda z}.$$

Here λ is viewed as a realisation of the latent variable Λ, and Z is exponentially distributed given that $\Lambda = \lambda$,

$$Z|\{\Lambda = \lambda\} \sim \text{Exponential}(\lambda).$$

Note that in this context, $\lambda > 0$ is an alternative skill measure, not a learning rate. The likelihood function for the ith archer is

$$\mathcal{L}_i(\lambda) = f_{\underline{Z}|\Lambda}(\underline{z}_i|\lambda) = \lambda^{n_i} \exp\left(-\lambda \sum_{j=1}^{n_i} z_{ij}\right),$$

where $\underline{z} = (z_{i1}, \ldots, z_{in_i})$ is a dataset of observations of the squared score. Since the gamma family is conjugate to the exponential likelihood, we'll approximate the prior on Λ with a gamma distribution, $\Lambda \sim \text{Gamma}(a, b)$. The posterior density for archer i is then

$$\Lambda|\{\underline{Z}_i = \underline{z}_i\} \sim \text{Gamma}\left(n_i + a, n_i\bar{z}_i + b\right),$$

as you will show in Exercise 10.14. This implies that the posterior mean is

$$\mathbb{E}\left(\Lambda|\underline{Z}_i = \underline{z}\right) = \frac{n_i + a}{n_i\bar{z}_i + b}.$$

It is instructive to express the posterior mean in terms of the prior mean μ_λ and variance σ_λ^2, which are related to the gamma parameters by

$$a = \frac{\mu_\lambda^2}{\sigma_\lambda^2}, \quad b = \frac{\mu_\lambda}{\sigma_\lambda^2}.$$

We find that

$$\mathbb{E}\left(\Lambda | \underline{Z}_i = \underline{z}\right) = \frac{1}{\bar{z}_i}\left(\frac{n_i\sigma_\lambda^2 + \mu_\lambda^2}{n_i\sigma_\lambda^2 + \mu_\lambda/\bar{z}_i}\right). \tag{10.17}$$

For large sample sizes, this expression has the following asymptotic behaviour[12]:

$$\mathbb{E}\left(\Lambda | \underline{Z}_i = \underline{z}\right) \sim \frac{1}{\bar{z}_i} + \frac{\mu_\lambda}{n_i\sigma_\lambda^2\bar{z}_i}\left(\mu_\lambda - \frac{1}{\bar{z}_i}\right) \quad \text{as } n_i \to \infty.$$

The first term, equal to the maximum likelihood estimate of λ, is purely data driven. The second term may be interpreted as a correction that shifts the posterior mean towards the prior mean. The magnitude of this correction term is inversely proportional both to the prior variance and the sample size. That is, a more concentrated prior *increases* the magnitude of the correction, whereas a larger dataset *decreases* its magnitude.

In order to evaluate expression (10.17), we need estimates for the prior mean and variance. One possibility is to calculate MAP estimates of each archer's skill parameter, based on a flat prior, and to approximate the mean and variance of the prior with the mean and variance of our estimates. We can then use the prior to update our parameter estimates, which will be collectively 'shrunk' towards the group mean. Using the dataset to obtain point estimates of the prior parameters, and then combining this prior with the data to approximate the posterior, is known as *empirical Bayes*. While it is a practically useful method, it is not fully Bayesian because the unknown prior parameters are not treated as latent variables.

■ ***Exercise*** 10.14 Suppose that conditional on $\Lambda = \lambda$, the random variables $Z_{i1}, Z_{i2}, \ldots, Z_{in_i}$ are independent and exponentially distributed with parameter λ. Assume the prior $\Lambda \sim \text{Gamma}(a, b)$. Show that given a dataset $\underline{z}_i = (z_{i1}, \ldots, z_{in_i})$, the posterior density of Λ satisfies

$$f_{\Lambda|\underline{Z}}(\lambda|\underline{z}_i) \propto \lambda^{n_i+a-1}\exp\left(-\lambda\left(\sum_{j=1}^{n_i} z_{ij} + b\right)\right).$$

Hence show that $\Lambda|\{\underline{Z}_i = \underline{z}_i\} \sim \text{Gamma}\left(n_i + a, n_i\bar{z}_i + b\right)$, where $\bar{z}_i$ is the sample mean of the data.

10.11.2 Hierarchical Models

We now return to our original parameterisation of the model and present a fully Bayesian approach to estimating priors. For simplicity, we'll assume that prior knowledge about the foreign archers' skill levels can be described with a gamma prior, Gamma(a, b). We can

[12] In this context, the symbol $\sim$ means 'tends asymptotically to' rather than 'is distributed as'.

represent the model for the ith archer with the following DAG:

(10.18)

In order to qualify as prior knowledge, the values of a and b must be known *before* observing the foreign archers. In the absence of such knowledge, the empirical Bayes approach was to fit a and b to a set of flat-prior (i.e. ML) point estimates of the skill levels of each archer, and then to compute corrected estimates using the resulting 'empirical' prior. An alternative, fully Bayesian approach is to view a and b as realisations of latent variables A and B. Since we view all skill levels as samples from the same prior, A and B will be realised just once. If there are three archers, the DAG representing this model is

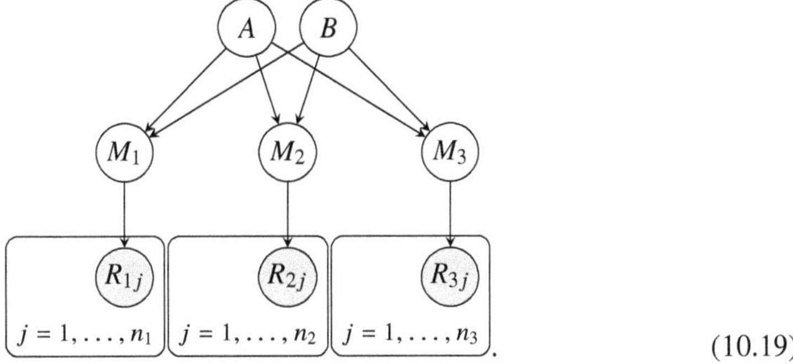

(10.19)

In order to incorporate the parameters of our prior into this model, we had to add an extra layer to our DAG, making it *hierarchical*. To complete the model, we need to specify a prior on A and B, known as a *hyperprior*. In the current context, we can think of the hyperprior as describing the range of possible skill distributions that different human societies are capable of developing – through different shooting techniques, training methods or bow technologies, for example.

The posterior distribution of the hierarchical model depends on all observations of all archers, which we view as a single a realisation D of the random dataset

$$\mathcal{D} = \{\underline{R}_i\}_{i=1}^{N},$$

where N is the number of archers. Writing the latent skill parameters of the individual archers as a vector $\boldsymbol{M} = (M_1, \ldots, M_N)$, the posterior is given by

$$f_{A,B,\boldsymbol{M}|\mathcal{D}}(a, b, \boldsymbol{\mu}|D) \propto \left(\prod_{i=1}^{N} f_{\underline{R}_i|M_i}(\underline{r}_i|\mu_i) f_{M|A,B}(\mu_i|a, b) \right) f_A(a) f_B(b), \tag{10.20}$$

where we have assumed the hyperprior factorises as $f_A(a)f_B(b)$. This posterior represents our state of knowledge about both the individual archers *and* the population from which they

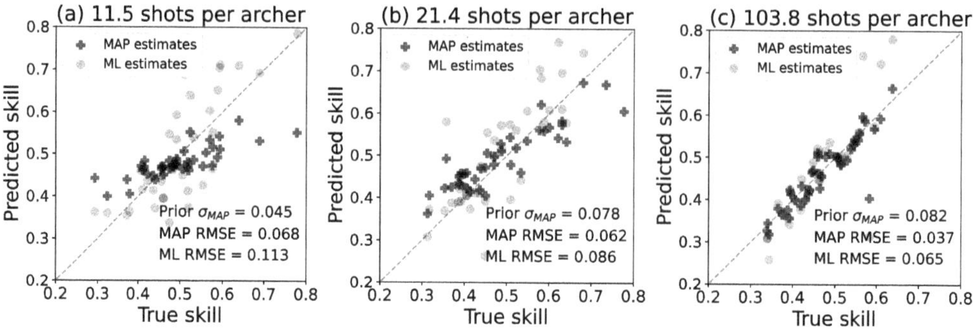

Figure 10.16 (a–c) Predicted versus true skill levels for a group of $N = 40$ archers using ML and hierarchical Bayes MAP estimates. Plots are annotated with the RMSEs in the skill predictions, and MAP estimates of the standard deviation of the inferred prior for each sample size.

were drawn. We have left it as an unnormalised density. It can be sampled from using MCMC or used to find MAP estimates via numerical optimisation.

Let us consider a specific example. Suppose that the true (but unknown to us) distribution of skill levels amongst all foreign archers is Gamma(25, 50) (mean $\mu = 0.5$, standard deviation $\sigma = 0.1$). Suppose also that there are $N = 40$ archers with the number of shots taken by the ith archer sampled from the Geometric(p) distribution (having mean p^{-1}). We consider three cases, letting p^{-1} take the values 10, 20 and 100. If individual skill levels are estimated without leveraging population-level information, then estimates for archers who have taken very few shots will be highly sensitive to noise. But if we simultaneously estimate both the individual skills *and* the distribution from which they are drawn, we can reduce this variance by shrinking individual estimates towards the population distribution. We set the hyperprior to $A \sim$ Halfnormal(100) and $B \sim$ Halfnormal(100), independently. Figure 10.16a shows maximum likelihood and hierarchical Bayes estimates for the case $p^{-1} = 10$ (sparse data). Because the data are sparse, there is not strong evidence for skill differences between individual archers. Accordingly, Bayesian hierarchical inference concludes that most of the archer-to-archer variation in the data is noise. Maximum likelihood estimates of individual skill levels, by contrast, are highly dispersed. As the number of observations per archer increases (Figure 10.16b and c), evidence for skill differences strengthens, and the Bayesian (MAP) estimates of individual skill levels become more similar to the maximum likelihood estimates. We can understand how this works by inspecting the form of the posterior density (10.20). Notice first that the factor $f_{M|A,B}(\mu_i|a,b)$ becomes large if a and b are tuned to give a gamma distribution that peaks sharply at μ_i. But the product $\prod_{i=1}^{N} f_{M|A,B}(\mu_i|a,b)$ can become large only in regions of parameter space where the μ_i are similar. If individual skill levels in fact differ substantially *and data is plentiful*, then the likelihood factors $f_{\underline{R}_i|M_i}(\underline{r}_i|\mu_i)$ prevent posterior concentration in such regions, because $f_{\underline{R}_i|M_i}(\underline{r}_i|\mu_i)$ will be tiny for any μ_i that deviates significantly from the mean of individual i's observed scores. But if data is sparse – just a few shots per archer – then the impact of the likelihood factors is greatly attenuated. Consequently, the posterior does indeed concentrate around the guess that all archers are roughly equally skilled. That is what we see in Figure 10.16a.

■ ***Exercise*** 10.15 Use d-separation to establish that in DAG (10.19) $\{\underline{R}_i, M_i\} \perp\!\!\!\perp \{\underline{R}_j, M_j\}|\{A, B\}$ whenever $i \neq j$.

Hierarchical models provide a way to exploit structure in naturally grouped data. In the example we presented, there were two kinds of groups. On the first level, we had the groups of scores obtained by each individual archer, and on the second level, we had the group of all archers. There are many situations where grouping is useful. Exercise 10.16 invites you to consider an example where there are three natural groupings.

■ ***Exercise*** 10.16 You want to estimate the reliability of different makes and models of car. There are three natural groupings: (1) individual cars of the same model, (2) models made by the same manufacturer and (3) all cars. Suppose you obtain reliability scores, X, for a wide range of individual cars. Draw the DAG for a hierarchical model which could be used to infer the reliability of different models and manufacturers.

10.12 Chapter Summary

In this chapter, we introduced a number of important new ideas and methods of Bayesian statistical modelling.

- The **posterior predictive distribution** is a model of the data-generating process which combines *epistemic* uncertainty in model parameters and *aleatory* uncertainty in future outputs conditional on those parameters.
- A **Bayesian p-value** gives the probability that a statistic computed from data output by a given model will be more extreme than the value of the same statistic computed from observed data. Bayesian p-values close to 0 or 1 suggest the model is questionable.
- **Markov Chain Monte Carlo** (MCMC) is a general-purpose method for sampling from arbitrary probability distributions which need not be normalised. It produces **dependent samples** so the **effective sample size** is usually smaller than the number of chain steps.
- **Informative priors** are useful when observational data leaves large posterior uncertainties in parameter values.
- **Empirical Bayes** allows us to combine information from multiple related datasets by computing a **frequentist** estimate of the distribution from which the parameters of the dataset-generating processes were drawn.
- **Hierarchical modelling** allows us to construct a single Bayesian model of multiple related datasets, with groupings expressed using a hierarchical graph.

10.13 Further Exercises

■ ***Exercise*** 10.17 Consider a random walker on the non-negative integers who, at each time step, moves left with probability $1/2$ (if a left move is possible), right with probability $p < 1/2$, and otherwise stays still. By writing down and solving detailed balance equations, find the equilibrium distribution of the walker.

11

Classification

In Chapter 1 we distinguished between the two main types of supervised learning: regression and classification. We've said plenty about the former; now it's time to look at the latter. In this chapter, we study binary logistic regression (which despite its name is a classification method), multinomial logistic regression (ditto) and k-nearest neighbours. The first two are especially important, because they serve as components of the neural-network-based classification methods we will meet in Chapter 13.

11.1 The Concept of Classification

Consider the biomedical dataset whose first few rows are shown in Table 11.1. Each row represents a patient. Columns represent variables – for example, fasting blood sugar and diabetes diagnosis. Now, suppose the patients are a random sample drawn from some large population. Could we predict whether a new patient sampled from this population has diabetes by observing the values of the other variables? Notice that this prediction problem is different in kind from those we have considered so far, because whether someone has diabetes is not a matter of degree: one either has the disease or not. The prediction target is *binary*.

Binary variables are a special case of categorical variables. A variable is categorical if it takes values in a finite set that lacks natural arithmetical structure. The possible values are known as 'categories', 'levels' or 'classes'. The set of possible values may still have a natural *order* – for example, the values might be 'low', 'mid' and 'high'. In that case, the categorical variable is *ordinal*. Otherwise, it is *nominal*. In our diabetes dataset, socioeconomic status

Table 11.1 *Data on diabetes and features possibly predictive of diabetes. Rows represent patients. There are 1,879 rows in the full dataset (El Kharoua (2024)); the first eight are shown here.*

Fasting blood sugar (mg/dL)	Haemoglobin A1C (%)	Cholesterol triglycerides (mg/dL)	Hypertension	Socioeconomic status	Diagnosis
120.2	7.93	235.4	No	High	No
75.1	4.50	232.9	No	Mid	No
187.3	8.62	180.4	No	Low	Yes
134.6	6.75	299.6	No	Mid	Yes
179.5	5.56	337.8	No	Mid	No
85.4	9.76	271.5	No	Mid	No
151.2	7.18	295.0	Yes	Mid	Yes
87.4	9.71	248.6	No	Low	No

is an ordinal variable. We don't have an ethnicity variable, but if we did it would be nominal (because there is no natural ordering of ethnicities). The question of natural order is largely uninteresting for binary variables, so these are often regarded as nominal by default.

Prediction problems in which the prediction target is categorical are known as classification problems. They are the topic of this chapter. We'll begin with binary classification, then generalise to multi-class targets. (We will restrict our attention to nominal targets, however.)

11.2 Probabilistic Binary Classification: Preliminaries

11.2.1 Why Probabilities?

We'd like to develop a *classifier* for the diabetes task sketched above. What should it do? It's clear it should implement a function of some sort. The inputs will be fasting blood sugar level, haemoglobin A1C percentage, cholesterol triglycerides level, hypertension classification and socioeconomic status. What about the output? One option is a 'best guess' as to whether the patient is diabetic or not: a binary prediction for a binary target. Some classifiers work this way, but we won't cover them. Instead, we will focus on tools whose output can be interpreted as a *probability* – in the diabetes case, a probability that the patient in question is diabetic. (Probabilities can be turned into best guesses if best guesses are needed.)

Probabilities will often be of interest to the users of classification tools. A 30% predicted probability that Joe is diabetic and a 0.01% predicted probability that Joe is diabetic may both correspond to the same 'best guess' (Joe probably isn't diabetic), but their practical implications are different. One warrants further investigation; the other doesn't. Moreover, probabilistic output is useful for fitting classifiers to data. When we make small adjustments to a classifier, the probabilities assigned to various observed outcomes will change, even if none of the best-guess binary judgements do. That's helpful for finding promising directions for further adjustment. We will make good on the directions metaphor in Section 11.3, when we consider classifiers whose behaviour is governed by continuous parameter vectors.

A good probabilistic diabetes classifier outputs values near 1 for diabetics, and values near 0 for non-diabetics. It therefore makes sense to recode the prediction target as an indicator variable: 1 for diabetes and 0 for no diabetes. The labels used to encode the prediction target in a classification problem are known as *class labels*; using 1 and 0 is standard for binary problems. We'll similarly recode hypertension (1 for presence, 0 for absence) and socioeconomic status (0 = low, 1 = mid, 2 = high). The first few rows of the recoded dataset are shown in Table 11.2. We can package the predictor variables for the ith patient into a single predictor vector $\boldsymbol{x}_i$. So if the first row in Table 11.2 represents patient 1, then

$$\boldsymbol{x}_1 = (120.2, 7.93, 235.4, 0, 2).$$

The dataset as a whole takes the form $D = \{(\boldsymbol{x}_i, y_i\}_{i=1}^n$, where y_i is the diabetes status (either 0 or 1) of the ith patient.

11.2.2 Likelihood and Cross-Entropy Loss

Let g be a probabilistic binary classifier intended to handle the classification problem sketched above. If $\boldsymbol{x}$ is the predictor vector of a patient drawn randomly from the relevant population,

Table 11.2 *Data from Table 11.1 with categorical variables recoded. Reminder: there are 1879 rows in the full dataset.*

Fasting blood sugar (mg/dL)	Haemoglobin A1C (%)	Cholesterol triglycerides (mg/dL)	Hypertension	Socioeconomic status	Diagnosis
120.2	7.93	235.4	0	2	0
75.1	4.50	232.9	0	1	0
187.3	8.62	180.4	0	0	1
134.6	6.75	299.6	0	1	1
179.5	5.56	337.8	0	1	0
85.4	9.76	271.5	0	1	0
151.2	7.18	295.0	1	1	1
87.4	9.71	248.6	0	0	0

then $g(\boldsymbol{x})$ is the classifier's confidence (expressed as a probability) that the patient is diabetic. We can also put this in frequentist terms. Think of a patient drawn randomly from the population as a single realisation of a random vector $(\boldsymbol{X}, Y)$. The random vector represents the datapoint-generating process. Our dataset, D, consists of n independent realisations of $(\boldsymbol{X}, Y)$. A classifier g corresponds to the following hypothesis about the conditional distribution of Y given $\boldsymbol{X} = \boldsymbol{x}$ (for arbitrary $\boldsymbol{x}$):

$$Y \mid \{\boldsymbol{X} = \boldsymbol{x}\} \sim \text{Bernoulli}(g(\boldsymbol{x})),$$

or, in words, the conditional probability that a patient is diabetic given that his or her predictor vector is $\boldsymbol{x}$ is $g(\boldsymbol{x})$. To avoid notational clutter, we use the same symbol, 'g', to denote the classifier, the function from predictor vectors to probabilities it implements, and the hypothesis to which it corresponds. Context will make clear which is meant.

The likelihood of hypothesis g is the conditional probability of all the observed class labels $(y_1, \ldots, y_n)$ in our dataset given the observed predictor vectors $(\boldsymbol{x}_1, \ldots, \boldsymbol{x}_n)$, according to the hypothesis. To calculate it, we first write the probability (under the hypothesis) that a patient's class label is y given that his or her predictor vector is $\boldsymbol{x}$. That is,

$$f_{Y|\boldsymbol{X}}(y|\boldsymbol{x}) = g(\boldsymbol{x})^y \, (1 - g(\boldsymbol{x}))^{1-y} \,. \tag{11.1}$$

The conditional probability (according to hypothesis g) of all the observed class labels given all the observed predictors is just the product of the conditional probabilities for the individual patients. So the likelihood function is

$$\mathcal{L}(g; D) = \prod_{(\boldsymbol{x},y)\in D} g(\boldsymbol{x})^y \, (1 - g(\boldsymbol{x}))^{1-y} \,. \tag{11.2}$$

We may sometimes wish to choose a probabilistic classifier by maximising the likelihood function over some family of candidates G, that is, to choose

$$\hat{g} = \arg\max_{g\in G} \mathcal{L}(g; D).$$

Maximising likelihood is equivalent to minimising the *cross-entropy loss*, which is just the negative of log-likelihood divided by the size of the dataset. It is defined by

$$\begin{aligned} l(g;D) &= -\frac{1}{n}\log\mathcal{L}(g;D) \\ &= -\frac{1}{n}\sum_{(\boldsymbol{x},y)\in D}(y\log g(\boldsymbol{x}) + (1-y)\log(1-g(\boldsymbol{x}))). \end{aligned} \tag{11.3}$$

The cross-entropy loss is so called because each term in the sum is the cross-entropy of predicted probabilities relative to a distribution placing all the mass on the true label. (See Section 10.8 for the general definition of cross-entropy.) It is convenient to work with because it is non-negative and because its expected value does not scale with dataset size.

■ ***Exercise*** 11.1 Show that $l(g;D)$ can never be negative. In what situations is it zero?

■ ***Exercise*** 11.2 Let random dataset $\mathcal{D}$ be a sequence of n i.i.d. copies of random vector $(\boldsymbol{X}, Y)$. Show that $\mathbb{E}(l(g;\mathcal{D}))$ does not depend on n.

Cross-entropy losses are ubiquitous in classification. The loss we have just defined is for binary classification problems (when we want to emphasise that, we call it binary cross-entropy loss), but generalising the idea to the multiclass case is straightforward, as we'll see in Section 11.4.2. The key thing to remember is that cross-entropy losses are lightly disguised measures of likelihood. If you have a family of classifiers to choose from and you pick the one with the smallest cross-entropy loss on your training data, you're doing maximum likelihood inference.

In Section 11.3.1 we will encounter *regularised* cross-entropy loss, that is, cross-entropy loss plus a regularisation penalty. This, as you would expect, is a lightly disguised measure of posterior probability, with the regularisation term encoding an implicit prior over classifiers.

11.3 Binary Logistic Regression

11.3.1 The Binary Logistic Regression Model

We have a likelihood-based loss function suitable for probabilistic binary classification (equation (11.3)), so we can now choose among candidate classifiers in a data-driven way. But which candidates should we consider? We could cast a wide net, considering all sorts of very complicated functions from predictor vectors to probabilities. We'll do that in Chapter 13, applying deep neural networks to classification. For now we will keep things simple.

It is natural to consider linear functions of the predictor vector. However, linear functions are unbounded. Given a constant scalar β_0 and a constant vector $\boldsymbol{\beta}$, you can make $\beta_0 + \boldsymbol{\beta}^T\boldsymbol{x}$ take any value you like by picking an appropriate $\boldsymbol{x}$ (unless $\boldsymbol{\beta}$ happens to be the zero vector, in which case there is no $\boldsymbol{x}$-dependence at all).[1] Since values outside $[0, 1]$ make no sense as probabilities, we shouldn't define classifiers by writing '$g(\boldsymbol{x};\beta_0,\boldsymbol{\beta}) = \beta_0 + \boldsymbol{\beta}^T\boldsymbol{x}$'. Instead,

[1] Due to the presence of the intercept term β_0, a mathematician would say we are talking about *affine* (not linear) functions from the predictor space to $\mathbb{R}$. In the machine learning literature, 'linear' usually means affine.

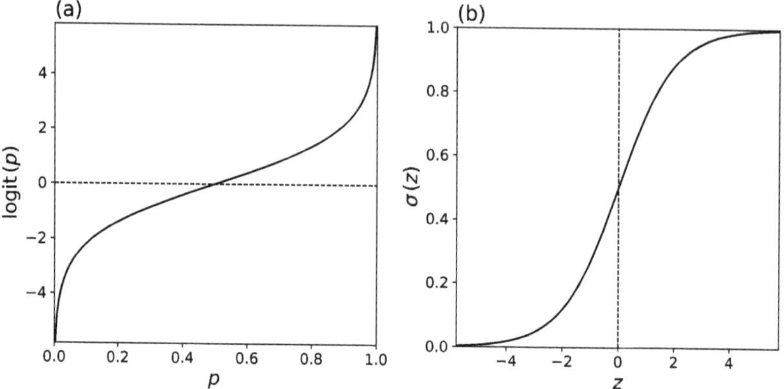

Figure 11.1 (a) The function $\text{logit}(p) = \log(p/(1-p))$ plotted against p. (b) The function $\sigma(z) = (1 + e^{-z})^{-1}$ plotted against z. Notice that these functions are mutual inverses.

we can postulate that some *transformed representation* of probability depends linearly on $\boldsymbol{x}$:

$$\psi(g(\boldsymbol{x}; \beta_0, \boldsymbol{\beta})) = \beta_0 + \boldsymbol{\beta}^T \boldsymbol{x}, \tag{11.4}$$

where $\psi : [0, 1] \to \mathbb{R}$ is a fixed invertible function. For then we have

$$g(\boldsymbol{x}; \beta_0, \boldsymbol{\beta}) = \psi^{-1}(\beta_0 + \boldsymbol{\beta}^T \boldsymbol{x}), \tag{11.5}$$

which is guaranteed to lie in the interval $[0, 1]$. The fixed invertible function ψ is called the *link function*. One common choice of link function is the *logit* function, defined by

$$\text{logit}(p) = \log\left(\frac{p}{1-p}\right). \tag{11.6}$$

Its inverse, $\sigma : \mathbb{R} \to [0, 1]$, is the *standard logistic function* or *standard sigmoid*, defined by

$$\sigma(z) = \frac{1}{1 + e^{-z}}. \tag{11.7}$$

Both functions are plotted in Figure 11.1. The value returned by the logit function – the quantity on the right-hand side of equation (11.6) – is sometimes called the *log odds*. So with $\psi = \text{logit}$, our postulate (11.4) says that the log odds depends linearly on $\boldsymbol{x}$.

■ ***Exercise*** 11.3 Compute the log odds corresponding to each of the following probabilities: 10^{-12}; 1%; 52%; 53%. Which of the following probability changes corresponds to the greater change in log odds:

- a change in probability from 10^{-12} to 1%;
- a change in probability from 52% to 53%?

With logit as the link function, equation (11.5) becomes

$$g(\boldsymbol{x}; \beta_0, \boldsymbol{\beta}) = \sigma(\beta_0 + \boldsymbol{\beta}^T \boldsymbol{x}). \tag{11.8}$$

This defines a family of functions mapping predictor vectors to probabilities. You specify a function in the family by picking values for the parameters β_0 and $\boldsymbol{\beta}$. If you do that by minimising (regularised or unregularised) cross-entropy loss on training data, you are doing *binary logistic regression*. We will focus for now on the unregularised case, in which the loss function is precisely as stated in equation (11.3). Expressed as a function of β_0 and $\boldsymbol{\beta}$, it is

$$l(\beta_0, \boldsymbol{\beta}; D) = \frac{1}{n} \sum_{(\boldsymbol{x},y)\in D} \left(y \log(1 + e^{-(\beta_0 + \boldsymbol{\beta}^T \boldsymbol{x})}) + (1-y)\log(1 + e^{\beta_0 + \boldsymbol{\beta}^T \boldsymbol{x}}) \right). \tag{11.9}$$

■ ***Exercise*** 11.4 Derive equation (11.9) by substituting equation (11.8) into equation (11.3).

There is no simple formula for the values of β_0 and $\boldsymbol{\beta}$ that minimise this loss, but there are fast algorithms (implemented by statistical software packages) for finding them computationally.

Before tackling the full diabetes classification task, let's experiment with some single-predictor models.

11.3.2 Simple Logistic Regression (One Predictor)

If there is just one scalar predictor x, equation (11.8) becomes

$$g(x; \beta_0, \beta_1) = \sigma(\beta_0 + \beta_1 x).$$

■ ***Exercise*** 11.5 Logistic regression with a single predictor is sometimes called simple logistic regression. This exercise will help you get acquainted with the general form of a simple logistic regression model.

(a) Sketch $g(x; \beta_0, \beta_1)$ as a function of x, first supposing $\beta_1 > 0$, and then supposing $\beta_1 < 0$. Annotate your sketches to show $x_{0.5}$, the value of x for which $g(x; \beta_0, \beta_1) = 0.5$. Write an expression for $x_{0.5}$ in terms of β_0 and β_1.
(b) What happens if $\beta_1 = 0$?
(c) Assuming $\beta_1 \neq 0$, what change in x is required to increase the log odds by one? Call this change δ_x.
(d) Still assuming $\beta_1 \neq 0$, what are the values of $g(x_{0.5} + \delta_x; \beta_0, \beta_1)$ and $g(x_{0.5} - \delta_x; \beta_0, \beta_1)$? Update your sketches in (a) to display this information.

■ ***Exercise*** 11.6 Rewrite equation (11.9) for the single-predictor case.

We will now fit five separate simple logistic regression models to the diabetes dataset, using each of the five potentially predictive features in Table 11.2 in turn as the predictor x. We'll use unregularised cross-entropy loss as the loss function, so these will be maximum likelihood fits. Figure 11.2 shows the resulting sigmoid curves. Each plot also includes a jittered scatter of individual datapoints, colour-coded by diabetes status, to give a rough visual sense of how the predictor values are distributed across the two response classes. The main takeaway from the figure is that fasting blood sugar and haemoglobin A1C are much

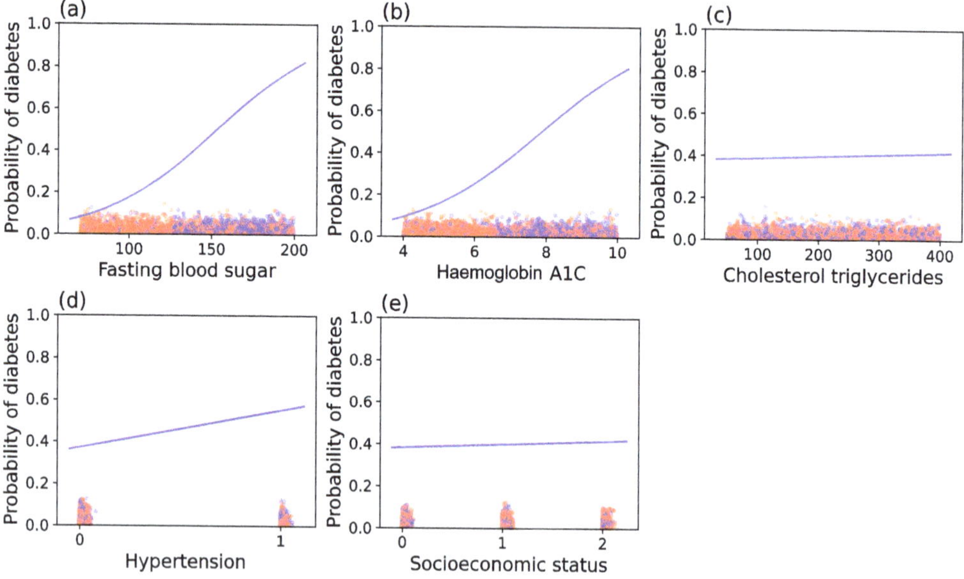

Figure 11.2 Simple logistic regression models fitted to each of the five predictors in the dataset previewed in Table 11.2. In (a)–(e), the curve shows predicted probability of diabetes as a function of the single predictor; training datapoints are plotted along the predictor axis, colour-coded by diagnosis – blue (darker) for diabetes and orange (lighter) for no diabetes. Datapoints are plotted with vertical jitter in all panels and with horizontal jitter in panels (d) and (e).

more strongly associated with diabetes (in this dataset) than the other three predictors. For example, according to the model plotted in Figure 11.2a, patients in the top fasting blood sugar decile have a dramatically higher conditional probability of diabetes (given their fasting blood sugar) than patients in the bottom decile: roughly 80% versus roughly 10%. By contrast, according to the model plotted in Figure 11.2c, patients in the top cholesterol triglycerides decile have only a slightly higher conditional probability of diabetes (given their cholesterol triglycerides) than patients in the bottom decile: just over 40% versus just under 40%. We see a similar pattern in Figure 11.2e. As you have probably guessed, the proportion of patients in the dataset who have diabetes is approximately 40%. When the predictor contains little information about the prediction target, the best a model can do is defer to the base rate.

What sort of out-of-sample prediction performance should we expect from these models? Naturally we turn to cross-validation to answer that question. We'll use five-fold cross-validation (see Section 7.4), with one minor tweak: when we divide the dataset into five equal-sized folds, we will make sure that each fold contains roughly the same number of diabetic patients. That's easy to do: first we share out the diabetic patients among the five folds, and then we share out the rest. This is called *stratified* cross-validation, and it is standard practice for classification problems. In all other respects, the cross-validation procedure is the same as it was for regression. We fit our model on four of the folds (lumped together into one training set) and validate it on the fifth, letting each of the five folds take its turn as the validation fold. Finally, we average over the five validation scores to get our performance

Table 11.3 *Coefficients, estimated standard errors in coefficients, cross-validated cross-entropy losses and cross-validated accuracies for single-predictor logistic regression models. Starred coefficients, β_1^*, are based on standardised predictors. Standard errors were estimated using the non-parametric bootstrap method with 1,000 replications.*

Predictor	$\hat{\beta}_1$	$\hat{\mathrm{se}}(\hat{\beta}_1)$	$\hat{\beta}_1^*$	$\hat{\mathrm{se}}(\hat{\beta}_1^*)$	cv ce loss	cv accuracy
Fasting blood sugar	0.0288	0.0016	1.080	0.059	0.563	0.680
Haemoglobin A1C	0.582	0.033	1.012	0.057	0.575	0.668
Cholesterol triglycerides	3.67×10^{-4}	4.51×10^{-4}	0.037	0.046	0.673	0.600
Hypertension	0.730	0.126	0.263	0.045	0.665	0.616
Socioeconomic status	0.0682	0.0628	0.0522	0.0480	0.673	0.600

measure. In fact, we will do this averaging twice, because there are two validation scores in common use and we want to illustrate both. The first is cross-entropy loss. For the avoidance of doubt, some version of cross-entropy loss is *always* used as the loss function for fitting logistic regression models. The issue here is how to evaluate the performance of fitted models on validation data, that is, on data that was not used in the fitting process. Cross-entropy loss can do that job too; we just compute it using the validation data rather than the training data. If we're working with validation set $D_{\mathrm{val}} = \{(x_i, y_i)\}_{i=1}^{n_{\mathrm{val}}}$, then the cross-entropy loss validation score for a single-predictor model with parameters (β_0, β_1) is

$$l_{\mathrm{val}}(\beta_0, \beta_1; D_{\mathrm{val}}) = -\frac{1}{n_{\mathrm{val}}} \sum_{(x,y)\in D_{\mathrm{val}}} (y \log g(x;\beta_0,\beta_1) + (1-y)\log(1-g(x;\beta_0,\beta_1))) . \tag{11.10}$$

This quantity is the average of the negative logarithm of the probability assigned by the classifier to the class label that was in fact correct, where the average is taken over the validation set. Lower cross-entropy loss validation scores indicate better performance. A classifier that classified every example in the validation set correctly with 100% confidence would achieve a perfect score: 0. (Logistic regression classifiers will never attain this ideal limit. Why not?)

The other score commonly used in binary classification problems is *accuracy*, which is the proportion of validation examples classified correctly. To compute an accuracy score for a probabilistic binary classifier, the classifier's probabilistic outputs need to be converted into best-guess class labels. We can do that by mapping probabilities greater than 0.5 to the guess $y = 1$ (patient is diabetic), and probabilities less than or equal to 0.5 to the guess $y = 0$ (patient is not diabetic). Higher accuracy scores indicate better performance. A classifier that classified every example in the validation set correctly would achieve a perfect accuracy score of 1.0, even if its classifications were not always very confident.

Cross-validation scores for the five simple logistic regression models are reported in the two rightmost columns of Table 11.3. Let's begin by considering the two least-informative predictors, cholesterol triglycerides and socioeconomic status. Logistic regression models based on either of these predictors achieve cross-validated accuracy of 0.600 and cross-validated cross-entropy loss of 0.673. Both scores make sense. Single-predictor models fitted to four-fifths of the available data should look very like single-predictor models fitted to all of the available data; and we saw those in Figure 11.2. When the predictor was cholesterol triglycerides or socioeconomic status, the predicted probability of diabetes was essentially constant at the base rate, 40%. If your best guess about every patient is that they are not diabetic, then if 40% of them *are* diabetic you will be right 60% of the time. Moreover, equation (11.10) implies that models that always predict probability π of diabetes, where π is the base rate of diabetes in the dataset, will achieve an average validation cross-entropy loss of

$$-\left(\pi \log \pi + (1-\pi)\log(1-\pi)\right).$$

When $\pi = 0.4$, this expression evaluates to approximately 0.673. Unsurprisingly, single-predictor models based on the more informative predictors do better, achieving higher cross-validated accuracies and lower cross-validated cross-entropy losses.

Table 11.3 also reports coefficient estimates for each model. As usual, these estimates are easier to interpret if we standardise predictors to have the same variance across the training data. It is common practice to do this, to pick unit variance, and to also shift the mean of each predictor to 0; the procedure is the familiar one described in Section 7.5.3.[2] Coefficient estimates based on standardised predictors are marked with asterisks in Table 11.3. Looking at the first row of the $\hat{\beta}_1^*$ column, we see that a one-standard-deviation difference in fasting blood sugar is associated (according to the maximum likelihood single-predictor model) with a difference in the log odds of diabetes of 1.08 – quite a strong effect. Scanning down the column we get an immediate sense of the predictive value of each predictor in isolation.

■ ***Exercise*** 11.7 Suppose you initially judge the probability that a particular patient is diabetic is 40%. What is your revised probability if your log odds increase by

(a) 1.08;
(b) 0.037?

The table also reports bootstrap standard errors in coefficient estimates (we used 1,000 resamples). For the single-predictor models based on cholesterol triglycerides and socioeconomic status, the bootstrap standard errors are comparable in magnitude to the coefficient estimates. There is therefore uncertainty about the *sign* of these coefficients.

11.3.3 Why Regularisation Is Normally Used in Practice

The single-predictor models we fitted in Section 11.3.2 were extremely simple, so using unregularised cross-entropy loss as our objective function – a pure maximum likelihood approach – seemed natural. It would certainly be odd to look at Figure 11.2 and worry about overfitting. Nevertheless, it is standard practice to include at least a small regularisation penalty in the logistic regression loss function, even in the single-predictor case. We shall explain the reasons in this subsection.

Figure 11.3 shows four simple logistic regression classification problems. In each panel, 200 points, each coloured either orange (lighter) or blue (darker), are plotted along the x-axis (with vertical jitter). Each set of 200 points is a random sample from a particular population. The population changes from panel to panel. In panel (a), the distribution of

[2] Shifting the mean does not affect the coefficient estimates, but it does make it easier to understand the data. For example, if you see a standardised predictor take an extreme value such as 9 or −12, you know at once that you are dealing with an extreme outlier.

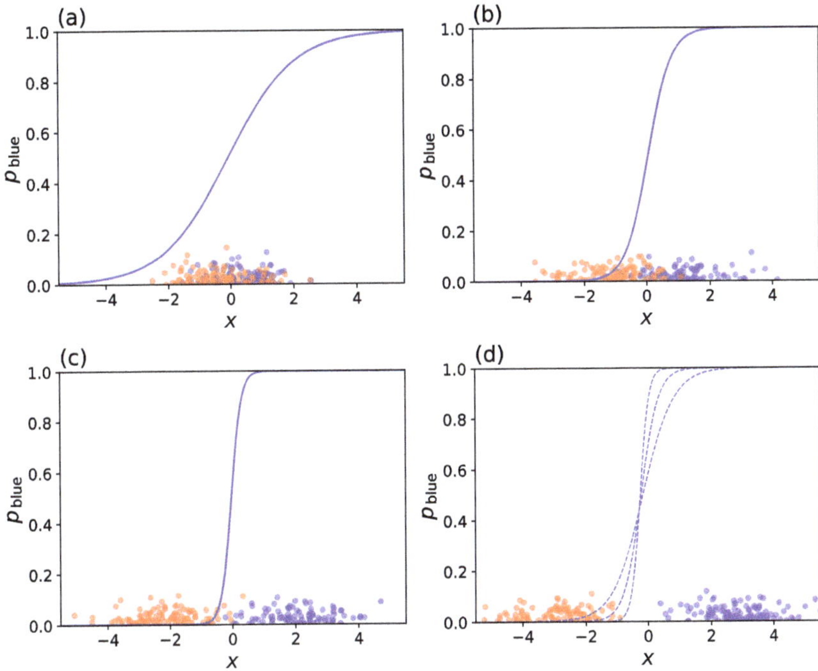

Figure 11.3 A sequence of simple logistic regression problems featuring simulated datasets with increasing class separation. Each solid sigmoid represents the probability that a point is blue (darker) (rather than orange (lighter)) as a function of predictor x, according to the logistic regression classifier that minimises unregularised cross-entropy loss on the training data (blue and orange circles, plotted with vertical jitter). In (a), the distributions of blue and of orange points are similar; the best-fitting sigmoid is therefore broad. In (b), the distributions are more clearly distinguishable, and the best-fitting sigmoid is sharper. It is sharper still in (c). In (d), the distributions are *linearly separable*, and there is no best-fitting sigmoid.

orange and blue points is similar, though blue becomes slightly more common as x increases. The distinction between the orange and blue classes is clearer in panel (b), clearer still in panel (c) and very clear indeed in panel (d). We treat each panel's set of points as a training dataset for logistic regression. In panels (a) through (c), we plot the maximum likelihood sigmoids. That is, we pick a member of the function family defined by equation (11.8) by minimising unregularised cross-entropy loss. As the orange and blue classes become more distinct, the likelihood-maximising sigmoid becomes sharper. But panel (d) is an awkward case for likelihood maximisation. Here, the orange and blue classes in the training data are *linearly separable*. That is, we can find a point x_{cut} on the x-axis such that every orange training point is to the left of x_{cut} and every blue training point is to the right of it. (For example, $x_{\text{cut}} = 0$ works.) One consequence of this situation is that there is no such thing as the likelihood-maximising sigmoid. So long as the location of the sigmoid's transition is appropriate – that is, so long as we snap from low probability to high probability in the region separating the orange class from the blue – sharp is good and sharper is better.

As we try sharper and sharper sigmoids, the likelihood just continues to increase, tending towards (but never quite attaining) a limit of 1. Equivalently, the cross-entropy loss continues to decrease, tending towards (but never quite attaining) a limit of zero. When we added regularisation penalties to likelihood-based loss functions in Chapter 9, it was because we didn't trust maximum likelihood estimates. Now we have a new motive: maximum likelihood (i.e. cross-entropy loss minimising) logistic regression fits *may not even exist*. By including a regularisation penalty – even a very small one – in the loss function, we ensure that our optimisation problems are well posed. With a ridge penalty, the loss function for simple logistic regression becomes

$$l^{\lambda}(\beta_0, \beta_1; D) = \frac{1}{n}\Big[\lambda \beta_1^2 - \sum_{(x,y)\in D} (y \log g(x;\beta_0,\beta_1) + (1-y)\log(1-g(x;\beta_0,\beta_1)))\Big]. \quad (11.11)$$

The second term on the right-hand side is the unregularised cross-entropy loss. The first term is the ridge penalty, with λ a hyperparameter controlling the strength of regularisation. So long as $\lambda > 0$, the regularised cross-entropy loss is sure to have a global minimum. In particular, when D is linearly separable, it is no longer possible to achieve lower and lower losses by sharpening sigmoids without bound. Sharpening means increasing $|\beta_1|$ (recall Exercise 11.5), and hence incurring larger ridge penalties.[3]

Of course, the classic motives for regularisation remain centrally important. Regularisation penalties are effective tools for controlling overfitting in general, not just when the training dataset happens to be linearly separable. Moreover, the Bayesian interpretation we stressed in Chapter 9 – regularisers as implicit priors – is still compelling. When we fit simple logistic regression models by minimising the loss defined in equation (11.11), we are doing maximum a posteriori (MAP) inference. We are imposing a prior that says that sigmoids with very sharp transitions are implausible: tiny changes in the predictor should not lead to large changes in predicted probability. Exercise 11.8 invites you to fill in the details.

■ ***Exercise*** 11.8 What prior precisely are we imposing on β_1 when we minimise the loss defined in equation (11.11)? (Hint: minimising l^{λ} is mathematically equivalent to maximising $\exp(-nl^{\lambda})$. If you substitute the right-hand side of equation (11.11) into the latter expression, you will see that it decomposes naturally into factors proportional to a prior and to a likelihood respectively.)

We saw in Section 9.3 that cross-validation can be used to tune regularisation strength hyperparameters. In Section 11.3.5, we will demonstrate essentially the same workflow in a multiple-predictor logistic regression context.[4]

11.3.4 Multiple Logistic Regression (Many Predictors)

Binary logistic regression with two or more predictors is sometimes called *multiple logistic regression* (though the word 'multiple' is often omitted). We have already presented the

[3] Since the ridge penalty is quadratic in β_1 and the (unregularised) cross-entropy loss is bounded below by zero, increasing $|\beta_1|$ without bound must eventually cause $l^{\lambda}(\beta_0, \beta_1; D)$ to increase.

[4] For the sake of concision, we'll skip a similar demonstration in the simple logistic regression context.

logistic regression model with an arbitrary number of predictors (Section 11.3.1). We restate the essentials in Box 11.1, updating the loss function to include a ridge penalty.

★ Box 11.1 Binary logistic regression

Binary logistic regression (sometimes just called logistic regression) is a probabilistic binary classification method defined by a model and a fitting strategy. The model postulates that the log odds depends linearly on the predictor vector, or equivalently that the conditional probability that an example with predictor vector $\boldsymbol{x}$ is a positive instance ($y = 1$ as opposed to $y = 0$) is

$$g(\boldsymbol{x}; \beta_0, \boldsymbol{\beta}) = \sigma(\beta_0 + \boldsymbol{\beta}^T \boldsymbol{x}) = \frac{1}{1 + e^{-(\beta_0 + \boldsymbol{\beta}^T \boldsymbol{x})}}, \tag{11.12}$$

where scalar β_0 and vector $\boldsymbol{\beta}$ are the parameters of the model. The fitting strategy is to choose parameters that minimise (regularised) cross-entropy loss on training data $D = \{(\boldsymbol{x}_i, y_i)\}_{i=1}^n$. With ridge regularisation, that loss is

$$l^\lambda(\beta_0, \boldsymbol{\beta}; D) = \frac{\lambda}{n} ||\boldsymbol{\beta}||^2 + l(\beta_0, \boldsymbol{\beta}; D), \tag{11.13}$$

where

$$l(\beta_0, \boldsymbol{\beta}; D) = -\frac{1}{n} \sum_{(\boldsymbol{x}, y) \in D} (y \log g(\boldsymbol{x}; \beta_0, \boldsymbol{\beta}) + (1 - y) \log(1 - g(\boldsymbol{x}; \beta_0, \boldsymbol{\beta})))$$

is the unregularised cross-entropy loss and $\lambda \geq 0$ is a hyperparameter controlling the strength of regularisation.

Before fitting a binary logistic regression model, it is good practice to standardise each predictor (predictor vector component) to have a mean of 0 and a variance of 1, where the mean and the variance are taken over the training data. This is especially important when regularisation is used, that is, when $\lambda > 0$. If one does not standardise, then components of $\boldsymbol{\beta}$ corresponding to predictors with smaller ranges of variation will be subject to stronger regularisation.

Understanding simple logistic regression is an excellent foundation for understanding multiple logistic regression, because the mapping from predictor vector to probability defined by equation (11.12) goes *via* a one-dimensional space, irrespective of the dimensionality of $\boldsymbol{x}$. That is because $\sigma(\beta_0 + \boldsymbol{\beta}^T \boldsymbol{x})$ depends on $\boldsymbol{x}$ only by way of the inner product $\boldsymbol{\beta}^T \boldsymbol{x}$. Indeed, if we write $x_\beta = \frac{\boldsymbol{\beta}^T \boldsymbol{x}}{||\boldsymbol{\beta}||}$ for the magnitude of $\boldsymbol{x}$'s component in the $\boldsymbol{\beta}$ direction, then we have

$$\sigma(\beta_0 + \boldsymbol{\beta}^T \boldsymbol{x}) = \sigma(\beta_0 + ||\boldsymbol{\beta}|| x_\beta).$$

Magnitude $||\boldsymbol{\beta}||$ controls the sharpness of the sigmoid. That is, it controls how abruptly $\sigma(\beta_0 + \boldsymbol{\beta}^T \boldsymbol{x})$ transitions from a value near 0 to a value near 1 as the magnitude of $\boldsymbol{x}$'s component in the the $\boldsymbol{\beta}$ direction increases. Magnitude $||\boldsymbol{\beta}||$ and parameter β_0 together control the location of the transition. The mathematics of this is structurally identical to the simple logistic regression case explored in Exercise 11.5.

■ ***Exercise*** 11.9 What would it mean for two classes to be linearly separable in a p-dimensional predictor vector space? (Hint: your answer should begin 'it would mean that there exists a $\boldsymbol{\beta}$ such that ...')

11.3.5 Multiple Logistic Regression on the Diabetes Dataset

It's time to fit a multiple logistic regression model to the diabetes dataset described in Section 11.1. But what value should we pick for hyperparameter λ? To answer that question, we need to explore how λ affects the out-of-sample classification performance of fitted models. Cross-validation is the natural strategy. We will use average cross-entropy loss and average accuracy on validation folds as our scores, just as we did when estimating the out-of-sample performance of simple logistic regression models in Section 11.3.2. Our earlier definition of validation cross-entropy loss (equation (11.10)) was for the simple logistic regression case; generalising in the obvious way, we have

$$l_{\text{val}}(\beta_0, \boldsymbol{\beta}; D_{\text{val}}) = -\frac{1}{n_{\text{val}}} \sum_{(\boldsymbol{x},y)\in D_{\text{val}}} (y \log g(\boldsymbol{x}; \beta_0, \boldsymbol{\beta}) + (1-y)\log(1 - g(\boldsymbol{x}; \beta_0, \boldsymbol{\beta}))) .$$

■ ***Exercise*** 11.10 Why is *regularised* cross-entropy loss never used as a validation score?

Cross-validation results for multiple logistic regression on the diabetes dataset are plotted in Figure 11.4. Panels (a) and (b) cover a wide range of regularisation strengths: they give us the big picture. At first glance, it seems that regularisation hurts out-of-sample performance. As λ increases, cross-validated cross-entropy loss increases, and cross-validated accuracy falls. We'll need to qualify that summary when we discuss panel (c), but the basic trend is clear. It is also unsurprising. Logistic regression is a simple technique, and our dataset is 'tall': we have many examples (1879 patients) but just five features (fasting blood sugar, haemoglobin A1C, cholesterol triglycerides, hypertension and socioeconomic status). Five features is more than one, but there is still little reason to worry about overfitting.

A logistic regression classifier fitted to four-fifths of the available data by minimising unregularised cross-entropy loss achieves a cross-validated accuracy of roughly 83%. That is considerably better than any of the simple logistic regression models we considered in Section 11.3.2, and much better than the 60% accuracy one would achieve if one just guessed that every patient was non-diabetic. (Recall that the base rate of diabetes in the dataset is 40%.) But in the strong regularisation ($\lambda \to \infty$) limit, cross-validated accuracy converges to precisely this rate, 60%. Moreover, in that same limit, cross-validated cross-entropy loss appears to be converging to the value achieved by simple logistic regression models fitted to uninformative predictors – approximately 0.67. Why? Well, since the fitting process minimises the right-hand side of equation (11.13), pushing λ to infinity drives every component of $\hat{\boldsymbol{\beta}}$ towards zero. From equation (11.12), we have

$$g(\boldsymbol{x}; \beta_0, \boldsymbol{0}) = \frac{1}{1 + e^{-\beta_0}},$$

which is independent of $\boldsymbol{x}$. So, zeroing out $\hat{\boldsymbol{\beta}}$ gives a classifier that ignores the predictor

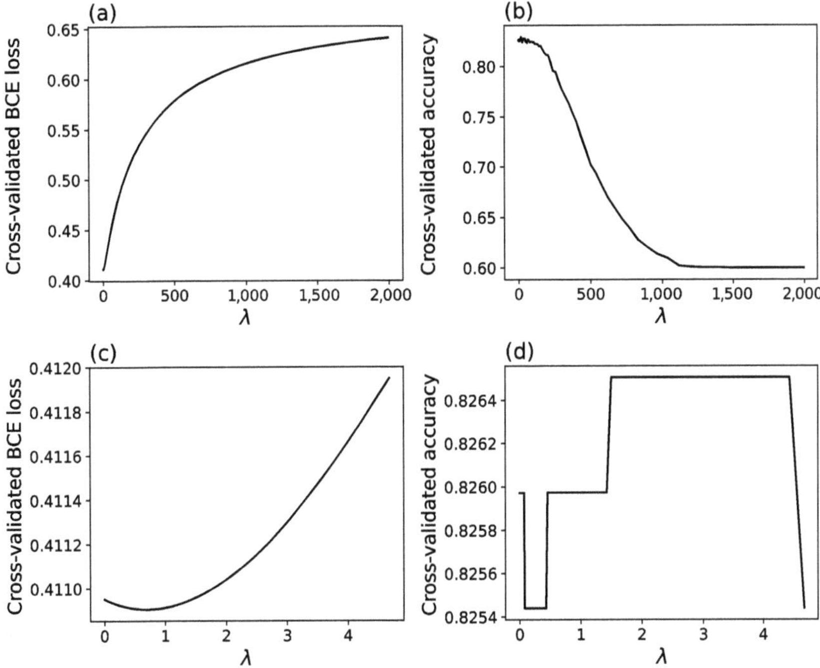

Figure 11.4 Cross-validated binary cross-entropy loss (a and c) and accuracy (b and d) of multiple logistic regression applied to the diabetes classification task described in Section 11.1, as a function of the regularisation strength λ. For each value of λ shown, each validation metric is averaged across all five folds in the cross-validation. Panels (a) and (b) cover a wide range of regularisation strengths. The initial portion of this range is replotted in panels (c) and (d).

vector and returns the same predicted probability $(1 + e^{-\beta_0})^{-1}$ for every input. When fitting such a classifier, the best one can do is tune that probability to match the base rate. Exercise 11.11 fills in the details.

■ ***Exercise*** 11.11 (a) Justify the claim that 'pushing λ to infinity drives every component of $\hat{\boldsymbol{\beta}}$ towards zero'. That is, explain why

$$\lim_{\lambda\to\infty} \operatorname*{arg\,min}_{(\beta_0,\boldsymbol{\beta})} \left[\frac{\lambda}{n}\|\boldsymbol{\beta}\|^2 + l(\beta_0,\boldsymbol{\beta};D)\right] = \left(\operatorname*{arg\,min}_{\beta_0} l(\beta_0,\mathbf{0};D),\ \mathbf{0}\right).$$

(b) Show that

$$g(\boldsymbol{x};\hat{\beta}_0,\mathbf{0}) = \pi,$$

where $\hat{\beta}_0 = \operatorname*{arg\,min}_{\beta_0} l(\beta_0,\mathbf{0};D)$, and $\pi = \frac{1}{n}\sum_{i=1}^{n} y_i$ is the base rate of the 1-coded class in dataset D.

So much for the big picture. Let's now zoom in for a closer look at the weak regularisation regime. Panels (c) and (d) of Figure 11.4 plot cross-validation scores against λ for very

Table 11.4 *Coefficients for binary logistic regression model ($\lambda = 0.7$) fitted to dataset previewed in Table 11.2. Predictor components were standardised prior to fitting. Standard errors were estimated using the non-parametric bootstrap method with 1,000 replications.*

Predictor	Predictor index, i	$\hat{\beta}_i$	$\widehat{\text{se}}(\hat{\beta}_i)$
Fasting blood sugar	1	1.584	0.090
Haemoglobin A1C	2	1.494	0.084
Cholesterol triglycerides	3	0.103	0.063
Hypertension	4	0.588	0.073
Socioeconomic status	5	0.102	0.065

small values of λ. Panel (c) reveals that the minimum cross-validated cross-entropy loss is not at $\lambda = 0$, but at $\lambda \approx 0.7$. It seems that even with our tall dataset and simple modelling strategy, a little regularisation is helpful. There is a striking contrast between panels (c) and (d), though. As λ varies, cross-validated cross-entropy loss changes smoothly, but cross-validated accuracy changes in discrete jumps. The jumps correspond to individual examples moving from one 'best-guess' predicted class to the other. In panel (d), we have zoomed in so far that the individual jumps are clearly visible. With just a handful of datapoints close to the decision boundary moving from one predicted class to the other as λ sweeps through its narrow window, the plot is noisy – and not much use for fine-tuning λ.

To fit our final logistic regression model to all the data, we'll go ahead and use $\lambda = 0.7$. Table 11.4 shows the coefficients of the fitted model, along with bootstrap standard errors. The table tells us that of the five predictors tracked in the dataset, fasting blood sugar and haemoglobin A1C are the most important. For example, holding all other predictors constant, a one-standard-deviation increase in fasting blood sugar implies an increase in the log odds of diabetes (according to the fitted model) of 1.58. By contrast, a one-standard-deviation increase in cholesterol triglycerides implies a log odds increase of only 0.1.

So far we have been working with a tall dataset in which examples greatly outnumber features. Regularisation didn't help much. What if we had fewer examples and more features? We would expect overfitting to become more of an issue, and we would expect regularisation to help more. In fact, the diabetes dataset previewed in Table 11.2 was obtained from a dataset with 43 features. We threw away 38 of these, leaving the five features used above and listed in Table 11.4. Let's now restore the full set of 43 features, but sample just 300 of the 1,879 examples.[5] Cross-validation results for logistic regression on the new dataset are plotted in Figure 11.5. We can see right away that our expectation was correct: regularisation does indeed help more here. Raising λ from 0 to 4 reduces cross-validated cross-entropy loss from around 0.6 to below 0.45. There are now also signs that a bit of regularisation improves cross-validated accuracy (raising it from 80% to 81%), but this effect is weak; moreover, the value of λ that is optimal for cross-validated accuracy is lower than the value that is optimal for cross-validated cross-entropy loss.

Why does light regularisation dramatically improve out-of-sample classification performance according to one score (cross-entropy loss) but only help a little according to another (accuracy)? It's important to remember that the two scores correspond to two different con-

[5] For a full list of the 43 features, see the online notebook for this chapter.

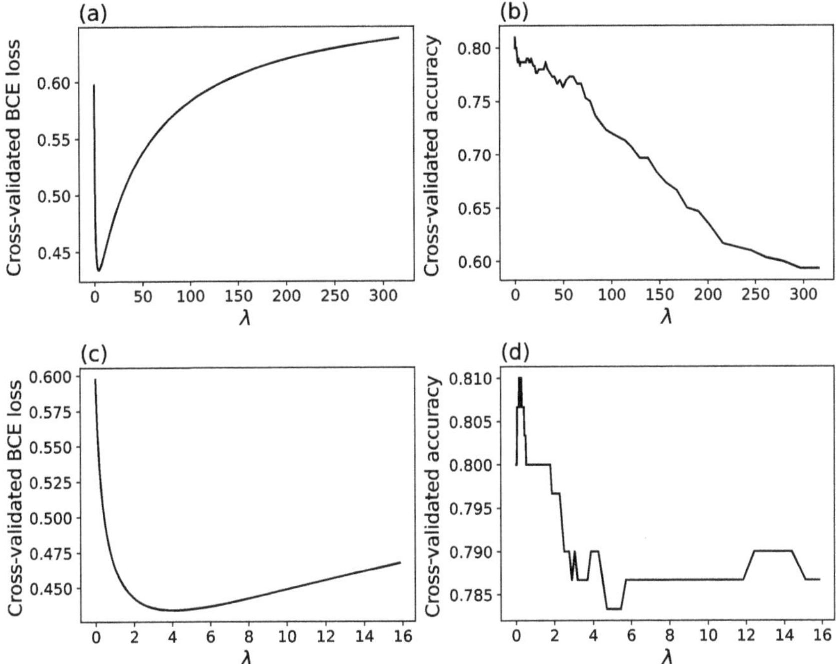

Figure 11.5 Cross-validated binary cross-entropy loss (a and c) and accuracy (b and d) of logistic regression applied to the diabetes classification task with respect to a wider dataset (43 features, 300 examples). As in Figure 11.4, λ is regularisation strength, test metrics are averaged across five cross-validation folds, and panels (c) and (d) replot an initial portion of the λ range plotted in panels (a) and (b).

cepts of success. Accuracy encodes a binary concept of success: on any given test example, a classifier either succeeds or fails; there is no such thing as an especially good or an especially bad performance. Cross-entropy loss, by contrast, encodes a graded concept of success. A classifier that assigns 10% probability to what turns out to be the correct class does badly ($-\log 0.1 \approx 2.3$); but if it assigns probability 10^{-9} to that class, it does very much worse ($-\log 10^{-9} \approx 20.7$). It is possible for a disastrous failure on a single example to wreck a classifier's cross-validated cross-entropy loss score. Now suppose that during cross-validation, for some particular choice of validation fold the training examples (the examples *not* in the validation fold) happen – by fluke – to be linearly separable, or close to linearly separable. Then if λ is too small, the sigmoid that minimises *training* loss will be very sharp. There is a danger that one or more of the validation fold examples will fall on the wrong side of the sharp transition; these will be examples that the classifier fails disastrously on. This is why the cross-validated cross-entropy loss shoots up as λ approaches zero in Figure 11.5. We saw the faintest hints of a similar phenomenon in Figure 11.4. It is much more apparent when working with 300 examples and 43 features than when working with 1,879 examples and 5 features because linear separability (or near linear separability) is more likely to arise by fluke with a small training set in a high-dimensional space than with a large training set in a low-dimensional space.[6]

Finally, we should note a 'big picture' contrast between Figures 11.5 and 11.4. In panels (a) and (b) of Figure 11.5, we see the convergence to the same strong regularisation limit we saw in Figure 11.4. However, the convergence is quicker: setting λ to 300 is now sufficient to drive cross-validated accuracy down to the 'always guess no' rate of 60%, whereas with the previous, taller dataset values in excess of 1,000 were required. Why? The answer is that priors are more influential when data is scarce. Equation (11.13) presents the loss function as a sum of two terms: a regularisation term $\frac{\lambda}{n}||\boldsymbol{\beta}||^2$ that is inversely proportional to n and the unregularised cross-entropy loss. The latter is an average over the training data, so its expected value is independent of n. Therefore, for any given value of λ, reducing the number of examples in the dataset increases the impact of the regularisation term.

11.4 Probabilistic Multiclass Classification: Preliminaries

11.4.1 Integer Encoding and Probability Vectors

So far we have focused on classification tasks in which the prediction target is a binary variable. It is now time to broaden our horizons, and consider *multiclass* classification tasks. These are classification tasks in which the prediction target can take many different values.[7] The prediction target remains a categorical variable, so you should not think of its possible 'values' as magnitudes. You should think of them as elements of a finite set that lacks any natural arithmetical structure. We will focus on prediction targets that are *nominal* categorical variables, meaning that the set of possible values also lacks any natural order. For example, identifying the language of a text snippet is a multiclass classification task (with a nominal target), so long as the set of possible values of the prediction target is well-defined and finite – e.g. {English, Spanish, French, German, Chinese}.

If g is a probabilistic *binary* classifier, then it maps a predictor vector $\boldsymbol{x}$ to a probability distribution over the set $\{0, 1\}$. We could represent that distribution as an ordered pair $(1 - g(\boldsymbol{x}), g(\boldsymbol{x}))$, where the first component is the predicted probability that the prediction target is 0 and the second component is the predicted probability that the prediction target is 1. Of course, there is really only one degree of freedom in that probability distribution: once you say that the probability of 1 is $g(\boldsymbol{x})$, normalisation forces you to say that the probability of 0 is $1 - g(\boldsymbol{x})$. That's why in Sections 11.2 and 11.3 we were able to present probabilistic binary classifiers as maps of the form $\boldsymbol{x} \mapsto g(\boldsymbol{x})$, with $g(\boldsymbol{x})$ understood to be the predicted probability that the prediction target takes value 1. Now that we are generalising to multiclass classification, we will need to distinguish notationally between probabilities assigned to the

[6] You might now be wondering why regularisation improves cross-validated accuracy *at all*. After all, a test point on the wrong side of the $g(\boldsymbol{x}; \hat{\beta}_0, \hat{\boldsymbol{\beta}}) = 0.5$ decision boundary hurts accuracy to precisely the same extent whether the sigmoid transition across that boundary is sharp (large $||\hat{\boldsymbol{\beta}}||$) or smooth (small $||\hat{\boldsymbol{\beta}}||$). Recall, however, that regularisation does have some effect on the location of the decision boundary. We have already seen that if λ is enormous (strong regularisation limit), the fitted model will predict approximately the same probability (the base rate) for all values of $\boldsymbol{x}$ that occur in the dataset. So unless that base-rate probability happens to be 0.5, pushing λ to infinity pushes the decision boundary away from all the datapoints.

[7] Binary classification can be viewed as a special case of multiclass classification, but when a task is described as 'multiclass classification', there will usually be at least three possible values of the prediction target.

different classes. Suppose the prediction target can take K possible values. It is common to use class labels $0, 1, \ldots, K-1$ to represent them; this is called integer encoding. (When the prediction target is nominal, the matching of integers to classes must be arbitrary.) Then if $\boldsymbol{g}$ is a probabilistic multiclass classifier, it maps predictor vectors to probability distributions over the set $\{0, 1, \ldots, K-1\}$. We can write the map implemented by the classifier as follows:

$$\boldsymbol{x} \mapsto (g_0(\boldsymbol{x}), g_1(\boldsymbol{x}), \ldots, g_{K-1}(\boldsymbol{x})),$$

where component $g_i(\boldsymbol{x})$ is the predicted probability of class label i. The whole K-tuple of probabilities, $\boldsymbol{g}(\boldsymbol{x}) = (g_0(\boldsymbol{x}), g_1(\boldsymbol{x}), \ldots, g_{K-1}(\boldsymbol{x}))$, is sometimes called a probability vector.[8]

11.4.2 Categorical Cross-Entropy Loss

As you would expect, a probabilistic multiclass classifier $\boldsymbol{g}$ corresponds to a hypothesis about a datapoint-generating process. We again represent that process by a random vector $(\boldsymbol{X}, Y)$, but Y is now an integer-valued random variable taking values in $\{0, 1, \ldots, K-1\}$. The hypothesis to which $\boldsymbol{g}$ corresponds is that the conditional distribution of Y given $\boldsymbol{X} = \boldsymbol{x}$ (for any $\boldsymbol{x}$) is

$$Y \mid \{\boldsymbol{X} = \boldsymbol{x}\} \sim \text{Categorical}\,(g_0(\boldsymbol{x}), g_1(\boldsymbol{x}), \ldots, g_{K-1}(\boldsymbol{x}))\,.$$

In other words, the hypothesis says for each i that the conditional probability that a datapoint has class label i given that its predictor vector is $\boldsymbol{x}$ is $g_i(\boldsymbol{x})$.

A multiclass classification dataset $D = \{(\boldsymbol{x}_i, y_i)\}_{i=1}^n$ consists of n independent realisations of $(\boldsymbol{X}, Y)$. The likelihood of $\boldsymbol{g}$ is the conditional probability (according to $\boldsymbol{g}$) of all the observed class labels $(y_1, \ldots, y_n)$ given the observed predictor vectors $(\boldsymbol{x}_i, \ldots, \boldsymbol{x}_n)$, so

$$\mathcal{L}(\boldsymbol{g}; D) = \prod_{(\boldsymbol{x}, y) \in D} g_y(\boldsymbol{x}).$$

(Compare this equation to equation (11.2) to convince yourself of the advantages of probability vector notation.) Notation aside, the story continues to parallel the binary case closely. The cross-entropy loss is

$$\begin{aligned} l(\boldsymbol{g}; D) &= -\frac{1}{n} \log \mathcal{L}(\boldsymbol{g}; D) \\ &= -\frac{1}{n} \sum_{(\boldsymbol{x}, y) \in D} \log g_y(\boldsymbol{x}), \end{aligned} \tag{11.14}$$

and we will see a regularised form soon. If we wish to stress that this loss function can handle any number of classes, we call it categorical (as opposed to binary) cross-entropy loss.

11.4.3 One-Hot Encoding

Integer encoding of nominal prediction targets is common, but it does have one drawback: integer class labels are arithmetically related, but these relationships are meaningless. For example, if the correct class label for a particular example is 6 and your classifier's best

[8] This terminology is slightly unfortunate, because probability vectors do not form a vector space. For example, the sum of two probability vectors is not itself a probability vector. (Why not?)

guess is that it is 5, that is not in any meaningful sense a 'near miss'.[9] If you use a sensible loss function when fitting your classifier, integer encoding won't introduce any distortions; but the potential for misunderstanding is there all the same. Relatedly, it is senseless to ask about the expected value of the random variable Y we used above to represent class labels. (Given a particular integer encoding scheme and a particular datapoint-generating process, the expectation of Y will be mathematically well defined, but it won't be *interesting*.) The bottom line is that integer encoding introduces surplus mathematical structure that we need to learn to ignore.

An alternative strategy is *one-hot encoding*. We have already used probability vectors to represent probabilistic predictions about categorical variables. The basic idea of one-hot encoding is to use them to represent ground truth too. We begin with an integer encoding of the categorical variable in question. Then we recode our class labels. For the ith class, instead of the integer i, we use a probability vector with a 1 in the ith slot and zeros everywhere else. (One component is 'hot' and the others are 'cold'.) For example, in the case of the language identification task mentioned above, we might begin by coding English as 0, Spanish as 1, French as 2, German as 3 and Chinese as 4. The corresponding one-hot encoding would represent English as $(1, 0, 0, 0, 0)$, Spanish as $(0, 1, 0, 0, 0)$ and so on. Our class labels are now vectors, so we write the random vector representing the datapoint-generating process as $(\boldsymbol{X}, \boldsymbol{Y})$; notice that $\boldsymbol{Y}$ is typeset in bold here because it is a random vector in its own right. This encoding has several advantages. First, it treats the classes symmetrically. The integer encoding misleadingly suggested that English is closer to Spanish than to German, because 0 is closer to 1 than to 3. The corresponding one-hot vectors, by contrast, are all equidistant from each other in $\mathbb{R}^5$. Second, expectations are now meaningful.

■ ***Exercise*** 11.12 How should we interpret the components of $\mathbb{E}(\boldsymbol{Y})$? What about the components of $\mathbb{E}(\boldsymbol{Y}|\boldsymbol{X} = \boldsymbol{x})$?

Third, one-hot encoding of the prediction target makes probabilistic classification look like a typical supervised task: when we feed a predictor vector $\boldsymbol{x}$ to our classifier, we'd like it to return a prediction $\boldsymbol{g}(\boldsymbol{x})$ that resembles the true label $\boldsymbol{y}$ as closely as possible. That familiar framing only makes sense if the true label is a probability vector. (Recall that $\boldsymbol{g}(\boldsymbol{x})$ is itself a probability vector. It cannot 'resemble' an integer!)

When class labels are probability vectors rather than integers, the formula for cross-entropy loss becomes

$$l(\boldsymbol{g}; D) = -\frac{1}{n} \sum_{(\boldsymbol{x},\boldsymbol{y}) \in D} \boldsymbol{y}^T \log \boldsymbol{g}(\boldsymbol{x}), \tag{11.15}$$

where the log function is applied component-wise.

■ ***Exercise*** 11.13 Assuming that the class labels in equation (11.15) are one-hot vectors, verify that this formula for l is equivalent to equation (11.14).

A nice feature of equation (11.15) is that it generalises gracefully to handle 'soft' class labels. That is, we could in principle allow the *ground truth* to be uncertain, in which case the label

[9] Or, to be more precise, if in some particular context this mislabelling *does* count as a near miss, that has nothing to do with the fact that 5 and 6 are consecutive integers.

vectors in equation (11.15) would be probability vectors with multiple non-zero components. The formula still defines a sensible loss function.

In practice, some software tools encode categorical variables as integers, and others encode them as one-hot vectors. It is important to be familiar with both coding schemes. When *talking* about multiclass classification problems, though, it is much more convenient to use integer class labels. That's what we'll do below.

11.5 Multinomial Logistic Regression

11.5.1 Theory

We will now generalise logistic regression to the multiclass case. Multiclass logistic regression is usually called *multinomial logistic regression* or *softmax regression*. Our basic task is to define a family of functions for mapping a predictor vector $\boldsymbol{x} \in \mathbb{R}^p$ to a probability vector over K classes. When there are more than two classes, we cannot speak of 'the' log odds. But we can define a separate score for each class. The essentials of multinomial logistic regression are as follows: (i) we stipulate that each class's score is a linear function of $\boldsymbol{x}$; (ii) we use the softmax function, defined below, to map scores to probabilities; and (iii) we fit our classifier to training data by minimising (regularised) cross-entropy loss.

Let's spell out the details. The score for the ith class, s_i, is a linear function of $\boldsymbol{x}$. That is,

$$s_i = \beta_0^{(i)} + \boldsymbol{\beta}^{(i)T}\boldsymbol{x}, \tag{11.16}$$

where scalar $\beta_0^{(i)}$ and p-dimensional vector $\boldsymbol{\beta}^{(i)}$ are parameters to be learned from data. There are K such equations ($i = 0, 1, ..., K-1$), but we can combine them into one by writing

$$\boldsymbol{s} = \boldsymbol{\beta}_0 + \underline{\boldsymbol{\beta}}\,\boldsymbol{x},$$

where $\boldsymbol{\beta}_0$ is the K-dimensional vector whose ith component is $\beta_0^{(i)}$, and $\underline{\boldsymbol{\beta}}$ is the $K \times p$ matrix whose ith row is $\boldsymbol{\beta}^{(i)}$. The score vector $\boldsymbol{s}$ now needs to be mapped to a probability vector – that is, to K non-negative numbers that sum to 1. This is done using the softmax function, defined component-wise by

$$\text{softmax}(\boldsymbol{s})_i = \frac{e^{s_i}}{\sum_{j=0}^{K-1} e^{s_j}}, \tag{11.17}$$

for $i = 1, 2, ..., K-1$.

■ ***Exercise*** 11.14 Verify that softmax($\boldsymbol{s}$) is a probability vector, that is, that its K components are non-negative numbers that sum to 1.

To see why this function is called 'softmax', consider a case in which one of the scores, say s_3, is very much greater (in the positive direction) than the others. It is then clear from equation (11.17) that the corresponding component of the vector returned by the softmax function, softmax$(\boldsymbol{s})_3$, will be very nearly 1, while every other component will be very nearly 0. Effectively, softmax($\boldsymbol{s}$) is a one-hot vector in this case, with its hot component indicating the location of the maximum in $\boldsymbol{s}$. When one component of $\boldsymbol{s}$ isn't overwhelmingly greater than all the others, softmax implements a 'soft' version of maximum detection. The largest scores are mapped to the largest probabilities, and in general, the probability a score gets mapped to reflects its value relative to the other scores.

■ ***Exercise*** 11.15 Show that adding a constant to each component of $\boldsymbol{s}$ (the same constant for each score) leaves $\text{softmax}(\boldsymbol{s})$ unchanged.

For numerical stability, it is common to subtract $\max_j s_j$ from each component of the input vector before computing $\text{softmax}(\boldsymbol{s})$. This transformation does not affect the output (see Exercise 11.14). Box 11.2 summarises multinomial logistic regression.

★ Box 11.2 Multinomial logistic regression

Multinomial logistic regression is a probabilistic multiclass classification method defined by a model and a fitting strategy. The model postulates that the vector of conditional class probabilities given predictor vector $\boldsymbol{x}$ is

$$\boldsymbol{g}(\boldsymbol{x}; \boldsymbol{\beta}_0, \underline{\boldsymbol{\beta}}) = \text{softmax}(\boldsymbol{\beta}_0 + \underline{\boldsymbol{\beta}}\,\boldsymbol{x}), \tag{11.18}$$

where K-dimensional vector $\boldsymbol{\beta}_0$ and $K \times p$ matrix $\underline{\boldsymbol{\beta}}$ are the parameters of the model, K is the number of classes, p is the dimensionality of the predictor vector and the softmax function is defined in equation (11.17). The fitting strategy is to choose parameters that minimise (regularised) cross-entropy loss on training data $D = \{(\boldsymbol{x}_i, y_i)\}_{i=1}^n$. With ridge regularisation, that loss is

$$l^\lambda(\boldsymbol{\beta}_0, \underline{\boldsymbol{\beta}}; D) = \frac{\lambda}{n}\|\underline{\boldsymbol{\beta}}\|^2 + l(\boldsymbol{\beta}_0, \underline{\boldsymbol{\beta}}; D), \tag{11.19}$$

where

$$l(\boldsymbol{\beta}_0, \underline{\boldsymbol{\beta}}; D) = -\frac{1}{n}\sum_{(\boldsymbol{x},y)\in D} \log g_y(\boldsymbol{x}; \boldsymbol{\beta}_0, \underline{\boldsymbol{\beta}})$$

is the unregularised cross-entropy loss, $\|\underline{\boldsymbol{\beta}}\|^2 = \sum_{i=1}^K \sum_{j=1}^p \beta_{ij}^2$ and $\lambda \geq 0$ is a hyperparameter controlling the strength of regularisation.

Before fitting a multinomial logistic regression model, it is good practice to standardise each predictor vector component to have a mean of 0 and a variance of 1 (where the mean and variance are taken over the training data).

When $K = 2$, multinomial logistic regression is equivalent to binary logistic regression. Let's verify that. The score vector in the $K = 2$ case has two components, s_0 and s_1. Applying the softmax function, we find that the predicted probability of class 1 is

$$\frac{e^{s_1}}{e^{s_0} + e^{s_1}} = \frac{1}{1 + e^{s_0 - s_1}} = \sigma(s_1 - s_0),$$

where σ is the standard sigmoid function defined in equation (11.7). We therefore recognise the difference $s_1 - s_0$ as the log odds. A multinomial logistic regression model represents both s_1 and s_0 as linear functions of $\boldsymbol{x}$. Hence, it implies that the log odds is also linear in $\boldsymbol{x}$; hence, it maps $\boldsymbol{x}$ to probability-of-class-1 in precisely the same way as a binary logistic regression model.

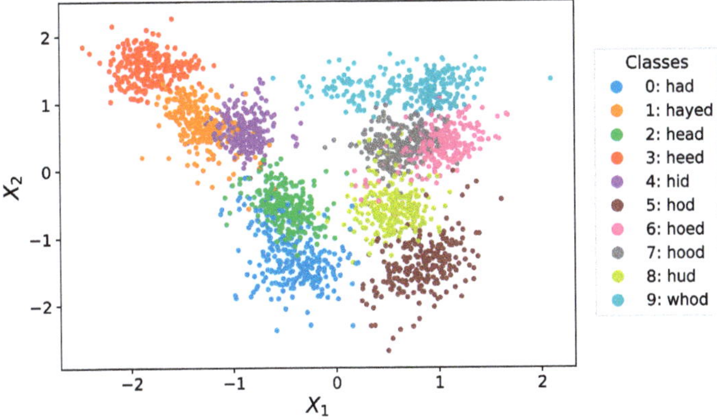

Figure 11.6 Backness (X_1) and height (X_2) measurements for 10 different syllables. Each point (n = 2,371) represents a single utterance of a syllable by a monolingual native speaker of American English. The plot aggregates data from 48 speakers. Data is drawn from Clopper et al. (2005).

■ ***Exercise*** 11.16 In general in the multinomial case, there is no single log odds, but we can define the log odds between classes i and j as

$$\log\left(\frac{g_i(\boldsymbol{x};\boldsymbol{\beta}_0,\underline{\boldsymbol{\beta}})}{g_j(\boldsymbol{x};\boldsymbol{\beta}_0,\underline{\boldsymbol{\beta}})}\right).$$

Show that increasing x_k by one unit, while holding all other components of the predictor vector constant, shifts the log odds between classes i and j by $\beta_{ik} - \beta_{jk}$.

11.5.2 Application to Human Speech Sounds

To illustrate multinomial logistic regression, we will take a first peek at a topic explored further in Chapter 12: human speech sounds. The specific task we'll focus on here is distinguishing between spoken syllables of American English based on vowel sounds. The sound of a vowel is largely determined by the position of the tongue – how *high* and how far *back* it is in the mouth. We can measure tongue 'height' and 'backness' using acoustic features that correspond to specific resonances ('formants') of the oral cavity. Precisely how this is done is not important for our purposes; what matters is that we have a two-dimensional feature space (p = 2), which makes it easy to visualise decision regions. (Of course, real-world classification problems typically involve much larger feature spaces.) To further simplify matters, we'll restrict our attention to just 10 possible syllables (K = 10). The training datapoints (n = 2,371) are plotted in Figure 11.6.

Let's now go ahead and fit a multinomial logistic regression model to this dataset. The workflow closely parallels that of Section 11.3.5, so we'll move briskly. Figure 11.7 plots cross-validated cross-entropy loss and accuracy against regularisation strength λ for a narrow range of λ values around the apparent sweet spot. For essentially the same reasons as applied in Section 11.3.5, very light regularisation is optimal here: we have a simple model (softmax

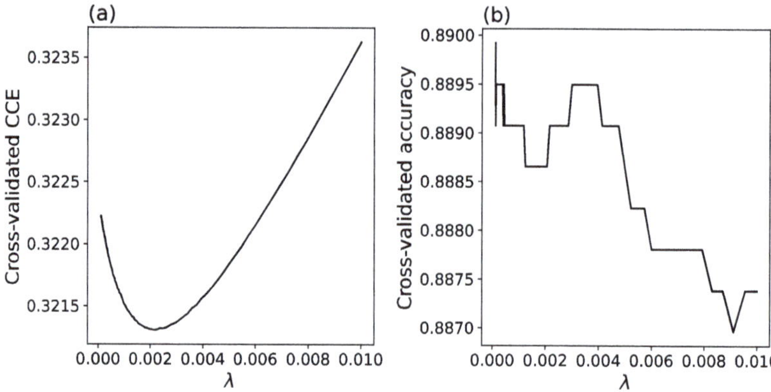

Figure 11.7 Cross-validated categorical cross-entropy (a) loss and (b) accuracy of multinomial logistic regression applied to the speech sound dataset (shown in Figure 11.6) as a function of the regularisation strength λ. For each value of λ, each test metric is averaged across five cross-validation folds.

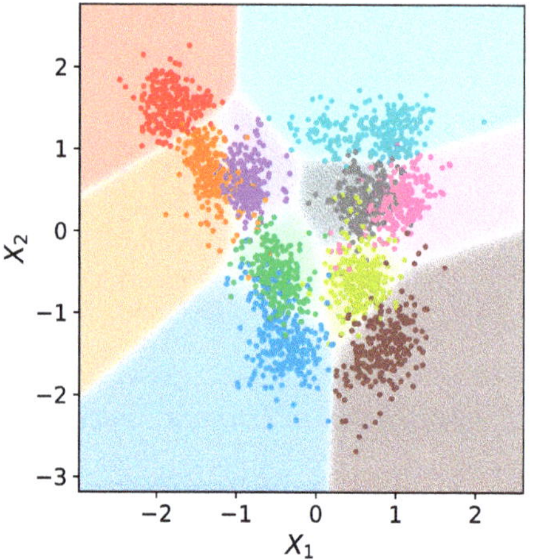

Figure 11.8 Decision regions of multinomial logistic regression model ($\lambda = 0.002$) fitted to the speech sound training data (overlaid scatter plot). Background colour represents the model's best-guess class label; background opacity represents the model's confidence in its best guess.

scores linear in $\boldsymbol{x}$), a small feature space ($p = 2$) and plenty of data ($n = 2371$). We'll take the lesson of Figure 11.7 to be that λ should be set to 0.002. With λ fixed, we can wrap up the cross-validation phase and fit our model to all the training data. Figure 11.8 visualises the decision regions of the resulting model and also replots the training data. Regions of feature space are colour-coded according to the model's 'best guess' class label. That is, the background colour at a point $\boldsymbol{x}$ is the colour associated with class label $\arg\max_i s_i(\boldsymbol{x})$, where

$s_i(\boldsymbol{x})$ is the ith component of the score vector assigned to point $\boldsymbol{x}$ by the fitted model. The region corresponding to best-guess class label i is the ith *decision region*. The opacity of the background colour is proportional to the model's confidence in its best guess. The colour fades near boundaries between decision regions because there, by definition, the model is not confident in its best guess. For points lying *precisely* on a boundary, there is no best guess; instead, there is a tie between two or more leading contenders.

■ ***Exercise*** 11.17 This exercise will help you understand the geometry of the decision regions in Figure 11.8, and of multinomial logistic regression decision regions in general.

(a) Show that logistic regression decision regions are *convex*. That is, show that if points $\boldsymbol{x}_1$ and $\boldsymbol{x}_2$ both lie in the ith decision region, then every point on the line segment connecting $\boldsymbol{x}_1$ and $\boldsymbol{x}_2$ also lies in the ith decision region.
(b) Show that in a two-dimensional feature space, the boundary between the ith and jth decision regions (if there is one) must be a line segment, a half line or a line. (Hint: use the fact that the ith and jth components of the score vector are equal at the boundary.)
(c) What is the generalisation of the result in (b) to feature spaces of arbitrary dimensionality?

■ ***Exercise*** 11.18 A multiclass classification dataset is said to be *simultaneously linearly separable* if for some choice of parameters $\boldsymbol{\beta}_0$ and $\underline{\boldsymbol{\beta}}$, the best-guess class label estimator

$$\begin{aligned}\hat{y} &= \arg\max_i s_i \\ &= \arg\max_i \left(\beta_0^{(i)} + \boldsymbol{\beta}^{(i)T}\boldsymbol{x}\right)\end{aligned}$$

achieves 100% accuracy on the dataset.

(a) Is the speech sounds dataset plotted in Figure 11.6 simultaneously linearly separable?
(b) What will happen if you attempt to fit a multinomial logistic regression model to a simultaneously linearly separable dataset by minimising *unregularised* ($\lambda = 0$) cross-entropy loss?

11.6 k-Nearest Neighbours

We'll conclude this chapter by looking at a very different probabilistic classification method: k-nearest neighbours (k-NN). We'll explain the theory (which is minimal), apply the method to the human speech sounds dataset and compare k-NN to multinomial logistic regression.

11.6.1 Theory

k-NN is a non-parametric method. There is no parameter vector to fit to the training data, and so no loss function to minimise. Given a training dataset $D = \{(\boldsymbol{x}_i, y_i)\}_{i=1}^n$ and an unlabelled point $\boldsymbol{x}$ to classify, the procedure is as follows: we find the k training examples whose predictor vectors lie closest to $\boldsymbol{x}$ in feature space.[10] Then the vector $\boldsymbol{g}(\boldsymbol{x})$ of predicted class label probabilities for point $\boldsymbol{x}$ is just the empirical distribution of class labels within the set of k nearest neighbours. So, for example, if half of those nearest neighbours have class label 3, then $g_3(\boldsymbol{x}) = 0.5$. This remarkably simple method often performs well in low-dimensional feature spaces. (We'll discuss the trouble that arises in high-dimensional feature spaces shortly.)

Although k-NN is non-parametric, there are some hyperparameters to worry about. One is obvious: k. In principle, k could be set to any integer value in $\{1, 2, ..., n\}$, where n is the size of the training set. If you set $k = n$, then the 'nearest' neighbours of any given point $\boldsymbol{x}$ will be *all* the points in the training set, so $\boldsymbol{g}(\boldsymbol{x})$ won't actually depend on $\boldsymbol{x}$ at all: it will just reproduce the base rates of the classes in the training data.[11] If you go to the other extreme and set $k = 1$, then $\boldsymbol{g}(\boldsymbol{x})$ will be a one-hot vector that predicts, with 100% confidence, that $\boldsymbol{x}$'s class matches the class of the training point that happens to lie closest to $\boldsymbol{x}$. Intuitively, this sort of prediction is too confident, and too sensitive to small changes in $\boldsymbol{x}$ or in the training data. Probabilistic predictions worth taking seriously ought to be based on more than one training example. A key part of the k-NN workflow is finding a sweet spot between these two extremes: a value of k that works well. We can do this by cross-validation. Cross-validated accuracy is an appropriate performance metric to use when tuning k, although it does not penalise confident incorrect guesses any more harshly than tentative ones. Unfortunately, cross-validated cross-entropy loss is not an appropriate performance metric in this context.

■ ***Exercise*** 11.19 Why is cross-validated cross-entropy loss not an appropriate performance metric when tuning k-NN hyperparameters? (Hint: consider what happens when a training point's class does not match any of its nearest neighbours.)

A less obvious hyperparameter is the distance metric used to identify nearest neighbours. *Euclidean distance* is a common default choice. The Euclidean distance between two points in p-dimensional feature space, $\boldsymbol{x}$ and $\boldsymbol{x}'$, is given by the familiar expression:

$$\sqrt{\sum_{i=1}^{p}(x_i - x_i')^2}.$$

We will explore some alternative distance metrics in Exercise 11.25, but we'll stick to Euclidean distance in the main text. When using this distance metric – and indeed most others – it is crucial to standardise features so that each predictor vector component displays a similar range of variation across the training data. To see the importance of this step, imagine a classification problem in a 10-dimensional feature space in which the first component of the predictor vector has a standard deviation across the training data of 1,000, while all the other components have standard deviations of 0.1 or less. Then the nearest neighbours of any given point will be determined almost entirely by that point's first component. If you implement k-NN classification for this problem without standardising features, you will effectively be ignoring all but one of the features.

[10] If ties (training examples that are precisely equidistant from $\boldsymbol{x}$) make it impossible to identify a unique set of k nearest neighbours, these ties are broken at random.

[11] A multinomial logistic regression classifier will approach – but not quite attain – the same constant function in the $\lambda \to \infty$ strong regularisation limit.

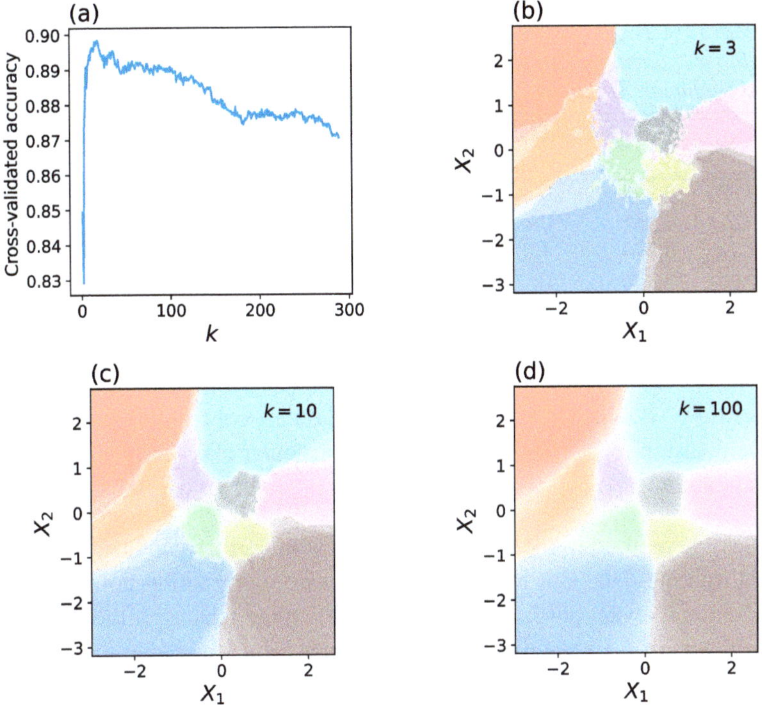

Figure 11.9 (a) Cross-validated accuracy of k-NN applied to the human speech sounds dataset (shown in Figure 11.6) as a function of k. For each value of k plotted, accuracy is averaged across five cross-validation folds. (b)–(d) Decision regions of k-NN fitted to the same dataset, for $k = 3, 10$ and 100. Background colour represents the model's best-guess class label; background opacity represents the model's confidence in its best guess.

11.6.2 Revisiting Human Speech Sounds

Figure 11.9 shows the results of applying k-NN to the human speech sounds dataset we met in Section 11.5.2. Cross-validated accuracy scores for a range of k values are shown in panel (a). As expected, very small and very large values of k perform suboptimally. There appears to be a sweet spot around $k = 18$.

■ ***Exercise*** 11.20 Suppose one picks a value of k that is too large. Does the resulting classifier *underfit* or *overfit* the training data?

Panels (b), (c) and (d) plot decision regions for k-NN classifiers fitted to the whole dataset with k set, in turn, to 3, 10 and 100, respectively; the colour-coding and opacity-coding are the same as in Figure 11.8, but to avoid clutter we have not replotted the data this time. Decision boundaries are jagged and complicated when $k = 3$, but they become smoother as k increases. Two contrasts with Figure 11.8 are worth stressing: in a two-dimensional feature space, k-NN's decision boundaries are not constrained to be straight line segments; and k-NN's decision regions are not constrained to be convex.

Table 11.5 *A sample of five rows drawn from an expanded version of the speech sounds dataset plotted in Figure 11.6. Two additional features are now recorded: duration of vowel and dialect of speaker. The latter is a categorical variable with six levels; it has been one-hot encoded in the penultimate six columns. There are 2,371 rows in the full dataset (Clopper et al. (2005)).*

X_1	X_2	Duration	Mid-Atlantic	Midland	New England	North	South	West	Syllable
0.47	−0.98	0.193	0	0	1	0	0	0	hud
−0.76	1.22	0.161	0	0	0	0	1	0	hid
−0.24	−1.14	0.174	1	0	0	0	0	0	had
−0.78	−0.95	0.320	0	0	0	1	0	0	had
−1.55	0.88	0.211	1	0	0	0	0	0	hayed

■ ***Exercise*** 11.21 Are k-NN's decision regions at least constrained to be *connected*? Or could the decision region corresponding to a particular class consist of two or more disconnected blobs?

k-NN is a flexible method in the sense that its decision regions can take almost any form, given the right training data. But for k-NN to work well, the training points do need to fill feature space (or the relevant portion of feature space) rather densely. If they do not, then the k-nearest neighbours of a given unlabelled point $\boldsymbol{x}$ may in fact be rather distant from $\boldsymbol{x}$; and the statistics of *distant* neighbours will usually not be a good basis for making probabilistic predictions about $\boldsymbol{x}$. Our current speech sounds task plays to k-NN's strengths. The feature space is small ($p = 2$) and there are plenty of training points ($n = 2371$). Indeed, comparison of Figure 11.9a with Figure 11.7b reveals that k-NN (with optimal k) narrowly outperforms multinomial logistic regression (with optimal λ).

11.7 Multinomial Logistic Regression versus k-NN in Higher Dimensional Spaces

The original dataset from which the data plotted in Figure 11.6 was taken (Clopper et al. (2005)) in fact tracks more than two features. We can explore the effect of passing to a higher-dimensional feature space by restoring some of the features we ignored above. Specifically, we'll restore the duration (in seconds) of the voiced vowel, and the regional dialect of the speaker. The latter is a categorical variable with six levels ('Mid-Atlantic', 'Midland', 'New England', 'North', 'South' and 'West'). We will one-hot encode dialect. That means we are adding seven new dimensions to the feature space: one for duration and six for dialect. A random sample of five rows from this expanded version of the human speech sounds dataset is shown in Table 11.5.

■ ***Exercise*** 11.22 We will be applying multinomial logistic regression and k-NN to the dataset sampled in Table 11.5 (with the Syllable column as the prediction target). Why would it not have made sense to use integer encoding for the dialect variable?

We now have a multiclass classification problem in a nine-dimensional feature space. It is no longer straightforward to visualise the dataset, but the multinomial logistic regression and k-NN workflows can be applied as before. Optimally regularised multinomial logistic regression achieves a mean cross-validated accuracy of 91.4% (and a mean cross-validated cross-entropy loss of 0.241) on the expanded dataset. It turns out that k-NN's mean cross-validated accuracy on this same dataset is highest when $k = 4$, and the score achieved – 90.0% – is worse than multinomial logistic regression. k-NN had a narrow advantage when the feature space was two-dimensional, but passing to nine dimensions has turned the tables in favour of the parametric method.

It is often the case that adding additional predictors (i.e. increasing the dimensionality of the feature space) while holding the number of training points constant favours parametric over distance-based non-parametric prediction methods. As the dimensionality of the feature space increases, the distance between a test point and a training point becomes a less and less informative measure of the predictive *relevance* of the training point. In a very high-dimensional space, all the training points will (probably) be roughly equally distant from the test point. Box 11.3 illustrates the general phenomenon with a mathematical thought experiment in the spirit of Exercise 1.5.

★ Box 11.3 Distance between random vectors in high-dimensional spaces

Consider two i.i.d. p-dimensional random vectors, $\boldsymbol{X}$ and $\boldsymbol{X}'$, with the following properties: their components $X_1, \ldots, X_p$ and $X'_1, \ldots, X'_p$ are i.i.d. random variables, and for each i the random variable $(X_i - X'_i)^2$ has well-defined expectation μ and variance σ^2. Now consider another random variable: the square of the distance between $\boldsymbol{X}$ and $\boldsymbol{X}'$. It is straightforward to compute its expectation and variance. We have

$$\begin{aligned}\mathbb{E}[\|\boldsymbol{X} - \boldsymbol{X}'\|^2] &= \mathbb{E}\left[\sum_{i=1}^{p}(X_i - X'_i)^2\right] \\ &= \mu p,\end{aligned}$$

$$\begin{aligned}\mathrm{Var}[\|\boldsymbol{X} - \boldsymbol{X}'\|^2] &= \mathrm{Var}\left[\sum_{i=1}^{p}(X_i - X'_i)^2\right] \\ &= \sigma^2 p.\end{aligned}$$

Consequently, the ratio of the standard deviation of $\|\boldsymbol{X} - \boldsymbol{X}'\|^2$ to its mean is

$$\frac{\sqrt{\mathrm{Var}[\|\boldsymbol{X} - \boldsymbol{X}'\|^2]}}{\mathbb{E}[\|\boldsymbol{X} - \boldsymbol{X}'\|^2]} = \frac{\sigma}{\mu\sqrt{p}}.$$

In the limit $p \longrightarrow \infty$, this ratio approaches zero.

High-dimensional feature spaces can cause trouble for parametric methods too. There will usually be more parameters to learn in higher-dimensional spaces (e.g. recall that matrix $\underline{\boldsymbol{\beta}}$ in equation (11.18) has p columns), which raises the danger of overfitting. But this danger can be controlled by regularisation. Multinomial logistic regression continues to be useful in very high-dimensional feature spaces, while k-NN usually does not. That is one reason why the deep neural networks we will apply to classification tasks in Chapter 13 implement fancy versions of multinomial logistic regression, not fancy versions of k-NN. (There is another reason – *differentiability* – but we'll defer discussion of that topic to Chapter 13.)

11.8 Chapter Summary

In this chapter, we introduced probabilistic models for supervised tasks in which the prediction target is a **categorical variable**.

- In **binary classification** tasks, the prediction target takes two possible values. Probabilistic models output the conditional probability of one of these values (given the predictors).
- Binary logistic regression models the **log odds** as a linear function of the predictors.
- Logistic regression models are fitted to data by minimising (regularised) **cross-entropy loss**. Minimising unregularised cross-entropy loss is equivalent to maximising likelihood.
- When training data is **linearly separable**, there is no maximum likelihood logistic regression model. Including a regularisation penalty in the loss function ensures that the loss-minimisation problem is well posed. Regularisation strength can be tuned by cross-validation to prevent overfitting or underfitting.
- In **multiclass classification** tasks, the prediction target takes $K > 2$ possible values. Probabilistic models output a K-dimensional vector representing the conditional probability of each class (given the predictors).
- **Multinomial logistic regression** models the K-dimensional score vector as a linear function of the predictors. The **softmax function** is used to map score vectors to valid probability vectors over class labels.
- The **k-nearest neighbours (k-NN)** classifier is a non-parametric method that estimates class probabilities based on the class distribution among the k closest training points.
- **High-dimensional predictor spaces** tend to favour parametric methods (such as logistic regression) over non-parametric ones (such as k-NN).

11.9 Further Exercises

■ ***Exercise*** 11.23 Your race horse, Harry Trotter, has an estimated odds of $\frac{1}{9}$ of winning each race he runs. The event that he wins any given race is assumed to be independent of the outcomes of previous races.

(a) What is the estimated probability that he wins each race?

(b) Closer examination of Harry Trotter's performance suggests that he is sensitive to the weather. When the air temperature is 18°C he has a 10% chance of winning, but when the temperature is 20°C he has a 12% chance of winning. Let $p(x)$ be the probability of winning when the temperature is x. Assuming a logistic model,

$$p(x) = \sigma(\beta_0 + \beta_1 x),$$

find the values of β_0 and β_1 consistent with the information above.

(c) How warm does it need to be for Harry Trotter to have a 15% chance of winning?

■ ***Exercise*** 11.24 We presented logistic regression with ridge regularisation in this chapter, but other regularisation schemes are possible. Rewrite equations (11.13) and

(11.19) with lasso regularisation instead of ridge. In which of the following two situations would lasso regularisation be a natural choice?

(a) You are trying to predict the value of categorical variable Y based on a high-dimensional predictor vector $\boldsymbol{X}$. You suspect that all the components of X are useful for predicting Y.
(b) You are trying to predict the value of categorical variable Y based on a high-dimensional predictor vector $\boldsymbol{X}$. You suspect that only a few of the components of X are useful for predicting Y, but you don't know which ones.

■ ***Exercise*** 11.25 The q-norm of a vector $\boldsymbol{x} \in \mathbb{R}^p$ is defined as

$$\|\boldsymbol{x}\|_q = \left(\sum_{i=1}^{p} |x_i|^q\right)^{\frac{1}{q}},$$

and the Minkowski distance of order q between vectors $\boldsymbol{x}$ and $\boldsymbol{x}'$ is defined as $\|\boldsymbol{x} - \boldsymbol{x}'\|_q$.[a]

(a) Verify that Minkowski distance of order 2 is Euclidean distance.
(b) Why is Minkowski distance of order 1 often called 'Manhattan' or 'taxicab' distance?
(c) Show that $\lim_{q \to \infty} \|\boldsymbol{x}\|_q = \max_i |x_i|$.
(d) Why are very high-order Minkowski distances rarely used for k-NN?
(e) For k-NN tasks in high-dimensional feature spaces, Manhattan distance sometimes performs better than Euclidean distance. Why might that be? In practice, how would you decide which metric to use for any particular task?

[a] The Minkowski distance of order q satisfies all the mathematical conditions required of distance metrics so long as $q \geq 1$. If $q < 1$, then the Minkowski distance no longer satisfies the triangle inequality (that is, given three vectors $\boldsymbol{x}$, $\boldsymbol{x}'$ and $\boldsymbol{x}''$, it is no longer guaranteed that $\|\boldsymbol{x} - \boldsymbol{x}''\|_q \leq \|\boldsymbol{x} - \boldsymbol{x}'\|_q + \|\boldsymbol{x}' - \boldsymbol{x}''\|_q$). It is possible to implement k-NN with a Minkowski distance of order less than 1, but it is not a good idea in practice, because efficient algorithms for finding nearest neighbours rely on the triangle inequality.

12

Unsupervised Learning: A Deeper Dive

The inference problems we have considered so far have taken one of two forms. Either we were attempting to model a response given one or more predictors, or we were attempting to model the density of a standalone random vector. Density estimation is an *unsupervised* problem because there is no 'supervisory signal' – the response – to guide us. In this chapter, we will build on our treasure hunt example from Chapter 3, showing how densities with unlimited flexibility can be learned. We will also present two new techniques: *clustering* which seeks to identify groups of similar observations within a dataset and *dimensionality reduction* which seeks to describe data using a reduced number of features. All three learning problems boil down to minimising a loss function which measures how well we have described our data. In quite a general sense, *learning* is optimisation. So far in this book, the process used for optimisation was largely taken for granted. As models become more complex and the number of parameters increases, devising efficient methods to learn their optimal values becomes more difficult. In this chapter and Chapter 13, we give these methods greater attention.

12.1 Clustering

Some datasets can be partitioned into groups of observations in a natural way, with observations in any one group more similar to each other than to observations in different groups. The process of automatically finding such groups within a given dataset is known as *clustering*. There are many applications of clustering. We will introduce it as a tool to automatically discover the discrete set of vowel sounds that American English speakers use to communicate. Our starting point is Figure 12.1, which shows scatter plots of standardised backness (X_1) and height (X_2) values for vowel sounds in a range of words spoken by speakers from three different dialect regions in the USA.[1] (We introduced backness and height in Section 11.5.2.) How many different vowel sounds can we identify in these plots?

12.1.1 The K-Means Algorithm

Let us take a step back from vowels and consider a more general clustering problem. Suppose we have a dataset $\underline{\boldsymbol{x}} = (\boldsymbol{x}_1, \ldots, \boldsymbol{x}_n)$ of n points in $\mathbb{R}^p$. We can think of these points as i.i.d. realisations of a random vector $\boldsymbol{X} = (X_1, \ldots, X_p)$ which contains p features. Suppose we wish to design a procedure that partitions our observations into $K \geq 1$ groups of *similar* points. We will assume that the similarity between any pair of points $\boldsymbol{x}_i$ and $\boldsymbol{x}_j$ can be measured

[1] Original acoustic data collected by Clopper et al. (2005).

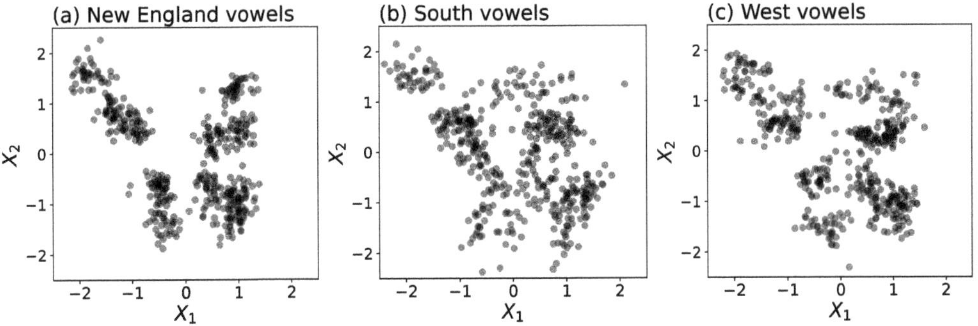

Figure 12.1 (a–c) Backness (X_1) and height (X_2) features extracted from the vowels in a range of words uttered by speakers in three different regions of the USA.

by the Euclidean distance $\|\boldsymbol{x}_i - \boldsymbol{x}_j\|$ between them, with smaller separations corresponding to more similar points. The result of our procedure will be a set of K *clusters* $C_1, \ldots, C_K$, where C_k is the set of indices of the points which belong to the kth cluster. We define the indicator function that point i belongs to cluster k,

$$z_{ik} = \begin{cases} 1 & \text{if } i \in C_k, \\ 0 & \text{otherwise.} \end{cases}$$

The *centroid* of cluster C_k is defined to be the mean of the points it contains,

$$\bar{\boldsymbol{x}}_k = \frac{1}{|C_k|} \sum_{i \in C_k} \boldsymbol{x}_i = \frac{\sum_{i=1}^n z_{ik} \boldsymbol{x}_i}{\sum_{i=1}^n z_{ik}},$$

where $|C_k|$ denotes the number of elements in C_k. We can represent the cluster membership of the ith point with a vector $z_i = (z_{i1}, \ldots, z_{iK})$. Since each point belongs to precisely one cluster, this is an example of *one-hot encoding* (Section 11.4.3). If point i belongs to cluster k, then $z_i = \boldsymbol{e}_k$, where $\boldsymbol{e}_k \in \mathbb{R}^K$ is the unit vector in coordinate direction k.

We want our clusters to be *dense* in the sense that points in the same cluster should be as close as possible to one another. We can measure this closeness using the within-cluster variation or *inertia*,

$$w(C_k) = \frac{1}{2|C_k|} \sum_{i,j \in C_k} \|\boldsymbol{x}_i - \boldsymbol{x}_j\|^2 = \sum_{i \in C_k} \|\boldsymbol{x}_i - \bar{\boldsymbol{x}}_k\|^2.$$

Note that the factor of 2 in the denominator compensates for the fact that the sum over $i, j \in C_k$ counts all pairs of points in the cluster twice.

■ ***Exercise*** 12.1 Establish that

$$\frac{1}{2|C_k|} \sum_{i,j \in C_k} \|\boldsymbol{x}_i - \boldsymbol{x}_j\|^2 = \sum_{i \in C_k} \|\boldsymbol{x}_i - \bar{\boldsymbol{x}}_k\|^2.$$

Hint: write $\|\boldsymbol{x}_i - \boldsymbol{x}_j\|^2 = \|(\boldsymbol{x}_i - \bar{\boldsymbol{x}}_k) - (\boldsymbol{x}_j - \bar{\boldsymbol{x}}_k)\|^2$.

The total within-cluster variation summed over all clusters – the total inertia – is then

$$W(\underline{z}) = \sum_{k=1}^{K} w(C_k) = \sum_{i=1}^{n} \sum_{k=1}^{K} z_{ik} \|\boldsymbol{x}_i - \bar{\boldsymbol{x}}_k\|^2,$$

where $\underline{z}$ is the matrix whose ith row is z_i. The optimal clusters are those that minimise this objective:

$$\hat{\underline{z}} = \arg\min_{\underline{z}} W(\underline{z}). \tag{12.1}$$

This optimisation problem is very difficult to solve exactly even for moderately large n because the number of possible realisations of $\underline{z}$ (each row with a single 1, and each column with at least one 1) is $\approx K^n$. Because of this we must use an approximate method.

To approximate the solution to (12.1), we first define a modified version of our objective function in which the cluster centroids are replaced with vectors $\boldsymbol{\mu}_1, \ldots, \boldsymbol{\mu}_K$ representing arbitrary cluster 'centres' which need not be equal to the centroids of the clusters. This new objective is

$$J = \sum_{i=1}^{n} \sum_{k=1}^{K} z_{ik} \|\boldsymbol{x}_i - \boldsymbol{\mu}_k\|^2.$$

The goal is now to find *both* the cluster centres *and* the cluster assignments which minimise J. The approach involves first optimising J over the assignments keeping the centres fixed, and then optimising over the centres, keeping the assignments fixed. These two optimisation problems are solved repeatedly until the cluster assignments stop changing. Keeping centres fixed, we can minimise J by assigning each point to the cluster whose centre it is closest to, so

$$z_{ik} = \begin{cases} 1 & \text{if } k = \arg\min_j \|\boldsymbol{x}_i - \boldsymbol{\mu}_j\|^2, \\ 0 & \text{otherwise.} \end{cases}$$

Keeping the assignments fixed, the gradient of the objective with respect to $\boldsymbol{\mu}_k$ is

$$\nabla_{\boldsymbol{\mu}_k} J = -2 \sum_{i=1}^{n} z_{ik} (\boldsymbol{x}_i - \boldsymbol{\mu}_k).$$

The minimum of J will lie at the fixed point where $\nabla_{\boldsymbol{\mu}_k} J = \mathbf{0}$, so

$$\boldsymbol{\mu}_k = \frac{\sum_{i=1}^{n} z_{ik} \boldsymbol{x}_i}{\sum_{i=1}^{n} z_{ik}} = \bar{\boldsymbol{x}}_k.$$

Therefore for a given cluster assignment, J is minimised when the centre of each cluster is equal to the centroid (mean) of the observations it contains. Each of these optimisation steps must either decrease J or leave the assignments unchanged, and since the second step sets centres equal to centroids, then their combined effect will be to either decrease $W(\underline{z})$ or leave it unchanged. Repeatedly applying these two steps is known as the K-means algorithm.

Once the assignments have stopped changing, then we will have found a *local minimum* of J. The minimum is *local* in the sense that although we cannot reduce the objective by either (infinitesimally) perturbing the centres or changing the clusters, there may be other assignments with different centroids which achieve a lower J. The fact that this optimisation

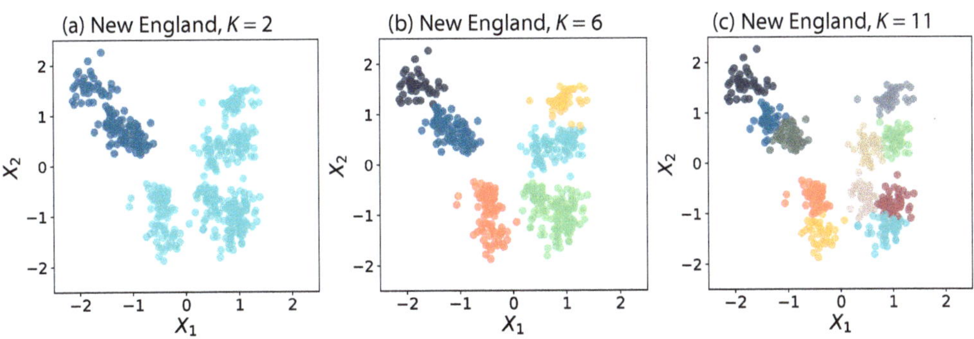

Figure 12.2 (a–c) K-means clusters of New England vowel features with three different K values.

scheme will not necessarily find the global minimum in J means that it will be sensitive to the cluster assignment used to initialise the process. For this reason, it is good practice to repeat the optimisation process multiple times starting from different randomised initial conditions each time and to select the cluster assignment which achieves the lowest objective value. Figure 12.2 shows the results of the K-means algorithm applied to the vowels of New England, for three different values of K. Each clustering is the best of 10 repeats with randomised initial conditions. The algorithm is summarised in Box 12.1.

★ Box 12.1 K-means algorithm

Given a dataset $D = \{\boldsymbol{x}_i\}_{i=1}^n$, assign each point a cluster label from $\{1, \ldots, K\}$ and compute the centroid, $\boldsymbol{\mu}_k$, of each cluster. Repeat the following two steps until assignments stop changing. The notation $a \leftarrow b$ means 'set the value of the variable a to be b'. The steps are labelled 'E' and 'M' for reasons that will become clear in Section 12.2.3.

(1) **E-step**. For each point, recalculate its cluster membership:

$$z_{ik} \leftarrow \begin{cases} 1 & \text{if } k = \arg\min_j \|\boldsymbol{x}_i - \boldsymbol{\mu}_j\|^2, \\ 0 & \text{otherwise.} \end{cases}$$

(2) **M-step**. For each cluster, recalculate its centroid:

$$\boldsymbol{\mu}_k \leftarrow \frac{\sum_{i=1}^n z_{ik}\boldsymbol{x}_i}{\sum_{i=1}^n z_{ik}}.$$

12.1.2 Choosing K

A crucial step in cluster analysis is to determine the most appropriate number of clusters. Applying our visual judgement to Figure 12.2, it seems that $K = 2$ is too few clusters while $K = 6$ and $K = 11$ are both somewhat plausible. In some applications of clustering, we might know what K should be in advance, but this won't be true in general. We therefore need a method to select K. We will present two closely related methods: the 'elbow' method and the 'gap' or 'null reference' method. Both are based on *Occam's razor*, the principle that we should accept the simplest explanation that fits the data.

Elbow Method

Clustering provides a simple summary of the complicated pattern of points in our dataset. Finding patterns of this kind is a way to *explain* the data in a simple way. The number of clusters is a measure of the complexity of our explanation. According to Occam's razor, we should make K just large enough to explain the patterns, but no larger.

K-means clustering works by minimising the total within-cluster variation, or inertia. Having found the clustering for a given value of K, the total inertia which results, $W(\hat{\underline{z}})$, can be used as a measure of how well that clustering explains the data. Unfortunately, the total inertia does not reward simple explanations; increasing K is guaranteed to reduce $W(\hat{\underline{z}})$. The idea behind the elbow method is that if K^* is the *true* number of clusters in our data, then when $K < K^*$, using one extra cluster should yield a substantial drop in inertia. However, when $K = K^*$, the result of using one extra cluster will be to artificially split one of the true clusters into two, bringing only a small reduction in inertia, with further increases in K bringing ever diminishing returns. A graph showing the dependence of $W(\hat{\underline{z}})$ on K, which will be monotonically decreasing, should contain a *kink* at $K = K^*$, where the rate of decrease of inertia with K drops suddenly. The graph will be shaped like a bent arm, with the kink located at the elbow. We should choose K by locating the elbow on the graph. Figure 12.3a shows the relationship between the logarithm of the inertia and K. Taking the logarithm allows us to see changes in inertia over a large range of magnitudes. In Figure 12.3a we have marked the point $K = 11$ as the elbow, based on our visual judgement. While this subjectivity may be seen as a deficiency of the method, it is important to realise that there is no definitively correct way to cluster data, so we should not be afraid to use common sense or domain knowledge to make and justify choices.

Null Reference Method

Finding an elbow on a graph is often difficult. We can make the K selection problem easier by turning it into one of identifying a *peak* rather than a *kink*. We do this by comparing the inertia of the real dataset to the inertia of a hypothetical dataset with no clusters. This approach, invented by Tibshirani et al. (2000), produces a *gap statistic* which measures how well the patterns in the data are explained by clustering. The 'official' gap statistic method involves simulation and K is selected algorithmically. We use a deterministic version of the method as an aid to identify the optimal number of clusters visually.

In the general clustering problem, we have a dataset of n points in $\mathbb{R}^p$. Let V be the volume of the smallest hypercube that contains all the points. If the data is naturally clustered then within the hypercube, there will be dense regions (clusters) and empty regions in between. Now imagine a 'null' dataset created by sampling n point locations uniformly at random from within the cube. Nothing can be learned about the nature of this data-generating process by clustering because any grouping together of points is purely due to random noise. The inertia values obtained by applying K-means to it therefore provide a baseline against which we can compare the inertia values obtained for the real data.

To estimate the inertia of the null data, we divide the hypercube into K clusters, each with

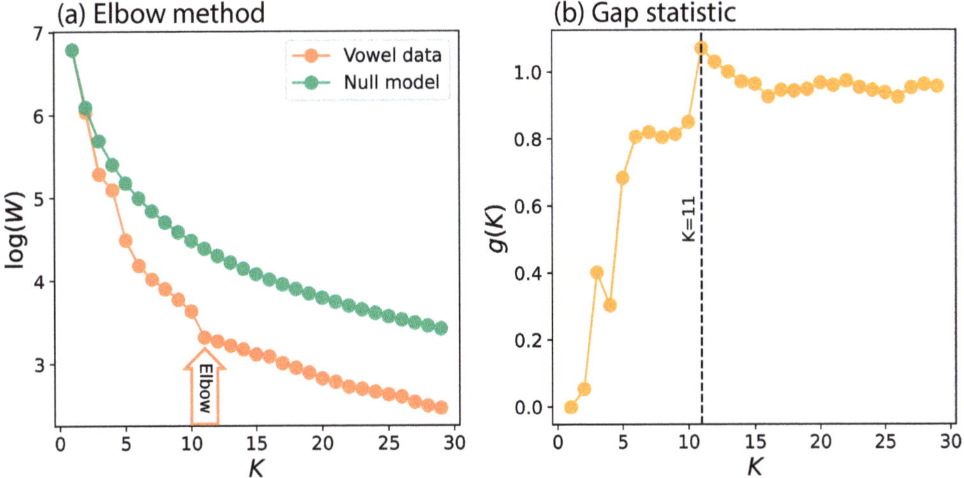

Figure 12.3 (a) Relationship between log inertia and K for the New England vowel data (lower series, orange) and a single cluster null dataset (upper series, green). (b) The corresponding gap statistic.

volume V/K. We approximate the inertia of each cluster by assuming the displacements of its observations with respect to its centroid are uniform random vectors $\boldsymbol{U} = (U_1, \ldots, U_p)$, where $U_i \sim \text{Uniform}\left(-\frac{s}{2}, \frac{s}{2}\right)$. The parameter s is chosen so that the volume of the hypercube in which these observations lies is equal to V/K. Therefore, $s = (V/K)^{1/p}$. According to these assumptions, the inertia of the null dataset will be

$$W_K^{\text{null}} = n\mathbb{E}\left(\|\boldsymbol{U}\|^2\right) = \frac{nps^2}{12} = \frac{np}{12}\left(\frac{V}{K}\right)^{\frac{2}{p}}.$$

■ ***Exercise*** 12.2 Show that if $U_i \sim \text{Uniform}\left(-\frac{s}{2}, \frac{s}{2}\right)$, then $\mathbb{E}(U_i^2) = \frac{s^2}{12}$. Hence, show that if $\boldsymbol{U} = (U_1, \ldots, U_p)$ then $\mathbb{E}\left(\|\boldsymbol{U}\|^2\right) = \frac{ps^2}{12}$.

If W_K is the inertia value of the true dataset partitioned into K clusters, then one way to measure how much has been explained by this partitioning is via the ratio

$$r_K = \frac{W_K^{\text{null}}}{W_K} \propto \frac{K^{-\frac{2}{p}}}{W_K}.$$

For given K, the larger the value of r_K the more we have explained relative to the null data, where clustering explains nothing. We define the logarithm of the scaled ratio $cK^{-2/p}/W_K$ to be the *gap statistic*[2]

$$g(K) := \log\left(\frac{W_1}{W_K}\right) - \frac{2}{p}\log K, \tag{12.2}$$

[2] This is a modified form of the *gap statistic* invented by Tibshirani et al. (2000).

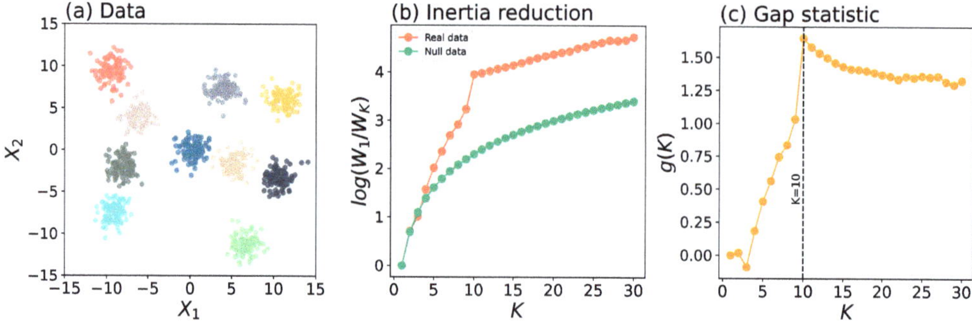

Figure 12.4 (a) A dataset consisting of samples from 10 randomly located Gaussian blobs. (b) Inertia reduction measures for the data (upper series, orange) and the null data (lower series, green). (c) The gap statistic and estimate $\hat{K} = 10$.

where we chose $c = W_1$ so that $g(1) = 0$. The gap statistic may be understood as comparing the inertia reduction factor $\log(W_1/W_k)$ due to clustering the real data with the inertia reduction factor we would have achieved in the null dataset using the same number of clusters. We can think of the $\log K$ term as a *complexity penalty* which compensates for the fact that using inertia alone as a performance measure does not reward simplicity of explanation. In this sense, the gap statistic is a quantitative formulation of Occam's Razor. The first term measures how well we have explained the patterns, and the second measures how complicated our explanation was. A large value of the gap statistic indicates that our clusters are explaining a lot of the patterns in the data in a simple way. Intuitively, a good choice of K should correspond to a significant peak in the gap statistic.

In Figure 12.4, we have applied the gap method to determine the number of clusters in an artificial dataset created by randomly generating $K^* = 10$ Gaussian blobs, shown in panel (a). Panel (b) shows the inertia reduction factors for the real data and the null data. We observe that for both datasets, increasing K reduces inertia, but in the clustered data this happens at a greater rate when $K < K^*$. Panel (c) shows the resulting gap statistic, which has a significant peak at $K = K^*$.

It is important to emphasise that, as with the elbow, using the gap statistic in the way we have defined it requires an element of judgement. We can make $g(K)$ artificially large by making K much larger than the natural number of clusters. In this case, many clusters will have only one point and zero inertia, and when $K = n$ the gap will be infinite. We therefore need to have *some* 'prior' sense of what range K should lie in, in advance. Whatever method or methods we use to choose K, we should always consider carefully whether the results we obtain make sense given our knowledge of the problem.

The result of applying the gap method to estimate the number of clusters in our vowel data is shown in Figure 12.3b. The result matches our earlier prediction using the elbow method. Figure 12.5b shows the New England data labelled according to the carrier word for each vowel. According to linguists, all but two of these words use different vowel phonemes, with 'frog' and 'logs' using the same sound.[3] These phoneme labels provide a ground truth against

[3] The vowel sounds in 'hood' and 'hoed' appear to be merging into a single sound in certain dialects.

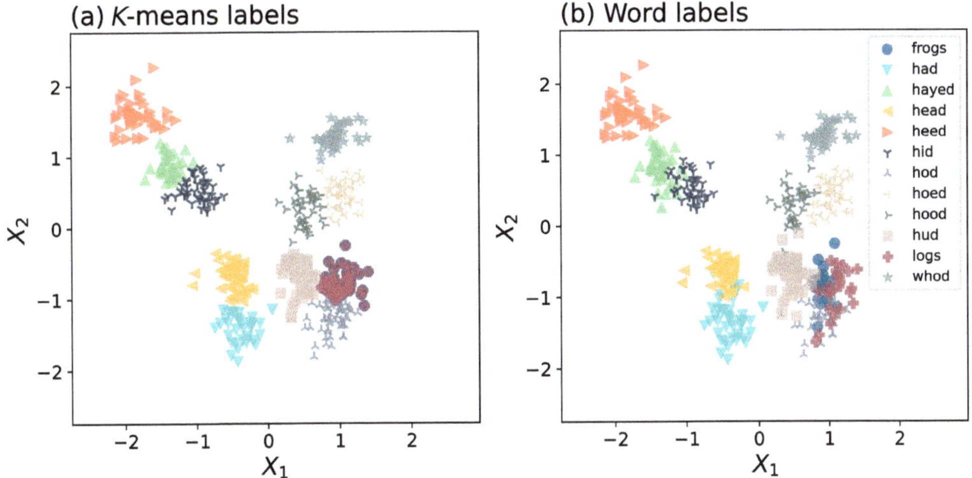

Figure 12.5 (a) New England vowel data labelled according to clusters automatically generated using K-means with $K = 11$. (b) New England vowel data labelled according to the carrier word of the vowel.

which we can compare the K-means predictions, shown in Figure 12.5a, which in this case are remarkably accurate.

12.2 Density Estimation

Imagine a datapoint-generating process which spits out p-dimensional vectors. Suppose we have collected a set of them, $D = \{\boldsymbol{x}_i\}_{i=1}^n$, and we want to build a model of the generating mechanism. We saw an example of this problem in Chapter 3, where the vectors were locations of buried treasure hoards. We have also seen several simple examples of the *probabilistic* approach to this problem, which is to view our observations as samples from a probability density (or mass function). In fact, the treasure hoard model considered in Chapter 3 was a probability density model in disguise – a Gaussian mixture model. Models in this family are capable of describing probability densities of arbitrary complexity in any number of dimensions. The goal of the current section is to put this family on a firm probabilistic footing.

12.2.1 Gaussian Mixtures

Let's start with a one-dimensional example. Suppose that we have a set of observations of a random variable, X, with an unknown distribution. We'll write this set

$$D_{\text{train}} = \{x_i\}_{i=1}^n.$$

Figure 12.6a shows D_{train}, along with a histogram computed from the sample. The histogram is a *non-parametric* approximation to the density of X. Non-parametric models are so called because they do not assume a particular parametrised mathematical model for the density

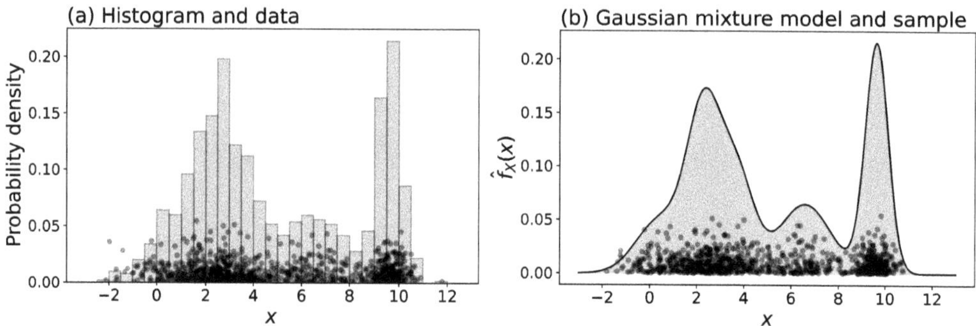

Figure 12.6 (a) Histogram computed from $n = 1{,}000$ realisations of a random variable X with unknown density. The individual datapoints are also shown, with vertical jitter applied to aid visualisation. (b) A $K = 5$ component Gaussian mixture model fitted to the same dataset, along with $n = 1{,}000$ samples from the model.

in advance of fitting to the data. Our aim is to construct a *parametric* approximation to the density. There are many ways we could approach this task, but a particularly flexible method is to assume that the density consists of a weighted sum of Gaussian (normal) density functions with different means and variances – a Gaussian mixture model. If our model has K *components* (individual Gaussian densities), its density may be written

$$f_X(x; \boldsymbol{\theta}) = \sum_{i=1}^{K} \pi_i \mathcal{N}(x|\mu_i, \sigma_i^2),$$

where the weights $\{\pi_i\}$ satisfy

$$0 \leq \pi_i \leq 1,$$
$$\sum_{i=1}^{K} \pi_i = 1.$$

These conditions ensure the density is positive and normalised. They also mean that the vector $\boldsymbol{\pi} = (\pi_1, \ldots, \pi_K)$ may be interpreted as a *probability mass function* over the components of the model. The parameter vector $\boldsymbol{\theta}$ contains the means, variances and weights of the individual Gaussians. The log-likelihood function is given by

$$\ell(\boldsymbol{\theta}) = \log \mathcal{L}(\boldsymbol{\theta}) = \log \left(\prod_{x \in D_{\text{train}}} f_X(x; \boldsymbol{\theta}) \right) = \sum_{x \in D_{\text{train}}} \log f_X(x; \boldsymbol{\theta}).$$

Notice that this is just the log-product measure of fit we used in Chapter 3. We can estimate these model parameters by the principle of maximum likelihood (Box 5.4) subject to the constraint, also discussed in Chapter 3, that we do not collapse any of the components onto single datapoints. Letting $\mathcal{S}$ be the set of 'feasible' parameter vectors – ones for which the component variances exceed a minimum threshold – our estimate of the parameter vector is

$$\hat{\boldsymbol{\theta}} = \arg\max_{\boldsymbol{\theta} \in \mathcal{S}} \ell(\boldsymbol{\theta}).$$

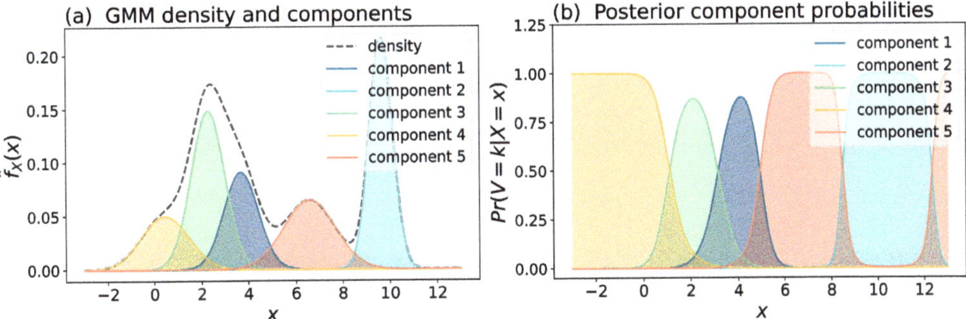

Figure 12.7 (a) The individual weighted components of the Gaussian mixture model shown in Figure 12.6b. (b) The posterior mass function over components.

This identifies the member of our constrained model family which makes the observed data *most probable*. Remember though, this is not necessarily the most plausible model given the data. Figure 12.6b shows the density of the maximum likelihood mixture model with $K = 5$ components, fitted to the data in Figure 12.6a. We will turn to the problem of finding $\hat{\theta}$ in Section 12.2.3. First, we consider the questions of how to generate new data from such a mixture model, and how to choose K.

One way to understand the Gaussian mixture model is as a probabilistic version of K-means, where the components play the role of clusters. Figure 12.7a shows the components of our mixture model. Generating data from the model can be accomplished by a two stage process. First, we select a component according to the probability mass function $\boldsymbol{\pi}$, then we sample from the density of this component. Following the same notation convention we used for K-means, we introduce the one-hot random vector $\mathbf{Z} = (Z_1, \ldots, Z_k)$, where Z_k is the indicator function that component k was selected, that is,

$$Z_k = \begin{cases} 1 & \text{if } k \text{ selected,} \\ 0 & \text{otherwise.} \end{cases}$$

We can write the probability mass function of $\mathbf{Z}$ as

$$f_{\mathbf{Z}}(z) = \prod_{k=1}^{K} \pi_k^{z_k}.$$

The probability density of X given $\mathbf{Z}$ is then

$$f_{X|\mathbf{Z}}(x|z) = \prod_{k=1}^{K} \mathcal{N}(x|\mu_k, \sigma_k^2)^{z_k},$$

and the joint distribution of X and $\mathbf{Z}$ is

$$f_{X,\mathbf{Z}}(x, z) = f_{X|\mathbf{Z}}(x|z) f_{\mathbf{Z}}(z). \tag{12.3}$$

The corresponding DAG representation of the datapoint-generating process is

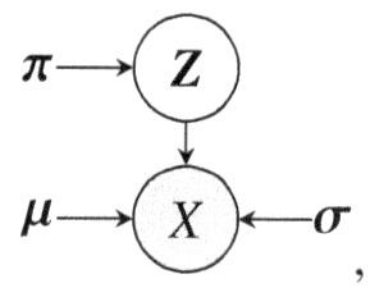

$$\begin{array}{ccccc} \pi & \longrightarrow & \boxed{Z} & & \\ & & \downarrow & & \\ \mu & \longrightarrow & \boxed{X} & \longleftarrow & \sigma \end{array}, \tag{12.4}$$

where $\boldsymbol{\mu} = (\mu_1, \ldots, \mu_K)$ and $\boldsymbol{\sigma} = (\sigma_1, \ldots, \sigma_K)$ are the parameters of the conditional density $f_{X|\mathbf{Z}}(x|z)$. The variable $\mathbf{Z}$ is *latent* because is not observable, except when we generate synthetic data using our model. Generating data in this way is known as *ancestral sampling*. Figure 12.6b shows a new dataset generated by this method, using mixture model parameters learned from the original dataset.

As well as describing the distribution of our data, the Gaussian mixture model provides us with a method for *soft* clustering. The goal of (hard) clustering is to assign every datapoint in our dataset a label which identifies its group such that points in the same group are as 'similar' as possible. In a Gaussian mixture model, the individual component densities play a role analogous to the clusters in K-means, and the latent variable z_i may be interpreted as a cluster assignment for x_i. Suppose we have fitted a Gaussian mixture model to a dataset sampled from some population. Let $(X, \mathbf{Z})$ be the observed and latent variables for a single entity selected at random from this population. The weight vector $\boldsymbol{\pi}$ of our model gives the *prior* mass function for $\mathbf{Z}$, *before* observing X. The *posterior* probability distribution of $\mathbf{Z}$, after observing X, may be found via Bayes' theorem:

$$f_{\mathbf{Z}|X}(z|x) = \frac{f_{X|\mathbf{Z}}(x|z) f_{\mathbf{Z}}(z)}{f_X(x)},$$

where $f_X(x) \equiv f_X(x; \boldsymbol{\theta})$. Writing this posterior in terms of the model parameters, we have

$$f_{\mathbf{Z}|X}(\mathbf{e}_k|x) = \mathbf{Pr}(\mathbf{Z} = \mathbf{e}_k | X = x) = \frac{\pi_k \mathcal{N}(x|\mu_k, \sigma_k^2)}{f_X(x; \boldsymbol{\theta})}.$$

This posterior mass function may be understood as a *soft* cluster assignment based on the observed value of each point. A *hard* cluster assignment may be obtained by computing the MAP estimate of $\mathbf{Z}$,

$$\hat{z} = \arg\max_k f_{\mathbf{Z}|X}(\mathbf{e}_k|x).$$

The dependence of posterior component probabilities on the observed value of X is shown in Figure 12.7b.

■ ***Exercise*** 12.3 Show that if $z_k = 1$ and $z_i = 0$ for all $i \neq k$ then

$$\frac{f_{X|\mathbf{Z}}(x|z) f_{\mathbf{Z}}(z)}{f_X(x)} = \frac{\pi_k \mathcal{N}(x|\mu_k, \sigma_k^2)}{f_X(x; \boldsymbol{\theta})}.$$

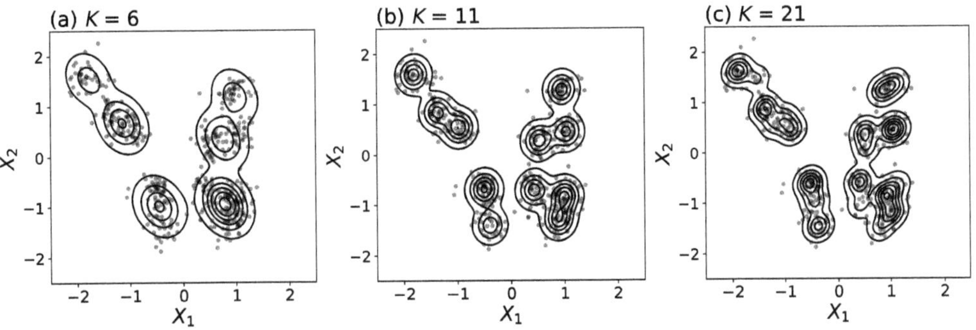

Figure 12.8 (a–c) Contour plots of spherical Gaussian mixture densities fitted to the New England vowel data, using three different K values. Contours are drawn at levels $\{0.05k | k \in \mathbb{N}\}$.

12.2.2 Choosing K

Let us estimate a probability density of normalised New England vowel formants using a mixture of bivariate normal densities,

$$f_X(\boldsymbol{x}; \boldsymbol{\theta}) = \sum_{i=1}^{K} \pi_i \mathcal{N}(\boldsymbol{x} | \boldsymbol{\mu}_i, \underline{\boldsymbol{\Sigma}}_i),$$

where $\boldsymbol{\mu}_i$ is the mean (location) of the ith component and $\underline{\boldsymbol{\Sigma}}_i$ is its covariance matrix. To keep things simple, we will use *spherical* component densities, which have covariance matrices of the form

$$\underline{\boldsymbol{\Sigma}}_i = \sigma_i^2 \underline{\boldsymbol{I}}, \tag{12.5}$$

where $\underline{\boldsymbol{I}}$ is the identity matrix. Figure 12.8 shows the result of fitting mixture models of this form to vowel data, using three different values of K. When $K = 11$ (Figure 12.8b), the locations of components coincide closely with the cluster locations found using K-means (Figure 12.5a).

For any given value of K, we can fit model (12.5) to training data by maximising likelihood. We cannot fit K to training data, because K is a hyperparameter controlling model flexibility. What we can do is apply our usual procedure for tuning hyperparameters. Rather than tuning K to maximise the probability (according to the model) of training data, we tune K to maximise the probability of *validation* data, that is, data not used to fit the model.

■ ***Exercise*** 12.4 What goes wrong if we try to fit K to training data by maximising likelihood without any constraints?

Let's spell out the details. We'll assume that the *true* datapoint-generating mechanism is described by some (unknown) density $\mathfrak{f}_X$. The training dataset D_{train} is a set of points drawn independently from $\mathfrak{f}_X$. To learn the parameters of our model (for any given K), we maximise the log probability density, according to our model, of D_{train}. Given a validation dataset,

$$D_{\text{val}} = \{\boldsymbol{x}_i\}_{i=1}^{n},$$

also drawn from the density $\mathfrak{f}_X$, we can measure how well our fitted model matches $\mathfrak{f}_X$ by evaluating the log probability of D_{val} according to our model. Writing $\hat{f}_X(\boldsymbol{x}) = f_X(\boldsymbol{x}; \hat{\boldsymbol{\theta}})$, the average log probability of the datapoints is

$$\frac{1}{|D_{\text{val}}|} \sum_{\boldsymbol{x} \in D_{\text{val}}} \log \hat{f}_X(\boldsymbol{x}). \tag{12.6}$$

In the limit of large numbers of validation datapoints, this sum converges to the expectation of the log density $\log \hat{f}_X(\boldsymbol{x})$ over the *true* distribution of $\boldsymbol{X}$. In Chapter 10 we used a Bayesian version of this measure – the expectation of the log posterior predictive density over the true data-generating distribution. The negative of the expectation approximated in (12.6) is the cross-entropy of $\hat{f}_X$ relative to $\mathfrak{f}_X$,

$$\mathbb{H}\left[\mathfrak{f}_X, \hat{f}_X\right] = -\int \mathfrak{f}_X(\boldsymbol{x}) \log \hat{f}_X(\boldsymbol{x}) d\boldsymbol{x}.$$

Here we place the arguments of $\mathbb{H}$ in square brackets to indicate that $\mathbb{H}$ is a *functional* – a function of one or more functions. The cross-entropy achieves its global minimum when $\hat{f}_X = \mathfrak{f}_X$ (see Appendix C). We want to get as close to this as we can.

We have one fitted model for each possible value of the hyperparameter K, and it is unlikely that this family will include the true density. Also, we cannot calculate the cross-entropy exactly, but must rely on sample approximations like expression (12.6). Our goal is to adjust K to make our estimate of the cross-entropy as small as possible, in the hope that this means our model is as close as it can be to the true density. To avoid permanently setting aside a validation set, we use k-fold cross-validation. We partition the dataset D that we are using for fitting and validation into k equal-sized folds, $\mathcal{F}_1, \ldots, \mathcal{F}_k$, and we let each fold take its turn as the validation set. That is, for each $i \in \{1, \ldots, k\}$, we compute the sample entropy

$$\hat{\mathbb{H}}_i = -\frac{1}{|\mathcal{F}_i|} \sum_{\boldsymbol{x} \in \mathcal{F}_i} \log \hat{f}_X(\boldsymbol{x}; \hat{\boldsymbol{\theta}}_{-i}),$$

where $\hat{\boldsymbol{\theta}}_{-i}$ is the parameter vector of the density model fitted to the data in all folds except $\mathcal{F}_i$. We think of $\hat{\mathbb{H}}_i$ as an estimate of $\mathbb{H}\left[\mathfrak{f}_X, \hat{f}_X(-; \hat{\boldsymbol{\theta}})\right]$, where $\hat{\boldsymbol{\theta}}$ is the parameter vector of the density model fitted to D. Our *best* estimate of this cross-entropy is then the average

$$\hat{\mathbb{H}}_{\text{CV}} = \frac{1}{k} \sum_{i=1}^{k} \hat{\mathbb{H}}_i.$$

Letting $\hat{\mathbb{H}}_{\text{CV}}(K)$ be our entropy estimate for a K-component mixture model, we select K as

$$\hat{K} = \arg\min_K \hat{\mathbb{H}}_{\text{CV}}(K).$$

Figure 12.9a shows estimates of the cross-entropy of K-component mixture models fitted to the New England vowel data. In this case, the minimum cross-entropy is achieved when $K = 11$, matching the optimal number of clusters for K means. This result is not surprising, given that the components of our mixture model play a similar role to the clusters in K-means. In Section 12.2.3 we will show how maximum likelihood estimates for a mixture model can be computed. We will see that learning procedures for K-means and Gaussian mixture models

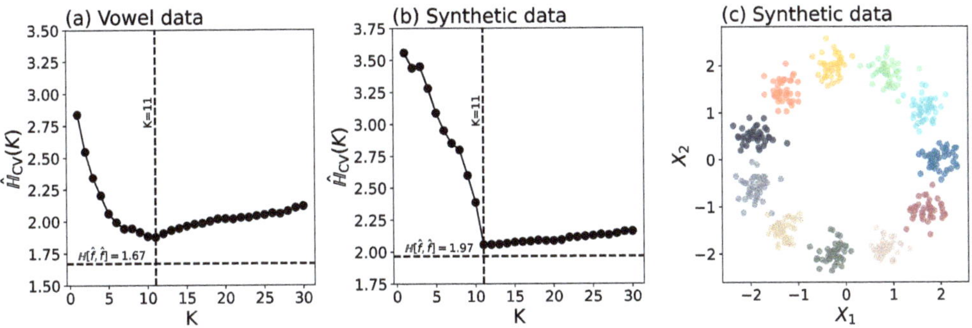

Figure 12.9 Cross-validation estimates of the cross-entropy of K-component mixture models fitted to (a) New England vowel data and (b) a synthetic dataset. The synthetic dataset itself is plotted in panel (c).

are very closely related and may be understood as examples of a general-purpose learning algorithm known as *expectation maximisation* (EM).

We noted above that whatever model family we use to approximate the true density $\mathfrak{f}_X$, the best we can hope for is a 'close match'. How close have we got in this case? The theoretical minimum value of the cross-entropy is $\mathbb{H}[\mathfrak{f}_X, \mathfrak{f}_X]$ – the *entropy* of $\mathfrak{f}_X$. If we knew this, and the cross-entropy $\mathbb{H}\left[\mathfrak{f}_X, \hat{f}_X\right]$, then we could measure how close we got as the gap between the two. This is the Kullback–Leibler (KL) divergence,

$$\mathbb{KL}\left[\mathfrak{f}_X \| \hat{f}_X\right] = \mathbb{H}\left[\mathfrak{f}_X, \hat{f}_X\right] - \mathbb{H}\left[\mathfrak{f}_X, \mathfrak{f}_X\right] = \int \mathfrak{f}_X(\boldsymbol{x}) \log\left(\frac{\mathfrak{f}_X(\boldsymbol{x})}{\hat{f}_X(\boldsymbol{x})}\right) d\boldsymbol{x},$$

which achieves its global minimum of zero when $\mathfrak{f}_X = \hat{f}_X$ (see Appendix C). In reality, we have only approximations to the two entropies. One way to approximate the entropy of the true distribution is using the entropy of our approximation $\hat{f}_X$, which is shown as a dotted line in Figure 12.9a. For comparison, Figure 12.9b shows entropy estimates based on a synthetic dataset with the same number of points as our vowel data, but containing 11 well-spaced Gaussian clusters, shown in Figure 12.9c. This synthetic data serves as a reference where the family of model densities contains the true density *by design*. In this case, our approximation to the KL divergence gets substantially closer to zero than it does for the vowel data.

12.2.3 The EM Algorithm

We now turn to the problem of fitting the parameters of our model. The Gaussian mixture model describes the joint distribution of *observed* random variables ($\boldsymbol{X}$) and *latent* (unobserved) random variables ($\mathbf{Z}$). As such it belongs to a very wide class of models known as *latent variable* models. From the Bayesian perspective, all models belong to this class, because their parameters are viewed as unobserved random variables. However, Bayesian parameters and unobserved outputs play distinct roles in the sense that *one* realisation of the latent parameters can be followed by *many* realisations of the latent outputs. In this chapter, we have treated the parameter vector of our mixture model from a frequentist perspective, so the only latent variables are the $\mathbf{Z}$s.

Our goal is to maximise the log-likelihood

$$\ell(\boldsymbol{\theta}) = \sum_{i=1}^{n} \log f_{X}(\boldsymbol{x}_i; \boldsymbol{\theta}), \tag{12.7}$$

subject to the constraint that $\boldsymbol{\theta}$ lies in our feasible set $\mathcal{S}$ where none of the components have collapsed onto single datapoints. Here, f_X is our Gaussian mixture density,

$$f_X(\boldsymbol{x}; \boldsymbol{\theta}) = \sum_{k=1}^{K} \pi_k \mathcal{N}\left(\boldsymbol{x} | \boldsymbol{\mu}_k, \underline{\boldsymbol{\Sigma}}_k\right). \tag{12.8}$$

The solution to the optimisation problem will be a local maximum of the log-likelihood, that is, one of its stationary points. Because the logarithms in (12.7) are applied to sums of Gaussians, the stationary point equations cannot be solved in closed form. An alternative approach to finding local maxima is to estimate gradients of $\ell(\boldsymbol{\theta})$ numerically and to perform *gradient ascent* – moving though parameter space in the direction that increases the likelihood fastest – until the gradient is approximately zero, meaning we are close to a local maximum. We will apply gradient methods when we come to train neural networks in Chapter 13. For the current problem, we will derive an alternative optimisation method known as EM. The EM algorithm iteratively updates both the parameter vector, $\boldsymbol{\theta}$, describing the component densities, and the 'responsibilities' that the components have for each datapoint, until a local maximum is found. An advantage of the method is that the iteration steps are expressible in closed form, and convergence is fast. We will present the EM algorithm as a general procedure for finding maximum likelihood (or MAP) estimates in latent variable models, before applying it to GMMs. We will also see that the K-means algorithm is EM in disguise.

Consider a probability model which contains both observable variables $\boldsymbol{X}$ and discrete latent variables $\boldsymbol{Z}$. The marginal density of the observable variables is

$$f_X(\boldsymbol{x}; \boldsymbol{\theta}) = \sum_{z} f_{X,Z}(\boldsymbol{x}, z; \boldsymbol{\theta}), \tag{12.9}$$

where the sum runs over all possible values of z. Examples in this class include the GMM and other mixture models of the form

$$f_X(\boldsymbol{x}; \boldsymbol{\theta}) = \sum_{k=1}^{K} \pi_k g(\boldsymbol{x}; \boldsymbol{\theta}_k),$$

where g is the density of some other distribution family. For a mixture model, the sum in equation (12.9) runs over all possible cluster assignments. *If* we were able to observe the latent variables, then we would know which component every observation $\boldsymbol{x}_i$ belonged to. Maximising the likelihood for a given set of component assignments would then boil down to fitting the individual components to the subsets of observations which they generated – a fairly straightforward task. However, the 'complete' dataset $D_{\text{comp}} = \{(\boldsymbol{x}_i, z_i)\}_{i=1}^{n}$ is not visible to us (the latent variables are a convenient fiction). We think of this dataset as a single realisation of the pair of random matrices $\underline{\boldsymbol{X}} = (\boldsymbol{X}_1, \ldots, \boldsymbol{X}_n)^T$ and $\underline{\boldsymbol{Z}} = (\boldsymbol{Z}_1, \ldots, \boldsymbol{Z}_n)^T$, where the $(\boldsymbol{X}_i, \boldsymbol{Z}_i)$ are i.i.d. copies of $(\boldsymbol{X}, \boldsymbol{Z})$. *Only* $\underline{\boldsymbol{X}}$ can be observed. Using this notation, the log-likelihood is

$$\ell(\boldsymbol{\theta}) = \log f_{\underline{\mathbf{X}}}(\underline{\boldsymbol{x}}; \boldsymbol{\theta}) = \log \sum_{\underline{z}} f_{\underline{\mathbf{X}},\underline{\mathbf{Z}}}(\underline{\boldsymbol{x}}, \underline{z}; \boldsymbol{\theta}), \tag{12.10}$$

where the joint density of the complete dataset is

$$f_{\underline{\mathbf{X}},\underline{\mathbf{Z}}}(\underline{\boldsymbol{x}}, \underline{z}; \boldsymbol{\theta}) = \prod_{i=1}^{n} f_{X,Z}(\boldsymbol{x}_i, z_i; \boldsymbol{\theta}).$$

The idea of EM is to use an approximation $q(\underline{z})$ to the posterior $f_{\underline{\mathbf{Z}}|\underline{\mathbf{X}}}(\underline{z}|\underline{\boldsymbol{x}})$ which makes it easy to compute an estimate for $\boldsymbol{\theta}$. This estimate is then used to improve approximation q, which is then used to re-estimate $\boldsymbol{\theta}$, and so on. Our starting point is the factorisation

$$f_{\underline{\mathbf{X}}}(\underline{\boldsymbol{x}}; \boldsymbol{\theta}) = \frac{f_{\underline{\mathbf{X}},\underline{\mathbf{Z}}}(\underline{\boldsymbol{x}}, \underline{z}; \boldsymbol{\theta})}{f_{\underline{\mathbf{Z}}|\underline{\mathbf{X}}}(\underline{z}|\underline{\boldsymbol{x}}; \boldsymbol{\theta})},$$

which holds for arbitrary $\underline{z}$. Multiplying by $1 \equiv q(\underline{z})/q(\underline{z})$ and taking logarithms, we obtain

$$\log f_{\underline{\mathbf{X}}}(\underline{\boldsymbol{x}}; \boldsymbol{\theta}) = \log\left(\frac{f_{\underline{\mathbf{X}},\underline{\mathbf{Z}}}(\underline{\boldsymbol{x}}, \underline{z}; \boldsymbol{\theta})}{q(\underline{z})}\right) - \log\left(\frac{f_{\underline{\mathbf{Z}}|\underline{\mathbf{X}}}(\underline{z}|\underline{\boldsymbol{x}}; \boldsymbol{\theta})}{q(\underline{z})}\right).$$

Since $\log f_{\underline{\mathbf{X}}}(\underline{\boldsymbol{x}}; \boldsymbol{\theta})$ does not depend on $\underline{z}$, multiplying by $q(\underline{z})$ and summing over $\underline{z}$ yields

$$\ell(\boldsymbol{\theta}) = \log f_{\underline{\mathbf{X}}}(\underline{\boldsymbol{x}}; \boldsymbol{\theta}) = \sum_{\underline{z}} q(\underline{z}) \log\left(\frac{f_{\underline{\mathbf{X}},\underline{\mathbf{Z}}}(\underline{\boldsymbol{x}}, \underline{z}; \boldsymbol{\theta})}{q(\underline{z})}\right) + \mathbb{KL}\left[q \| f_{\underline{\mathbf{Z}}|\underline{\mathbf{X}}}\right](\boldsymbol{\theta}),$$

where the KL divergence depends on $\boldsymbol{\theta}$ via the posterior $f_{\underline{\mathbf{Z}}|\underline{\mathbf{X}}}$. This is an important identity. The KL divergence is a measure of the distance between q and $f_{\underline{\mathbf{Z}}|\underline{\mathbf{X}}}$. It is always non-negative, and zero when $q = f_{\underline{\mathbf{Z}}|\underline{\mathbf{X}}}$ (see Appendix C). This non-negativity implies the following inequality:

$$\ell(\boldsymbol{\theta}) \geq \sum_{\underline{z}} q(\underline{z}) \log\left(\frac{f_{\underline{\mathbf{X}},\underline{\mathbf{Z}}}(\underline{\boldsymbol{x}}, \underline{z}; \boldsymbol{\theta})}{q(\underline{z})}\right) = \mathrm{ELBO}[q](\boldsymbol{\theta}).$$

From this we see that the functional we have called ELBO – the *Evidence Lower BOund* – is a lower bound on the log-likelihood based on the observable data (after marginalising out the latent variables).[4]

The EM algorithm exploits the decomposition

$$\ell(\boldsymbol{\theta}) = \mathrm{ELBO}[q](\boldsymbol{\theta}) + \mathbb{KL}\left[q \| f_{\underline{\mathbf{Z}}|\underline{\mathbf{X}}}\right](\boldsymbol{\theta})$$

to alternately adjust q and $\boldsymbol{\theta}$ so as to maximise the ELBO, until no further increases can be achieved. Suppose our current guess at the maximum likelihood parameter vector is $\boldsymbol{\theta}_t$. Given $\boldsymbol{\theta}_t$, the ELBO is maximised when the KL divergence is zero, which is achieved by setting $q(\underline{z}) = f_{\underline{\mathbf{Z}}|\underline{\mathbf{X}}}(\underline{z}|\underline{\boldsymbol{x}}; \boldsymbol{\theta}_t)$. Viewed as a function of $\boldsymbol{\theta}$, the ELBO is then

$$\mathrm{ELBO}[q](\boldsymbol{\theta}) = \sum_{\underline{z}} f_{\underline{\mathbf{Z}}|\underline{\mathbf{X}}}(\underline{z}|\underline{\boldsymbol{x}}; \boldsymbol{\theta}_t) \log f_{\underline{\mathbf{X}},\underline{\mathbf{Z}}}(\underline{\boldsymbol{x}}, \underline{z}; \boldsymbol{\theta}) + \mathbb{H}[f_{\underline{\mathbf{Z}}|\underline{\mathbf{X}}}, f_{\underline{\mathbf{Z}}|\underline{\mathbf{X}}}](\boldsymbol{\theta}_t). \tag{12.11}$$

4 The term 'evidence' is used loosely here. In a Bayesian model, where the parameters are latent variables, the marginal likelihood may be interpreted as the evidence (see Box 10.3), but in the current model the parameters are treated as constants.

We let the first term on the right define the function

$$Q(\boldsymbol{\theta}, \boldsymbol{\theta}_t) = \sum_{\underline{z}} f_{\underline{Z}|\underline{X}}(\underline{z}|\underline{x}; \boldsymbol{\theta}_t) \log f_{\underline{X},\underline{Z}}(\underline{x}, \underline{z}; \boldsymbol{\theta}).$$

Let $(\underline{\boldsymbol{X}}_t, \underline{\boldsymbol{Z}}_t)$ be the random data matrices with density $f_{\underline{X},\underline{Z}}(\underline{x}, \underline{z}; \boldsymbol{\theta}_t)$ – our current guess at the density of the complete dataset. We can then write

$$Q(\boldsymbol{\theta}, \boldsymbol{\theta}_t) = \mathbb{E}\left(\log f_{\underline{X},\underline{Z}}(\underline{\boldsymbol{X}}_t, \underline{\boldsymbol{Z}}_t; \boldsymbol{\theta}) \middle| \underline{\boldsymbol{X}}_t = \underline{x}\right).$$

That is, Q is the expectation of the log-likelihood function based on the *complete* dataset, over our current guess at the posterior distribution of $\underline{\boldsymbol{Z}}$. The entropy of our current guess at the posterior (the second term on the right-hand side of equation (12.11)) depends only on $\boldsymbol{\theta}_t$, so it may be treated as a $\boldsymbol{\theta}$-independent constant. Therefore,

$$\text{ELBO}[q](\boldsymbol{\theta}) = Q(\boldsymbol{\theta}, \boldsymbol{\theta}_t) + \text{constant}.$$

The act of computing Q is the *expectation* or E-step of the algorithm. The *maximisation* or M-step is to maximise the ELBO with respect to $\boldsymbol{\theta}$ for given q, giving

$$\boldsymbol{\theta}_{t+1} = \arg\max_{\boldsymbol{\theta}} Q(\boldsymbol{\theta}, \boldsymbol{\theta}_t).$$

The E-step of EM leaves the log-likelihood unchanged because it does not change the value of $\boldsymbol{\theta}$; it simply makes $\text{ELBO}[q](\boldsymbol{\theta}_t) = \ell(\boldsymbol{\theta}_t)$ by choosing q to make the KL divergence zero. The M-step must either increase the log-likelihood or leave it unchanged. We repeatedly apply these steps to generate the sequence $\boldsymbol{\theta}_0, \boldsymbol{\theta}_1, \ldots$ which in practice is terminated once the successive log-likelihood values meet some convergence condition.

Let us apply EM to the Gaussian mixture model with general (as opposed to spherical) covariance matrices. For a single observation, we have the following densities:

$$f_Z(z; \boldsymbol{\theta}) = \prod_{k=1}^{K} \pi_k^{z_k},$$

$$f_{X|Z}(\boldsymbol{x}|z; \boldsymbol{\theta}) = \prod_{k=1}^{K} \mathcal{N}(\boldsymbol{x}|\boldsymbol{\mu}_k, \underline{\boldsymbol{\Sigma}}_k)^{z_k}.$$

The joint distributions of a single complete datapoint $(\boldsymbol{X}, \boldsymbol{Z})$ and of the complete dataset $(\underline{\boldsymbol{X}}, \underline{\boldsymbol{Z}})$ are therefore

$$f_{X,Z}(\boldsymbol{x}, z; \boldsymbol{\theta}) = \prod_{k=1}^{K} \left(\pi_k \mathcal{N}(\boldsymbol{x}|\boldsymbol{\mu}_k, \underline{\boldsymbol{\Sigma}}_k)\right)^{z_k},$$

$$f_{\underline{X},\underline{Z}}(\underline{x}, \underline{z}; \boldsymbol{\theta}) = \prod_{i=1}^{n} \prod_{k=1}^{K} \left(\pi_k \mathcal{N}(\boldsymbol{x}_i|\boldsymbol{\mu}_k, \underline{\boldsymbol{\Sigma}}_k)\right)^{z_{ik}},$$

and the log of the density of the complete dataset is

$$\log f_{\underline{X},\underline{Z}}(\underline{x}, \underline{z}; \boldsymbol{\theta}) = \sum_{i=1}^{n} \sum_{k=1}^{K} z_{ik} \left(\log \pi_k + \log \mathcal{N}(\boldsymbol{x}_i|\boldsymbol{\mu}_k, \underline{\boldsymbol{\Sigma}}_k)\right).$$

This expression has the useful property of being *linear* in the individual components of the latent variables, allowing us to exploit the linearity of the expectation operator to compute $Q(\boldsymbol{\theta}, \boldsymbol{\theta}_t)$ very efficiently. We first define the *responsibility* r_{ik} of component k for observation i to be the probability that $\boldsymbol{x}_i$ came from cluster k,

$$r_{ik} = f_{\mathbf{Z}|\mathbf{X}}(\boldsymbol{e}_k|\boldsymbol{x}_i; \boldsymbol{\theta}).$$

This probability is also the expectation of random variable Z_{ik}, taken with respect to the density $f_{\underline{\mathbf{Z}}|\underline{\mathbf{X}}}(-|\underline{\boldsymbol{x}}_i; \boldsymbol{\theta})$. Since $f_{\mathbf{Z}|\mathbf{X}}(z|\boldsymbol{x}; \boldsymbol{\theta}) = f_{\mathbf{X},\mathbf{Z}}(\boldsymbol{x}, z; \boldsymbol{\theta})/f_{\mathbf{X}}(\boldsymbol{x}; \boldsymbol{\theta})$, it follows that

$$r_{ik} = \frac{\pi_k \mathcal{N}(\boldsymbol{x}|\boldsymbol{\mu}_k, \underline{\boldsymbol{\Sigma}}_k)}{\sum_{j=1}^{K} \pi_j \mathcal{N}(\boldsymbol{x}|\boldsymbol{\mu}_j, \underline{\boldsymbol{\Sigma}}_j)}.$$

Writing $r_{ik}(t)$ for the responsibilities computed using parameter estimate $\boldsymbol{\theta}_t$ and $Z_{ik}(t)$ for the kth component of the ith vector in $\underline{\mathbf{Z}}_t$, we have

$$\begin{aligned} Q(\boldsymbol{\theta}, \boldsymbol{\theta}_t) &= \mathbb{E}\left(\sum_{i=1}^{n} \sum_{k=1}^{K} Z_{ik}(t) \left(\log \pi_k + \log \mathcal{N}(\boldsymbol{x}_i|\boldsymbol{\mu}_k, \underline{\boldsymbol{\Sigma}}_k) \right) \middle| \underline{\mathbf{X}}_t = \underline{\boldsymbol{x}} \right) \\ &= \sum_{i=1}^{n} \sum_{k=1}^{K} r_{ik}(t) \left(\log \pi_k + \log \mathcal{N}(\boldsymbol{x}_i|\boldsymbol{\mu}_k, \underline{\boldsymbol{\Sigma}}_k) \right). \end{aligned}$$

Finally, we compute $\boldsymbol{\theta}_{t+1}$ as the value of $\boldsymbol{\theta}$ which maximises $Q(\boldsymbol{\theta}, \boldsymbol{\theta}_t)$. The details of this calculation are given in Box 12.2. The complete algorithm, including formulae for updating the parameters, is given in Box 12.3.

★ Box 12.2 Maximising $Q(\boldsymbol{\theta}, \boldsymbol{\theta}_t)$ with respect to $\boldsymbol{\theta}$

We use the following notation for derivatives of functions with respect to vectors and matrices. If f is a scalar function of the vector $\boldsymbol{x}$, then $\frac{\partial f}{\partial \boldsymbol{x}}$ denotes the vector with components $[\frac{\partial f}{\partial \boldsymbol{x}}]_i = \frac{\partial f}{\partial x_i}$. If f is a scalar function of the matrix $\underline{\boldsymbol{A}}$, then $\frac{\partial f}{\partial \underline{\boldsymbol{A}}}$ denotes the matrix with components $[\frac{\partial f}{\partial \underline{\boldsymbol{A}}}]_{ij} = \frac{\partial f}{\partial A_{ij}}$. We will use the following identities, in which $\underline{\boldsymbol{A}}$ is a symmetric invertible matrix with positive determinant:

$$\frac{\partial}{\partial \boldsymbol{x}}(\boldsymbol{x}^T \underline{\boldsymbol{A}} \boldsymbol{x}) = 2\underline{\boldsymbol{A}}\boldsymbol{x}, \tag{12.12}$$

$$\frac{\partial}{\partial \underline{\boldsymbol{A}}} \boldsymbol{x}^T \underline{\boldsymbol{A}}^{-1} \boldsymbol{x} = -\underline{\boldsymbol{A}}^{-1} \boldsymbol{x}\boldsymbol{x}^T \underline{\boldsymbol{A}}^{-1}, \tag{12.13}$$

$$\frac{\partial}{\partial \underline{\boldsymbol{A}}} \log(\det(\underline{\boldsymbol{A}})) = \underline{\boldsymbol{A}}^{-1}. \tag{12.14}$$

Using the definition of the multivariate normal density (Box 9.1), we have

$$Q(\boldsymbol{\theta}, \boldsymbol{\theta}_t) = \sum_{i=1}^{n} \sum_{k=1}^{K} r_{ik} \left(\log \pi_k - \frac{1}{2}(\boldsymbol{x}_i - \boldsymbol{\mu}_k)^T \underline{\boldsymbol{\Sigma}}_k^{-1} (\boldsymbol{x}_i - \boldsymbol{\mu}_k) - \frac{1}{2} \log(\det(\underline{\boldsymbol{\Sigma}}_k)) - \frac{p}{2} \log(2\pi) \right). \tag{12.15}$$

Applying identity (12.12), we have the stationary point condition

$$\frac{\partial Q}{\partial \boldsymbol{\mu}_k} = -\underline{\boldsymbol{\Sigma}}^{-1} \sum_{i=1}^{n} r_{ik}(\boldsymbol{x}_i - \boldsymbol{\mu}_k) = 0,$$

so, defining $n_k = \sum_{i=1}^{n} r_{ik}$, at a stationary point we have

$$\boldsymbol{\mu}_k = \frac{1}{n_k} \sum_{i=1}^{k} r_{ik}\boldsymbol{x}_i.$$

Applying identities (12.13) and (12.14), we have a second stationary point condition,

$$\frac{\partial Q}{\partial \underline{\boldsymbol{\Sigma}}_k} = \frac{1}{2}\underline{\boldsymbol{\Sigma}}_k^{-1} \left(\sum_{i=1}^{n} r_{ik}(\boldsymbol{x}_i - \boldsymbol{\mu}_k)(\boldsymbol{x}_i - \boldsymbol{\mu}_k)^T \underline{\boldsymbol{\Sigma}}_k^{-1} - n_k \underline{\boldsymbol{I}} \right) = 0,$$

implying

$$\underline{\boldsymbol{\Sigma}}_k = \frac{1}{n_k} \sum_{i=1}^{n} r_{ik}(\boldsymbol{x}_i - \boldsymbol{\mu}_k)(\boldsymbol{x}_i - \boldsymbol{\mu}_k)^T.$$

To find the stationary weights, we need to maximise Q subject to the constraint that $\sum_{k=1}^{K} \pi_k = 1$, which is achieved by finding the stationary point of $J = Q + \lambda(\sum_{j=1}^{K} \pi_j - 1)$ with respect to the weights and λ (the method of *Lagrange multipliers*). We have

$$\frac{\partial J}{\partial \pi_k} = \frac{n_k}{\pi_k} + \lambda = 0, \qquad \frac{\partial J}{\partial \lambda} = \sum_{k=1}^{K} \pi_k - 1 = 0.$$

The first equation yields $\pi_k = -n_k/\lambda$. Combining with the second, we obtain

$$\pi_k = \frac{n_k}{\sum_{j=1}^{K} n_j}.$$

★ Box 12.3 EM algorithm for Gaussian mixtures

A K-component Gaussian mixture model is defined by the density

$$f_{\boldsymbol{X}}(\boldsymbol{x}; \boldsymbol{\theta}) = \sum_{k=1}^{K} \pi_k \mathcal{N}(\boldsymbol{x} | \boldsymbol{\mu}_k, \underline{\boldsymbol{\Sigma}}_k),$$

where $\boldsymbol{\theta}$ is a parameter vector containing $\{\pi_k, \boldsymbol{\mu}_k, \underline{\boldsymbol{\Sigma}}_k\}_{k=1}^{K}$. Given a dataset $D = \{\boldsymbol{x}_i\}_{i=1}^{n}$, the log-likelihood is $\ell(\boldsymbol{\theta}) = \sum_{i=1}^{n} \log f_{\boldsymbol{X}}(\boldsymbol{x}_i; \boldsymbol{\theta})$. The procedure below generates a sequence of parameter vector estimates $\boldsymbol{\theta}_0, \boldsymbol{\theta}_1, \ldots$ that is non-decreasing in likelihood and usually converges to a local maximum of the likelihood function.

Initialise $\boldsymbol{\theta}_0$ and repeat the following two steps until $|\ell(\boldsymbol{\theta}_{t+1}) - \ell(\boldsymbol{\theta}_t)| < \epsilon$, where $\epsilon > 0$ is a small tolerance parameter.

(1) **E-step**. Recalculate the responsibilities:

$$r_{ik} \leftarrow \frac{\pi_k \mathcal{N}(\boldsymbol{x}_i|\boldsymbol{\mu}_k, \underline{\boldsymbol{\Sigma}}_k)}{\sum_{j=1}^K \pi_j \mathcal{N}(\boldsymbol{x}_i|\boldsymbol{\mu}_j, \underline{\boldsymbol{\Sigma}}_j)},$$

$$n_k \leftarrow \sum_{i=1}^n r_{ik}.$$

(2) **M-step**. Recalculate parameters (see Box 12.2):

$$\boldsymbol{\mu}_k \leftarrow \frac{1}{n_k} \sum_{i=1}^n r_{ik} \boldsymbol{x}_i,$$

$$\underline{\boldsymbol{\Sigma}}_k \leftarrow \frac{1}{n_k} \sum_{i=1}^n r_{ik}(t)(\boldsymbol{x}_i - \boldsymbol{\mu}_k)(\boldsymbol{x}_i - \boldsymbol{\mu}_k)^T,$$

$$\pi_k \leftarrow \frac{n_k}{\sum_{j=1}^K n_j}.$$

To avoid components collapsing onto datapoints, we set a minimum component covariance determinant designed to detect excessive concentration. A component that meets this condition can be replaced with a randomly located component with high variance. Component collapse may also be dealt with by placing a Bayesian prior on the parameters. Details of this approach may be found in Bishop (2006).

When introducing the Gaussian mixture model, we noted that the posterior distribution of the latent variables provides a soft version of the cluster assignments generated by K-means. In particular, the vector of responsibilities $\boldsymbol{r}_i = (r_{i1}, \ldots, r_{iK})^T$ of the components for datapoint $\boldsymbol{x}_i$ is the soft version of the K-means one-hot vector z_i. Both $\boldsymbol{r}_i$ and z_i may be understood as probability mass functions over component or cluster membership. Having derived the EM algorithm for Gaussian mixture models, we are now able to explain why the steps of the K-means algorithm (Box 12.1) were also labelled 'E' and 'M'. In both cases, the E-step computes assignments (soft for mixtures, hard for clustering) and the M-step updates cluster means (maximising likelihood for mixtures and minimising inertia for clustering). In fact, K-means may be understood as a limiting case of the EM algorithm. If we assume all clusters are spherical with fixed variance σ, then in the limit $\sigma \longrightarrow 0$ the responsibility r_{ik} will equal 1 if k is the nearest component to $\boldsymbol{x}_i$ and 0 otherwise – exactly the definition of z_{ik} in K-means. Component means and cluster assignments will therefore match in the two algorithms. We can exploit this connection by using the cluster assignments output by K-means to initialise the responsibilities in the mixture model, from which the initial values of the parameters $\{\pi_k, \boldsymbol{\mu}_k, \underline{\boldsymbol{\Sigma}}_k\}_{k=1}^K$ can be calculated.

12.3 Dimensionality Reduction

Our final example of unsupervised learning is *dimensionality reduction*. Many interesting datasets are high dimensional. For example, black and white digital images may contain hundreds or thousands of individual pixels, stored as numerical 'whiteness' values. If these values lie in $[0, 1]$, then a single image containing p pixels can be represented as a vector $\boldsymbol{x}_i$ belonging to the p-dimensional unit hypercube $[0, 1]^p \subset \mathbb{R}^p$. Figure 12.10 shows four

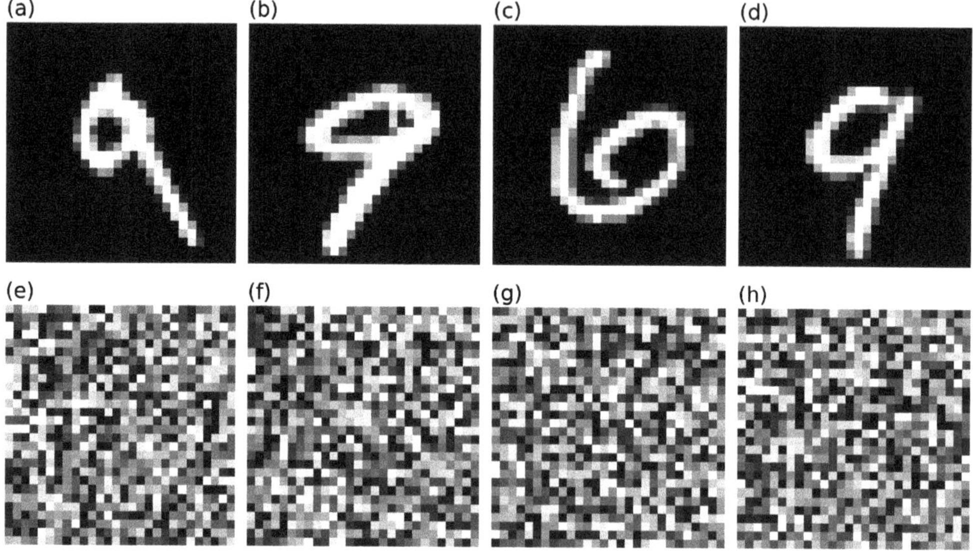

Figure 12.10 Panels (a)–(d) show $28 \times 28 = 784$ pixel images of handwritten digits. The pixels in panels (e)–(h) are i.i.d. Uniform$(0, 1)$.

28×28 pixel images of handwritten squiggles (digits in this example), each corresponding to a point in a 784-dimensional hypercube. Also shown are four images generated by sampling a single point uniformly at random from the same hypercube. The chance of creating anything resembling a handwritten squiggle via this process is vanishingly small. That is, the fraction of all possible pixel configurations which depict squiggles is tiny.

We can think of a squiggle as a realisation of a random vector $\boldsymbol{X} = (X_1, \ldots, X_p)$. The density $f_{\boldsymbol{X}}$ is a mathematical representation of the squiggle-generating process. The tiny region of the hypercube in which $f_{\boldsymbol{X}}$ is not zero we will call 'squiggle space', S.

$$S = \{\boldsymbol{x} \in [0, 1]^p | f_{\boldsymbol{X}}(\boldsymbol{x}) > 0\}.$$

Squiggle space is a *subset* of the hypercube. That is, $S \subset [0, 1]^p$. In order for realisations of $\boldsymbol{X}$ to look like squiggles, we require a complex set of dependencies between its elements. Without such dependencies, realisations of $\boldsymbol{X}$ would look like meaningless noise, as in Figure 12.10e–h. Moving away from any particular point in squiggle space will produce a distortion of the squiggle, with different directions producing different kinds of distortion. The dependencies between pixel values restrict the range of directions which keep us in S; most directions will simply add random noise to the image, taking us out of S. Because of this, it should be possible to describe S, approximately at least, using a coordinate system of *lower dimension* than the cube itself. Finding a way to do this is called *dimensionality reduction.*

12.3.1 Loss Function for Dimensionality Reduction

Let us formulate the dimensionality reduction problem in general terms. Our starting point is a dataset of points in $\mathbb{R}^p$,

$$D = \{\boldsymbol{x}_i\}_{i=1}^n.$$

We wish to find a lower dimensional coordinate system which is sufficient to accurately describe these points. To do this, we seek a mapping of each *observable* point $\boldsymbol{x} \in \mathbb{R}^p$ to a point z in an M-dimensional space (typically $\mathbb{R}^M$) where $M < p$. We may write this mapping as a function,

$$z = \text{encode}(\boldsymbol{x}; \boldsymbol{\theta}).$$

The components of z represent *latent* features which can be used to describe the observations in D in simple terms. Here we use the term 'latent' simply to refer to the fact that we do not directly observe z values. We call the function which maps from $\boldsymbol{x}$ to z an *encoder* because z encodes $\boldsymbol{x}$ in a shortened form. Given any point z' in latent space, we find the corresponding $\boldsymbol{x}'$ in real space using a *decoder* function,

$$\boldsymbol{x}' = \text{decode}(z'; \boldsymbol{\theta}).$$

The subset of $\mathbb{R}^p$ described by our latent vectors is the set of points that can be reached by decoding a latent point,

$$S = \left\{\text{decode}(z; \boldsymbol{\theta}) \middle| z \in \mathbb{R}^M\right\}.$$

Encoding a point and then decoding its latent representation produces a *reconstruction* of the original point,

$$\tilde{\boldsymbol{x}} = \text{decode}(\text{encode}(\boldsymbol{x}; \boldsymbol{\theta}); \boldsymbol{\theta}).$$

If a latent representation does not provide a complete description of the point it represents, it will not be possible to perfectly recover that point. We will see below that this may be understood as a consequence of datapoints lying *close* to S, but not inside it. We measure the performance of an encoder–decoder pair by computing a *reconstruction loss*. A common choice is the mean squared deviation of the reconstructions from the original points,

$$J(\boldsymbol{\theta}) = \frac{1}{n} \sum_{i=1}^{n} \|\boldsymbol{x}_i - \tilde{\boldsymbol{x}}_i\|^2. \tag{12.16}$$

We can learn the parameters of an encoder–decoder pair by minimising this loss.

12.3.2 Principal Component Analysis

We now consider the important case where the encoder and decoder are *linear transformations* between $\mathbb{R}^p$ and $\mathbb{R}^M$ (see Appendix B). In order to derive these transformations, we will need to *centre* our data. We assume that our raw (uncentred) data is a set of observations of a random vector $\boldsymbol{X} = (X_1, \ldots, X_p)^T$,

$$D^* = \{\boldsymbol{x}_i^*\}_{i=1}^n.$$

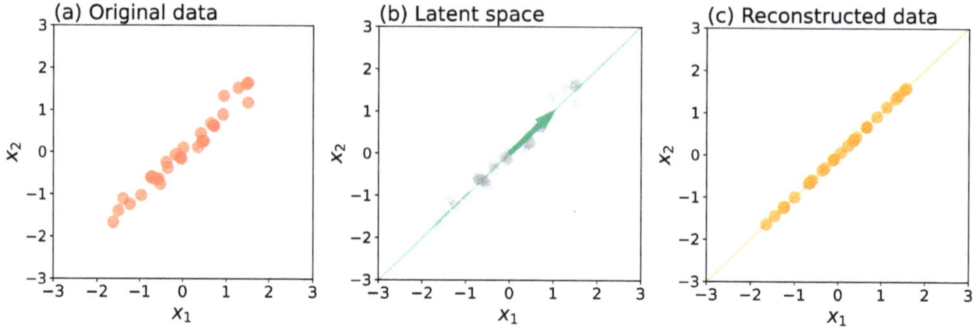

Figure 12.11 (a) A two-dimensional centred dataset. (b) A possible choice of one-dimensional latent space. (c) Reconstruction of the original data after encoding and decoding.

Here, we have used * to denote uncentred data, which has sample mean $\boldsymbol{\mu} = \frac{1}{n}\sum_{i=1}^{n} \boldsymbol{x}_i^*$. We centre the data by subtracting $\boldsymbol{\mu}$ from every point. The ith centred observation is then

$$\boldsymbol{x}_i = \boldsymbol{x}_i^* - \boldsymbol{\mu}.$$

Acting on the centred data, our linear encoder and decoder will take the form

$$z = \text{encode}(\boldsymbol{x}) = \underline{\boldsymbol{A}}_{\text{e}}\boldsymbol{x},$$
$$\tilde{\boldsymbol{x}} = \text{decode}(z) = \underline{\boldsymbol{A}}_{\text{d}}z,$$

where $\underline{\boldsymbol{A}}_{\text{e}}$ is an $M \times p$ matrix which defines the encoder, and $\underline{\boldsymbol{A}}_{\text{d}}$ is a $p \times M$ matrix which defines the decoder. We will discover that the optimal linear encoder and decoder are closely related to the empirical covariance matrix of the data (Box 9.5), defined elementwise by

$$\left[\underline{\boldsymbol{C}}\right]_{ij} = \frac{1}{n}\sum_{k=1}^{n}(x_{ki}^* - \mu_i)(x_{kj}^* - \mu_j) = \frac{1}{n}\sum_{k=1}^{n} x_{ki}x_{kj},$$

where $\boldsymbol{x}_k = (x_{k1}, \ldots, x_{kp})$ and $\boldsymbol{\mu} = (\mu_1, \ldots, \mu_p)$. More compactly,

$$\underline{\boldsymbol{C}} = \frac{1}{n}\underline{\boldsymbol{x}}^T\underline{\boldsymbol{x}}, \tag{12.17}$$

where $\underline{\boldsymbol{x}} = (\boldsymbol{x}_1, \ldots, \boldsymbol{x}_n)^T$ is the *centred* design matrix.

■ ***Exercise*** 12.5 Show that equation (12.17) is equivalent to the elementwise definition.

The estimated variance of the kth component, X_k, of $\boldsymbol{X}$ is the kth diagonal element of $\underline{\boldsymbol{C}}$,

$$\hat{\sigma}_k^2 = \left[\underline{\boldsymbol{C}}\right]_{kk} = \frac{1}{n}\sum_{i=1}^{n} x_{ik}^2,$$

and the *total variance* of the data is the sum of these individual variances,

$$\sum_{k=1}^{p}\hat{\sigma}_k^2 = \frac{1}{n}\sum_{i=1}^{n}\sum_{k=1}^{p} x_{ik}^2 = \frac{1}{n}\sum_{i=1}^{n}\|\boldsymbol{x}_i\|^2.$$

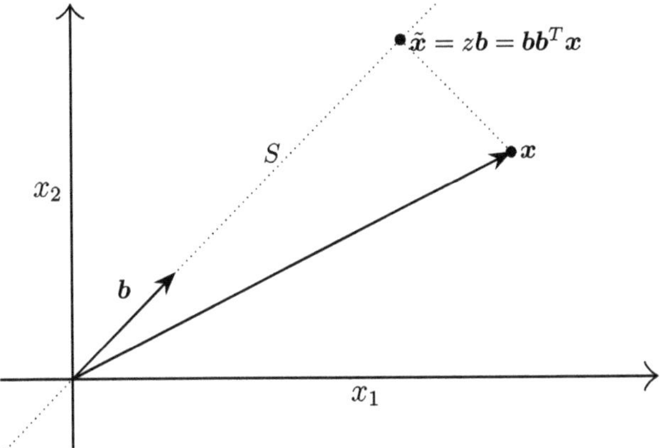

Figure 12.12 Orthogonal projection of the point $\boldsymbol{x}$ onto the one-dimensional subspace spanned by the unit vector $\boldsymbol{b}$.

The total variance of the data, and the total variance of its latent representation, will play central roles in determining the optimal encoder–decoder pair.

We begin with an illustration of linear encoding and decoding in action, using a simple example. Consider the set of two-dimensional observations shown in Figure 12.11a. The observations lie roughly along the thin green line shown in Figure 12.11b, which passes through $\mathbf{0}$. This line is a subspace, S, of $\mathbb{R}^2$. Letting $\boldsymbol{b}$ be the unit vector in the direction of the line (the green arrow in Figure 12.11b), we can write S as the set of all possible scalar multiples of this vector (its *span*),

$$S = \{z\boldsymbol{b} | z \in \mathbb{R}\}.$$

The unit vector $\boldsymbol{b}$ is a *basis vector* for S, allowing us to write any point $\boldsymbol{x}' \in S$ as $\boldsymbol{x}' = z'\boldsymbol{b}$. The number z' is said to be the coordinate of $\boldsymbol{x}'$ with respect to the basis vector. We can *approximately* encode any $\boldsymbol{x} \in \mathbb{R}^2$ using the coordinate of the point in S which is closest to $\boldsymbol{x}$. This point is the *orthogonal projection* of $\boldsymbol{x}$ onto the line, given by $\tilde{\boldsymbol{x}} = z\boldsymbol{b}$, where

$$z = \boldsymbol{b}^T\boldsymbol{x} = \text{encode}(\boldsymbol{x}).$$

Figure 12.12 shows the datapoint, $\boldsymbol{x}$, the subspace, S, and the projection of the point onto the subspace. The projection is *orthogonal* because the vector connecting $\boldsymbol{x}$ to its projection is orthogonal (perpendicular) to S. That is,

$$\boldsymbol{b}^T(\boldsymbol{x} - \tilde{\boldsymbol{x}}) = 0.$$

The number z is our latent representation of $\boldsymbol{x}$. The mapping from latent to observation space $z \mapsto z\boldsymbol{b}$ may be understood as the action of a *decoder* which *reconstructs* $\boldsymbol{x}$ in approximate form. The reconstruction is just the point on the line that the original point was projected to,

$$\tilde{\boldsymbol{x}} = z\boldsymbol{b} = \text{decode}(z).$$

We can view $\tilde{\boldsymbol{x}}$ as the result of encoding and then decoding $\boldsymbol{x}$:

$$\tilde{\boldsymbol{x}} = \text{decode}(\text{encode}(\boldsymbol{x})) = \boldsymbol{b}\boldsymbol{b}^T\boldsymbol{x}.$$

Here, $\boldsymbol{b}\boldsymbol{b}^T$ represents a (2×2) *projection matrix* which projects $\boldsymbol{x}$ to S. Figure 12.11c shows the result of encoding and decoding every point in the dataset. Since the original points do not lie exactly in S, our reconstructions are imperfect. However, the closer our dataset is to S, the better our reconstructions will be.

■ ***Exercise*** 12.6 Let $\boldsymbol{b} = \left(\frac{4}{5}, \frac{3}{5}\right)^T$, and $\boldsymbol{x} = (1, 2)^T$.

(a) Draw the subspace $S = \{z\boldsymbol{b} | z \in \mathbb{R}\}$ on a pair of axes.
(b) Find the projection matrix $\underline{\boldsymbol{P}} = \boldsymbol{b}\boldsymbol{b}^T$.
(c) Find the projection of $\boldsymbol{x}$ onto S, given by $\tilde{\boldsymbol{x}} = \underline{\boldsymbol{P}}\boldsymbol{x}$, and plot $\boldsymbol{x}$ and $\tilde{\boldsymbol{x}}$ on your axes.

The above procedure is a simple example of a more general strategy which will allow us to map p-dimensional data to an M-dimensional representation and back again using a linear encoder–decoder pair. We will derive a procedure for doing this – *principal component analysis* (PCA) – which minimises reconstruction loss. As with our introductory example, we will find that the loss can be minimised using *orthogonal projection*. In the general case, this projection is to an M-dimensional subspace. This concept is summarised in Box 12.4. Appendix B contains a brief review of the essential linear algebra which underpins our derivation of the PCA algorithm.

★ Box 12.4 Orthogonal projection

Suppose we have a set of M linearly independent vectors $\{\boldsymbol{v}_1, \ldots, \boldsymbol{v}_M\}$ in $\mathbb{R}^p$, where $M \leq p$. The set of all points which can be expressed as linear combinations of these vectors is a subspace S of $\mathbb{R}^p$ called their *span*,

$$S = \operatorname{span}(\{\boldsymbol{v}_1, \ldots, \boldsymbol{v}_M\}) = \left\{ \sum_{i=1}^{M} \alpha_i \boldsymbol{v}_i \,\middle|\, (\alpha_1, \ldots, \alpha_M) \in \mathbb{R}^M \right\}.$$

The vectors are said to form a *basis* for S. Any point $\boldsymbol{y} \in S$ may be written in the form

$$\boldsymbol{y} = \sum_{i=1}^{M} \alpha_i \boldsymbol{v}_i,$$

where $\alpha_1, \ldots, \alpha_M$ are the *coordinates* of $\boldsymbol{y}$ with respect to the basis $\boldsymbol{v}_1, \ldots, \boldsymbol{v}_M$. The *orthogonal projection* of any vector $\boldsymbol{x} \in \mathbb{R}^p$ to S is the point $\tilde{\boldsymbol{x}}$ in S which is *closest* to $\boldsymbol{x}$ in the sense that $\|\boldsymbol{x} - \tilde{\boldsymbol{x}}\|$ is as small as possible. Given a subspace S, there exists a projection matrix $\underline{\boldsymbol{P}}$ such that for any $\boldsymbol{x} \in \mathbb{R}^p$ its orthogonal projection to S is $\tilde{\boldsymbol{x}} = \underline{\boldsymbol{P}}\boldsymbol{x}$. The fact that the vector connecting $\boldsymbol{x}$ to $\tilde{\boldsymbol{x}}$ represents the shortest route from $\boldsymbol{x}$ to S means that it will be orthogonal to every vector in S, or equivalently,

$$\boldsymbol{v}_i^T(\boldsymbol{x} - \underline{\boldsymbol{P}}\boldsymbol{x}) = 0$$

for all $i \in \{1, \ldots, M\}$. We say that the vector $\boldsymbol{x} - \underline{\boldsymbol{P}}\boldsymbol{x}$ is orthogonal to S. If $\underline{\boldsymbol{V}}$ is the matrix with columns given by the basis vectors, and if that basis is *orthonormal*, meaning that $\boldsymbol{v}_i^T \boldsymbol{v}_j = \delta_{ij}$ for all $i, j \in \{1, \ldots, M\}$, then $\underline{\boldsymbol{P}} = \underline{\boldsymbol{V}}\,\underline{\boldsymbol{V}}^T$.

We will derive the PCA algorithm by finding a subspace and a mapping to that subspace which minimise the reconstruction loss. Let us start by considering an arbitrary M-dimensional subspace, S, spanned by linearly independent vectors $\boldsymbol{b}_1, \ldots, \boldsymbol{b}_M$ in $\mathbb{R}^p$, so

$$S = \operatorname{span}(\{\boldsymbol{b}_1, \ldots, \boldsymbol{b}_M\}) = \left\{ \sum_{m=1}^{M} z_m \boldsymbol{b}_m \middle| (z_1, \ldots, z_M) \in \mathbb{R}^M \right\}.$$

Without loss of generality, we can assume the vectors form an *orthonormal basis* (ONB).[5] That is, they are mutually orthogonal and of unit length, so $\boldsymbol{b}_i^T \boldsymbol{b}_j = \delta_{ij}$. Suppose we have a centred dataset of points in $\mathbb{R}^p$, $D = \{\boldsymbol{x}_i\}_{i=1}^n$. We can approximate any point $\boldsymbol{x}_i \in \mathbb{R}^p$ with a point $\tilde{\boldsymbol{x}}_i \in S$ using the basis expansion

$$\tilde{\boldsymbol{x}}_i = \sum_{m=1}^{M} z_{im} \boldsymbol{b}_m.$$

For now, we won't assume that $\tilde{\boldsymbol{x}}$ is an orthogonal projection to S. Writing the expansion coefficients as a vector $z_i = (z_{i1}, \ldots, z_{iM})^T$ and letting $\underline{\boldsymbol{B}}$ be the $p \times M$ matrix with columns equal to the basis vectors, we can write the basis expansion in the compact form:

$$\tilde{\boldsymbol{x}}_i = \underline{\boldsymbol{B}} z_i.$$

The vector $z_i \in \mathbb{R}^M$ is an M-dimensional latent encoding of $\boldsymbol{x}$, so we view it as the output of an encoder function,

$$z_i = \operatorname{encode}(\boldsymbol{x}_i; \underline{\boldsymbol{B}}).$$

The point $\tilde{\boldsymbol{x}}_i$ may be viewed as the output of a decoder which maps from $\mathbb{R}^M$ to $\mathbb{R}^p$,

$$\tilde{\boldsymbol{x}}_i = \operatorname{decode}(z_i; \underline{\boldsymbol{B}}) = \underline{\boldsymbol{B}} z_i.$$

Since $\tilde{\boldsymbol{x}}_i$ can be interpreted as the result of encoding and then decoding $\boldsymbol{x}_i$, we call it a *reconstruction* of $\boldsymbol{x}_i$.

As it stands the subspace S and the expansion coefficients z_i are arbitrary. We now show how to choose them so as to minimise the overall reconstruction loss (12.16). We first find the optimal latent coordinates for a given basis. The most direct route is to use the properties of orthogonal projections given in Box 12.4. We observe that $\|\boldsymbol{x}_i - \tilde{\boldsymbol{x}}_i\|$ is minimised if $\tilde{\boldsymbol{x}}_i$ is the orthogonal projection of $\boldsymbol{x}_i$ to S, given by $\tilde{\boldsymbol{x}}_i = \underline{\boldsymbol{B}}\,\underline{\boldsymbol{B}}^T \boldsymbol{x}_i$. Since $\tilde{\boldsymbol{x}}_i = \underline{\boldsymbol{B}} z_i$ then the optimal latent coordinates are

$$z_i = \underline{\boldsymbol{B}}^T \boldsymbol{x}_i = \operatorname{encode}(\boldsymbol{x}_i; \underline{\boldsymbol{B}}).$$

Figure 12.13 shows the projection process. It remains to find the optimal subspace.

■ ***Exercise*** 12.7 Let $\tilde{\boldsymbol{x}} = \underline{\boldsymbol{B}}\,\underline{\boldsymbol{B}}^T \boldsymbol{x}$ where $\underline{\boldsymbol{B}}$ is a $p \times M$ matrix with columns given by the basis vectors of an ONB.

(a) Show that $\underline{\boldsymbol{B}}^T \underline{\boldsymbol{B}} = \underline{\boldsymbol{I}}$, where $\underline{\boldsymbol{I}}$ is the $M \times M$ identity matrix.
(b) Show that $\tilde{\boldsymbol{x}}^T (\boldsymbol{x} - \tilde{\boldsymbol{x}}) = \boldsymbol{x}^T \underline{\boldsymbol{B}}\,\underline{\boldsymbol{B}}^T (\boldsymbol{x} - \underline{\boldsymbol{B}}\,\underline{\boldsymbol{B}}^T \boldsymbol{x}) = \boldsymbol{0}$.

[5] Given a linearly independent set of vectors, we can always find an orthonormal set with the same span.

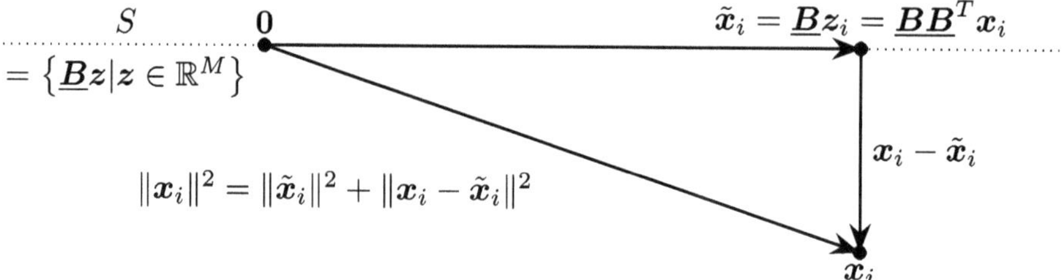

Figure 12.13 The vector $\tilde{x}_i$ is the orthogonal projection of x_i onto S. The vectors x_i, $\tilde{x}_i$ and $x_i - \tilde{x}_i$ form a right-angle triangle in $\mathbb{R}^p$. Note: S is represented as a line to aid visualisation.

The fact that the projection $x_i \mapsto \tilde{x}_i$ is orthogonal implies that the vectors $x_i, \tilde{x}_i$ and $x_i - \tilde{x}_i$ form three sides of a right-angle triangle (Figure 12.13), so by Pythagoras' theorem

$$\|x_i - \tilde{x}_i\|^2 = \|x_i\|^2 - \|\tilde{x}_i\|^2. \tag{12.18}$$

This identity may also be derived algebraically (Exercise 12.8) using $\tilde{x}_i^T(x_i - \tilde{x}_i) = 0$.

■ ***Exercise*** 12.8 The Pythagorean identity (12.18) is a special case of the general identity

$$\|u - v\|^2 = \|u\|^2 - \|v\|^2 + 2v^T(v - u). \tag{†}$$

Establish identity (†) and hence derive (12.18) by setting $u = x_i$ and $v = \tilde{x}_i$, where $\tilde{x}_i$ is the orthogonal projection of x_i onto a subspace S.

■ ***Exercise*** 12.9 If $\underline{B}$ is a matrix with columns given by ONB vectors, and $\tilde{x} = \underline{B}z$, show that $\|\tilde{x}\|^2 = \|z\|^2$.

Using the Pythagorean identity (12.18), we can write the reconstruction loss as

$$\begin{aligned}\frac{1}{n}\sum_{i=1}^{n}\|x_i - \tilde{x}_i\|^2 &= \frac{1}{n}\sum_{i=1}^{n}\|x_i\|^2 - \frac{1}{n}\sum_{i=1}^{n}\|\tilde{x}_i\|^2\\ &= \text{const.} - \frac{1}{n}\sum_{i=1}^{n}\|z_i\|^2.\end{aligned}$$

Here we see that the loss is the difference between the total variance of the data (a constant) and the total variance of the reconstructions. The latter is equal to the total variance of the latent coordinates (see Exercise 12.9). The optimal subspace should therefore be chosen so that the total latent variance is maximised. Intuitively, maximum variation in the latent features can be achieved by choosing a subspace spanned by basis vectors which are aligned with the directions along which we see the greatest variation in the data. Let us formalise this intuition.

Our first step is to recognise that the total variance of the latent vectors may be written in terms of the empirical covariance matrix as follows:

$$\frac{1}{n}\sum_{i=1}^{n}\|z_i\|^2 = \frac{1}{n}\sum_{i=1}\sum_{m=1}^{M}(\boldsymbol{b}_m^T\boldsymbol{x}_i)^2 = \frac{1}{n}\sum_{i=1}\sum_{m=1}^{M}\boldsymbol{b}_m^T\boldsymbol{x}_i\boldsymbol{x}_i^T\boldsymbol{b}_m$$

$$= \sum_{m=1}^{M}\boldsymbol{b}_m^T\left(\frac{1}{n}\sum_{i=1}\boldsymbol{x}_i\boldsymbol{x}_i^T\right)\boldsymbol{b}_m = \sum_{m=1}^{M}\boldsymbol{b}_m^T\underline{\boldsymbol{C}}\boldsymbol{b}_m.$$

Since $\underline{\boldsymbol{C}}$ is a real, symmetric, positive semi-definite $p \times p$ matrix it has p orthogonal eigenvectors which may be normalised to form an ONB $\{\boldsymbol{u}_1, \ldots, \boldsymbol{u}_p\}$ with corresponding non-negative eigenvalues $\{\lambda_1, \ldots, \lambda_p\}$. Letting $\underline{\boldsymbol{U}}$ be the matrix with columns equal to the eigenvectors and $\underline{\boldsymbol{\Lambda}}$ be the diagonal matrix with diagonal elements given by the corresponding eigenvalues, $\underline{\boldsymbol{C}}$ has eigen decomposition (see Appendix B)

$$\underline{\boldsymbol{C}} = \underline{\boldsymbol{U}}\underline{\boldsymbol{\Lambda}}\underline{\boldsymbol{U}}^T.$$

Since the eigenvectors form a basis for $\mathbb{R}^p$, we can write the basis vectors of S in terms of them. That is,

$$\boldsymbol{b}_m = \sum_{k=1}^{p}\pi_{mk}\boldsymbol{u}_k,$$

where $\sum_{k=1}^{p}\pi_{mk}^2 = 1$ since $\|\boldsymbol{b}_m\| = 1$. Using this representation, we can write the total variance of the latent vectors as

$$\sum_{m=1}^{M}\boldsymbol{b}_m^T\underline{\boldsymbol{C}}\boldsymbol{b}_m = \sum_{m=1}^{M}\boldsymbol{b}_m^T\underline{\boldsymbol{U}}\underline{\boldsymbol{\Lambda}}\underline{\boldsymbol{U}}^T\boldsymbol{b}_m = \sum_{m=1}^{M}\sum_{k=1}^{p}\pi_{mk}^2\lambda_k.$$

We will assume, without loss of generality, that the eigenvalues are ordered by size so $\lambda_{k+1} \leq \lambda_k$ for all $k \in \{1, \ldots, p-1\}$. Suppose we want to find the best one-dimensional subspace (the best one-dimensional latent representation). In that case, we wish to maximise

$$\boldsymbol{b}_1^T\underline{\boldsymbol{C}}\boldsymbol{b}_1 = \sum_{k=1}^{p}\pi_{1k}^2\lambda_k = \pi_{11}^2\lambda_1 + \cdots + \pi_{1p}^2\lambda_p.$$

Since λ_1 is the largest eigenvalue, we should set $\pi_{1k} = \delta_{1k}$ so $\boldsymbol{b}_1 = \boldsymbol{u}_1$ and

$$\boldsymbol{b}_1^T\underline{\boldsymbol{C}}\boldsymbol{b}_1 = \lambda_1.$$

Now suppose we wish to increase the dimension of S to $M = 2$ by adding a second basis vector $\boldsymbol{b}_2$, orthogonal to $\boldsymbol{b}_1$ so $\pi_{21} = 0$. The additional contribution to the total variance due to $\boldsymbol{b}_2$ will be

$$\boldsymbol{b}_2^T\underline{\boldsymbol{C}}\boldsymbol{b}_2 = \sum_{k=2}^{p}\pi_{2k}^2\lambda_k = \pi_{22}^2\lambda_2 + \cdots + \pi_{2p}^2\lambda_p.$$

Since λ_2 is the largest remaining eigenvalue, we should set $\pi_{2k} = \delta_{2k}$ so $\boldsymbol{b}_2 = \boldsymbol{u}_2$ and

$$\boldsymbol{b}_2^T\underline{\boldsymbol{C}}\boldsymbol{b}_2 = \lambda_2.$$

Continuing this line of reasoning, we find that the optimal M-dimensional subspace is spanned by the first M eigenvectors – or *principal components* – of $\boldsymbol{C}$, and that the total variance of the latent features in this case is

$$\frac{1}{n}\sum_{i=1}^{n}\|z_i\|^2 = \sum_{m=1}^{M}\lambda_m.$$

Orthogonal projection of the centred data to the principal components of $\underline{\boldsymbol{C}}$ is *principal component analysis* (PCA). We can generalise PCA to the case where the data has not been centred by applying our encoder to $\boldsymbol{x} - \boldsymbol{\mu}$, so $z = \text{encode}(\boldsymbol{x} - \boldsymbol{\mu})$, and by adding $\boldsymbol{\mu}$ to the output of decoder, so $\tilde{\boldsymbol{x}} = \text{decode}(z) + \boldsymbol{\mu}$. We measure the information retained by our lower-dimensional encoding as the *proportion of variance explained*, $\sum_{m=1}^{M}\lambda_m / \sum_{k=1}^{p}\lambda_k$. The complete algorithm is summarised in Box 12.5.

★ Box 12.5 PCA algorithm

Suppose we have a dataset $D = \{\boldsymbol{x}_i\}_{i=1}^{n}$ with each $\boldsymbol{x}_i \in \mathbb{R}^p$. The sample mean vector and empirical covariance matrix are

$$\boldsymbol{\mu} = \frac{1}{n}\sum_{i=1}^{n}\boldsymbol{x}_i, \quad [\underline{\boldsymbol{C}}]_{ij} = \frac{1}{n}\sum_{k=1}^{n}(x_{ki} - \mu_i)(x_{kj} - \mu_j).$$

Let $\lambda_1 \geq \lambda_2 \geq \cdots \geq \lambda_p$ be the eigenvalues of $\underline{\boldsymbol{C}}$ with corresponding normalised eigenvectors $\{\boldsymbol{b}_1, \ldots, \boldsymbol{b}_p\}$ known as *principal components*. Let $\underline{\boldsymbol{B}}$ be the $p \times M$ matrix with columns consisting of the first M principal components

$$\underline{\boldsymbol{B}} = \begin{pmatrix} | & | & \cdots & | \\ \boldsymbol{b}_1 & \boldsymbol{b}_2 & \cdots & \boldsymbol{b}_M \\ | & | & \cdots & | \end{pmatrix}.$$

We define

$$z = \text{encode}(\boldsymbol{x}) = \underline{\boldsymbol{B}}^T(\boldsymbol{x} - \boldsymbol{\mu}),$$
$$\tilde{\boldsymbol{x}} = \text{decode}(z) = \underline{\boldsymbol{B}}z + \boldsymbol{\mu}.$$

The proportion of the total variance of the original data which is retained in the reconstructions is $\sum_{m=1}^{M}\lambda_m / \sum_{k=1}^{p}\lambda_k$. No linear encoder–decoder pair achieves lower mean-squared-deviation reconstruction error.

Figure 12.14 shows the principal components and corresponding eigenvalues for three artificial datasets. In Figure 12.14a, the scatter of datapoints is approximately spherical, meaning that the eigenvalues are of similar size; only about half the variation would be explained by projecting to the first principal component. In Figure 12.14b, the scatter is elongated, with most of the variation occurring in the direction of the first principal component (i.e. the first eigenvalue is much larger than the second). In this case, approximately 90% of the variance would be explained by projecting to the first principal component. The data in Figure 12.14c is also elongated, but with the datapoints lying on a spiral curve. In this case, the proportion of variance explained by projecting to the first principal component is approximately 85%.

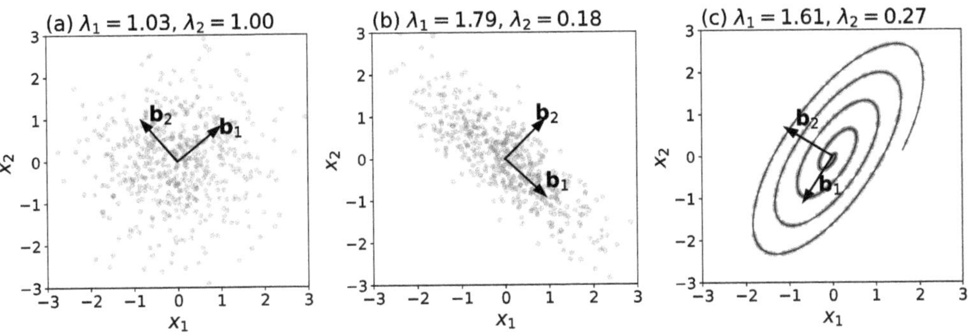

Figure 12.14 (a–c) The principal components of the covariance matrix of three datasets, with eigenvalues given in the plot titles.

However, suppose we had the curve in parameterised form,

$$(x_1, x_2) = (g_1(z), g_2(z)),$$

or more succinctly $\boldsymbol{x} = \boldsymbol{g}(z)$. We could then *perfectly* encode every datapoint $\boldsymbol{x}_i$ as $z_i = \boldsymbol{g}^{-1}(\boldsymbol{x}_i)$. This encoder is a non-linear function of the data. Learning optimal non-linear encoder–decoder pairs is a challenging problem.

12.3.3 PCA Applied to Digits

We will illustrate PCA in action by using it to encode a dataset of 1,967 digital images of the handwritten digits six and nine. Each image consists of 28×28 pixels, each represented as a number between 0 (black) and 1 (white). Four examples are shown in Figures 12.10a–d. The 28 rows of 28 pixels which make up each image may be placed end-to-end into a 784-dimensional vector. Having computed the covariance matrix of these vectors, we can display its principal components by reforming them into 28×28 squares. The first four principal components are shown in Figure 12.15a–d.

Suppose we project to the subspace spanned by the first component, $\boldsymbol{b}_1$. Letting $\boldsymbol{\mu}$ be the mean pixel vector, the reconstruction of image $\boldsymbol{x}_i$ is

$$\tilde{\boldsymbol{x}}_i = z_i \boldsymbol{b}_1 + \boldsymbol{\mu},$$

where $z_i = \boldsymbol{b}_1^T(\boldsymbol{x}_i - \boldsymbol{\mu}) = \text{encode}(\boldsymbol{x}_i)$. Figure 12.16a shows the distribution of z_i values for all images of the two different digits. Here we see that the latent features for handwritten sixes and nines form two distinct clusters. These clusters are centred at $z \approx \pm 3$. Figures 12.17a and c show the image reconstructions produced by these two z values, which are clearly recognisable as 6 and 9. Figure 12.17b shows the image produced by mid-point latent coordinate ($z = 0$). By projecting to the first principal component, we have produced a single feature which efficiently measures how 'nine-like' or 'six-like' each image is. One application of latent features is as replacements for high-dimensional predictor vectors in regression or classification, with the dimension M serving as a regularising hyperparameter. For example, using logistic regression to classify each digit image as a six or a nine based on its one-dimensional latent representation produces a cross-validated accuracy of 94%.

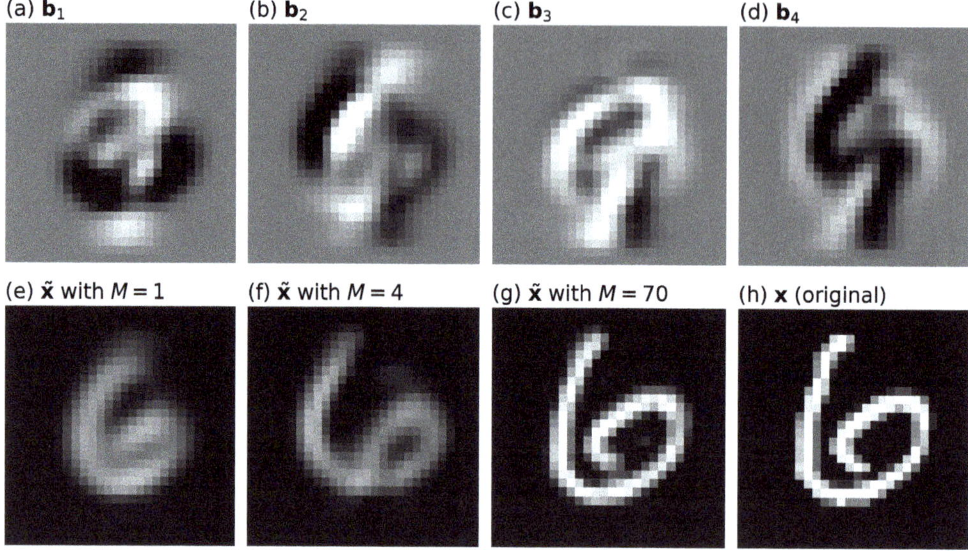

Figure 12.15 Panels (a)–(d) show the first four principal components for the handwritten digits 6 and 9. Panels (e)–(g) show reconstructions of an original image (h) using $M = 1, 4$ and 70 components.

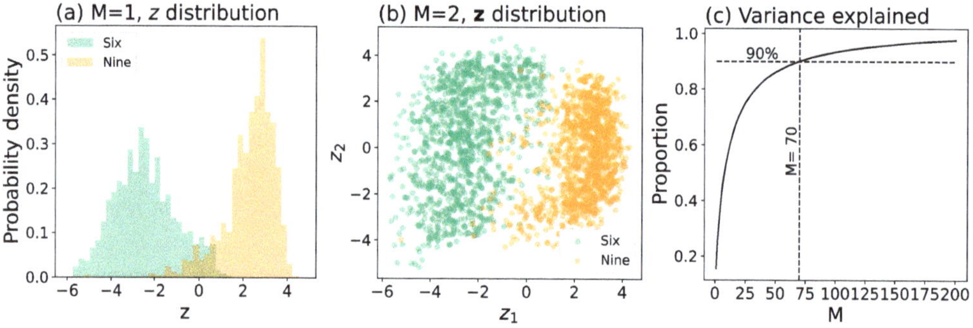

Figure 12.16 Distribution of latent coordinates of sixes and nines when projecting to (a) $M = 1$ dimension and (b) $M = 2$ dimensions. Panel (c) shows relationship between proportion of variance explained and dimension of representation.

Using a two-dimensional representation this jumps to 97.5%, reflecting the fact that the latent clusters in two dimensions, shown in Figure 12.16b, are more clearly separated than in one dimension.

Figure 12.16c shows how the explained variance increases with the number of principal components used. Here we see that the explained variance increases very rapidly at first, because early components encode the main variations in the data. To retain 90% of the variance, we requires $M = 70$ components. Panels (e)–(g) in Figure 12.15 show the recon-

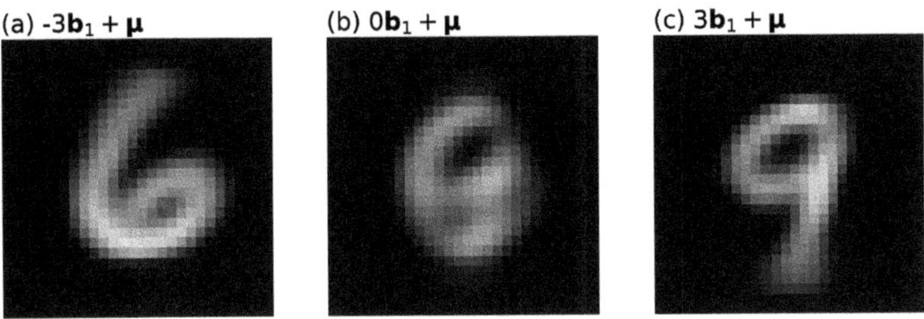

Figure 12.17 (a–c) Image reconstructions using the first principal component.

structions obtained using $M = 1, 4$ and 70 components. The $M = 70$ reconstruction is almost indistinguishable from the original, shown in panel (h).

12.4 Chapter Summary

Unsupervised learning – learning patterns and structure in data without labelled responses – is a very broad category of methods and models. In this chapter, we have seen three important examples: cluster analysis using K-means, density estimation using Gaussian mixture models and dimensionality reduction using principal component analysis.

- **Cluster analysis** seeks to divide data into groups within which all observations are similar to one another.
- The **K-means** algorithm achieves this by iteratively reassigning each observation to the group it was nearest to at the previous iteration. At each step, this reduces the **inertia** of the clustering, or leaves it unchanged.
- **Density estimation** seeks to learn a probabilistic model of a data-generating process based on a finite number of outputs from the process.
- **Gaussian mixture models** are probability densities consisting of a weighted sum of K multivariate normal densities known as **components**.
- The **EM algorithm** for fitting Gaussian mixtures iteratively assigns each component a *responsibility* for each observation. Component locations are then updated to the responsibility-weighted average over all observations.
- The **cross-entropy** of a density $q\colon \mathbb{R}^p \to [0, \infty)$ relative to the density f_X of a random variable $X \in \mathbb{R}^p$ is defined $\mathbb{H}[f_X, q] = \mathbb{E}[\log q(X)]$. The cross-entropy is minimised when the two densities are equal. $\mathbb{H}[\mathfrak{f}_X, \hat{f}]$ measures the performance of $\hat{f}$ as model of $\mathfrak{f}_X$.
- **Dimensionality reduction** seeks a mapping – the **encoder** – from observation space to a lower-dimensional latent space. An approximate inverse map – the **decoder** – **reconstructs** datapoints from their encoded forms.
- **Principal component analysis** (PCA) uses an encoder–decoder pair consisting of **linear transformations** which minimise **reconstruction error**.

12.5 Further Exercises

■ ***Exercise*** 12.10 Consider the following table of six observations in two-dimensional feature space. The ith observation is $\boldsymbol{x}_i = (x_{i1}, x_{i2})$, where x_{ij} is feature $j \in \{1, 2\}$ of observation number $i \in \{1, \ldots, 6\}$.

i	1	2	3	4	5	6
x_{i1}	1	1	0	5	6	4
x_{i2}	4	3	4	1	2	0

(12.19)

(a) Plot the observations on an (x_1, x_2) graph.

(b) Suppose we want to group these observations into $K = 2$ clusters. Starting from the grouping of observations $C_1 = \{1, 3, 4\}$ and $C_2 = \{2, 5, 6\}$, apply the K-means algorithm to determine the optimal assignments.

■ ***Exercise*** 12.11 The standard normal density function is

$$\phi(x) = \mathcal{N}(x|0, 1) = \frac{\exp\left(-\frac{x^2}{2}\right)}{\sqrt{2\pi}}.$$

Consider the two-component mixture $f_X(x; \boldsymbol{\mu}) = (\phi(x - \mu_1) + \phi(x - \mu_2))/2$, where $\boldsymbol{\mu} = (\mu_1, \mu_2)^T$. Use the EM algorithm to fit the parameters of this model (the locations of the densities) to the dataset $D = \{-3, 3\}$ starting from an initial guess $\boldsymbol{\mu}_0 = (-1, 1)^T$. Note: you will not need to update component weights or variances because the model assumes they are fixed.

■ ***Exercise*** 12.12 Use PCA to find the optimal linear encoder–decoder pair to reduce the dimension of the dataset $D = \{(1, 1), (4, 3), (3, 4)\}$ to 1. Plot the original points, the low-dimensional subspace and the reconstructed points.

13

Neural Networks and Deep Learning

Many inference techniques in this book amount to some version of *searching within a specific function family for functions that fit the data well.* So far, we've described such families using standard mathematical notation or shorthand, e.g. 'the family of 3rd-order polynomials'. That works for simple cases, but it becomes unwieldy as complexity grows. In this chapter, we take a different approach: specifying function families by describing neural network architectures.

We begin with very simple networks – single neurons and single layers – and show how familiar methods like linear and logistic regression fit naturally into this framework (Sections 13.2–13.3). We then introduce hidden layers and deep networks (Sections 13.4–13.5); we motivate but do not prove the universal approximation theorem. Deep networks are always trained via gradient-based methods. Roughly speaking, that means starting from a random parameter vector and repeatedly taking small loss-reducing steps in parameter space. Section 13.6 fleshes out that sketch, and Section 13.7 walks through a concrete training workflow. A central theme in deep learning research is the design of architectures that 'bake in' structural assumptions about the problem domain. We explore this in Section 13.8 through the example of convolutional neural networks (CNNs), designed for images, and touch briefly on other architectural ideas (residual blocks, transformers) in Section 13.9. Finally, we present the *backpropagation* algorithm for computing loss gradients, and explain the challenge of vanishing and exploding gradients in deep networks (Sections 13.10–13.11).[1]

13.1 Introduction

Biological neurons, the building blocks of animal nervous systems, are living cells that specialise in information processing. A typical neuron receives many input signals but generates a single output signal. Neural signals usually consist of trains of electrical pulses. Precisely how these trains of pulses carry information varies from context to context. In some cases it seems that the timing of individual pulses is important ('temporal coding'); in other cases it seems that average pulse *rates* are what matters ('rate coding'). The very simplest models of biological neurons adopt the rate-coding perspective, and also assume that a neuron's output (at any given time) is a function of its inputs (at that time).

Artificial neurons are mathematical abstractions loosely inspired by the simplest biological models. They are the building blocks of artificial neural networks – henceforth just 'neural networks' – and from now on, when we speak of neurons we will mean the artificial sort. A

[1] This is a short chapter on a big topic. For more detailed introductions, see Prince (2023) and Bishop and Bishop (2024).

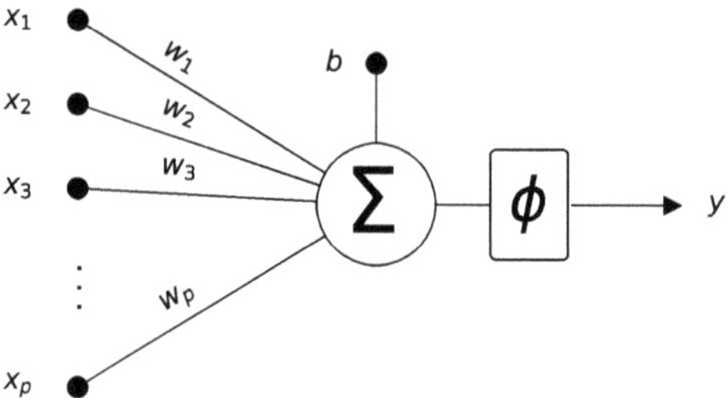

Figure 13.1 A single neuron with inputs $x_1, ..., x_p$ and output y. The neuron computes a weighted sum $\sum_{i=1}^{p} w_i x_i$ of its inputs, adds a bias b, and passes the result through an activation function ϕ.

neuron for us, then, is a mathematical function that maps a vector of inputs $(x_1, ..., x_p)$ to a scalar output y defined by

$$y = \phi\left(b + \sum_{i=1}^{p} w_i x_i\right), \tag{13.1}$$

as shown graphically in Figure 13.1. Parameters $w_1, ..., w_p$ are *weights*, with the ith weight controlling the neuron's sensitivity to its ith input. Parameter b is the *bias*, and ϕ is a fixed function known as the *activation function*. A neuron computes a weighted sum of its inputs, adds a bias, and passes the result of that computation – a single real number known as the *pre-activation*, often denoted by z – through an activation function. The output y is the neuron's *activation*. (So the activation function maps pre-activation to activation.) Notice that this recipe is rather constraining: y is allowed to depend on each x_i only by way of the pre-activation $z = b + \sum_{i=1}^{p} w_i x_i$. Even if we have the freedom to pick any activation function we like, only a narrow class of functions can be computed by a single neuron. But *networks* of neurons are much more flexible.

■ ***Exercise*** 13.1 Give an example of a function from $\mathbb{R}^2$ to $\mathbb{R}$ that cannot be computed by a single neuron.

Neurons can be wired together in arbitrarily complicated ways. We'll say that neuron A is *connected* to neuron B if A's output serves as one of B's inputs. A neuron can be connected to many other neurons.[2] If we want our neural network to implement a function – that is, a deterministic map from input to output – there is one restriction on the wiring diagram: it must be acyclic.[3] So, for example, if neuron A connects to neuron B and neuron B connects to neuron C, then neuron C cannot connect back to neuron A again. Apart from this 'no looping' rule, anything goes. Figure 13.2 shows two admissible (acyclic) neural network structures in panels (a) and (b), and one inadmissible (cyclic) structure in panel (c). From now on we will only be considering acyclic neural networks. These are also known as *feedforward* networks.

[2] Each neuron has a single output in the sense that it outputs (at any given moment) a single scalar value; but there is no restriction on how widely this value may be communicated.

[3] *Recurrent neural networks* break the no looping rule. As a result, they implement not functions but *dynamical systems*. This topic lies beyond the scope of the present chapter.

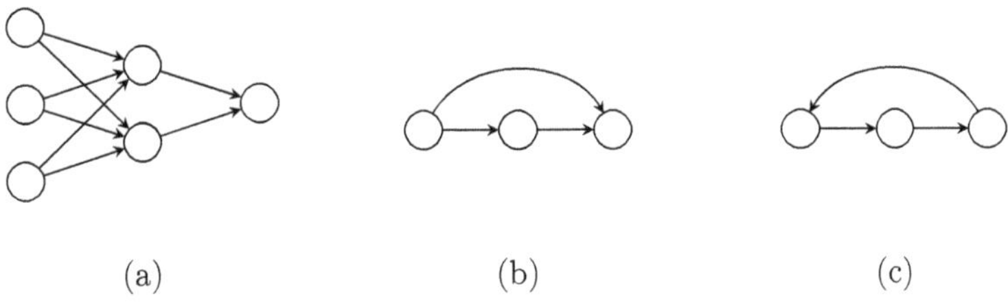
(a) (b) (c)

Figure 13.2 Illustrative neural network wiring diagrams. Examples (a) and (b) are acyclic; example (c) is cyclic. Neural networks that compute functions of their input must have acyclic wiring diagrams.

When drawing them, we will adopt the convention that connections always run from left to right, so in future we won't need to put direction-marking arrows on our wires.

It will be helpful to have some terminology for distinguishing between different kinds of feedforward networks. A *layered* network is one in which the neurons are partitioned into a sequence of layers. A layered network will always have at least one non-neural layer too – its input, usually regarded as the zeroth layer. It may also have some additional non-neural layers performing simple transformations such as normalisation or pooling. (We will encounter pooling layers in Section 13.8.) The word 'unit' covers both neurons and non-neural nodes such as inputs. Thus we can say that in a layered network, for each $k > 0$, units in layer k receive connections only from units in layer $k - 1$. Figure 13.2a is a layered network. Figure 13.2b is feedforward but not layered. A layered network is said to be *fully connected* if for every neural layer, each neuron in the layer receives a connection from *every* unit in the previous layer. Fully connected networks with at least two layers of neurons after the input layer are for historical reasons known as *multilayer perceptrons*. Figure 13.2a is a simple example. We will encounter layered architectures that are not fully connected when we study CNNs in Section 13.8.

Picking suitable activation functions is an important part of neural network design, but in practice a short menu of options suffices. The activation function we will see most often is ReLU (rectified linear unit), defined by

$$\mathrm{ReLU}(x) = \begin{cases} x & \text{if } x \geq 0, \\ 0 & \text{otherwise.} \end{cases}$$

ReLU acts like an identity function on non-negative inputs (passing them through unchanged), but maps negative inputs to zero. It never returns a negative value. Neurons with this activation function typically make up the guts of modern neural networks. They are responsible for computing intermediate signals, that is, signals that get fed to other neurons in the network for further processing. The fact that ReLU returns exactly zero for a wide range of inputs is fine in this context: ON/OFF switching behaviour is quite useful for intermediate processing. ReLU's non-negativity is also acceptable, because intermediate signals get multiplied by (possibly negative) weights en route to downstream neurons. Neurons responsible for computing a neural network's final output, by contrast, tend not to be ReLUs. Common choices of

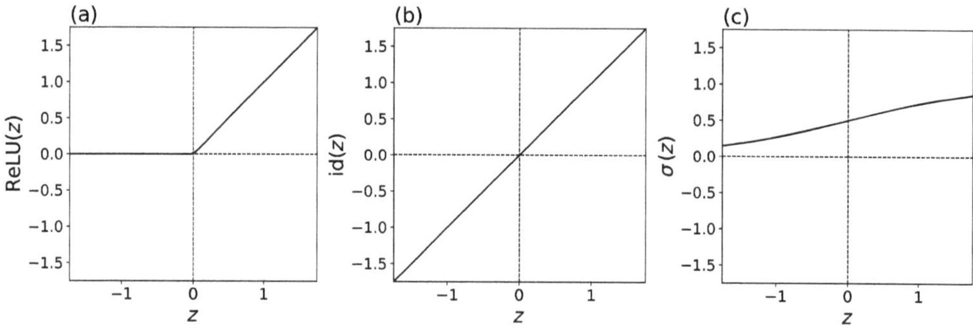

Figure 13.3 (a) ReLU activation, (b) linear activation, (c) sigmoid activation.

activation function for these neurons include identity ('linear activation'),

$$\mathrm{id}(x) = x,$$

and the sigmoid function,

$$\sigma(x) = \frac{1}{1 + e^{-x}}.$$

The sigmoid returns values in the interval $(0, 1)$, so is especially useful if the output is supposed to be a probability. Figure 13.3 plots ReLU, linear and sigmoid activation functions.

13.2 Single-Neuron Networks

The simplest possible neural networks consist of just a single neuron. We'll consider two examples: a single neuron with linear activation and a single neuron with sigmoid activation.

13.2.1 One Linear Neuron: Linear Regression

A neural network consisting of a single linear neuron maps a p-dimensional input vector $\boldsymbol{x} = (x_1, ..., x_p)$ to the scalar output

$$\begin{aligned} y &= b + \sum_{i=1}^{p} w_i x_i \\ &= b + \boldsymbol{w}^T \boldsymbol{x}, \end{aligned}$$

where $\boldsymbol{w} = (w_1, ..., w_p)$ is the vector of weights attached to the neuron's inputs and b is the neuron's bias. If we fix the value of p but leave $\boldsymbol{w}$ and b undefined, then we have pinned down the network's *architecture* but not its parameters. Equivalently, we have specified a particular function family – the family of linear functions from $\mathbb{R}^p$ to $\mathbb{R}$ – without picking out an individual function within that family.[4] To pick out a function, we must pick values for the weights and the bias.

[4] As already noted in Section 11.3.1, 'linear' usually means affine in the machine learning literature.

Suppose we wish to fit a linear prediction function to a dataset $D = \{(\boldsymbol{x}_k, y_k)\}_{k=1}^n$. For each k, vector $\boldsymbol{x}_k$ is a p-dimensional predictor and y_k is a scalar response. We can think of our task as the task of picking weights $\boldsymbol{w}$ and bias b for a single linear neuron. We want the neuron to map each predictor vector in D to a value that lies close to the corresponding response. In other words, we want $b + \boldsymbol{w}^T\boldsymbol{x}_k$ to lie close to y_k, for every k. If we use the mean squared error (MSE) as our loss function, then we will pick $\boldsymbol{w}$ and b to minimise

$$\text{MSE}(b, \boldsymbol{w}; D) = \frac{1}{n}\sum_{k=1}^{n}(y_k - b - \boldsymbol{w}^T\boldsymbol{x}_k)^2. \tag{13.2}$$

This, of course, is just least-squares linear regression.[5] In practice, no one solving a least-squares linear regression problem describes what they are doing as 'training a one-neuron neural network' – but the description would not be inaccurate.

So, can a single neuron 'do' linear regression? Yes and no. As we have just seen, a single linear neuron with p inputs can express any linear function from $\mathbb{R}^p$ to $\mathbb{R}$. But the neuron cannot choose which function to express. *We* define and implement a data-driven procedure for picking one function from the family selected by the neural architecture. In the one-neuron least-squares regression case the procedure is simple, because the right-hand side of equation (13.2) is quadratic in $\boldsymbol{w}$ and b (and is therefore easy to minimise). But finding good weights and biases for larger, more complicated neural networks is often challenging. We will return to this important point in Section 13.6.

13.2.2 One Sigmoid Neuron: Binary Logistic Regression

A neural network consisting of a single sigmoid neuron maps a p-dimensional input vector $\boldsymbol{x}$ to the scalar output

$$\begin{aligned} y &= \sigma(\boldsymbol{w}^T\boldsymbol{x} + b) \\ &= (1 + e^{-b-\boldsymbol{w}^T\boldsymbol{x}})^{-1}, \end{aligned}$$

where $\boldsymbol{w}$ is a p-dimensional vector of weights and b is a bias. Once again, if we fix the value of p but leave $\boldsymbol{w}$ and b undefined, we are picking out a function family but not a specific function. What is this family? You should recognise it from Section 11.3.1 as the function family that arises in binary logistic regression.

Let's reimagine dataset $D = \{(\boldsymbol{x}_k, y_k)\}_{k=1}^n$. For each k, we'll now have y_k be a binary response, taking the value 0 or 1 (while $\boldsymbol{x}_k$ remains a p-dimensional predictor, the components of which can take any real values). Suppose we wish to fit a binary logistic regression model to this dataset. We can think of our task as that of picking weights $\boldsymbol{w}$ and bias b for a single sigmoid neuron. If we use unregularised cross-entropy loss (equation (11.3)) as our loss function, then we will pick $\boldsymbol{w}$ and b to minimise

$$l(b, \boldsymbol{w}; D) = -\frac{1}{n}\sum_{k=1}^{n}\left(y_k \log\left(\sigma(b + \boldsymbol{w}^T\boldsymbol{x}_k)\right) + (1 - y_k)\log\left(1 - \sigma(b + \boldsymbol{w}^T\boldsymbol{x}_k)\right)\right).$$

[5] Minimising MSE is equivalent to minimising RSS, since RSS = $n \times$ MSE. The convention in the neural network literature is to express loss functions as averages of pointwise losses (adding regularisation penalties separately, if required). The advantages of this convention will become clear in Section 13.6.3.

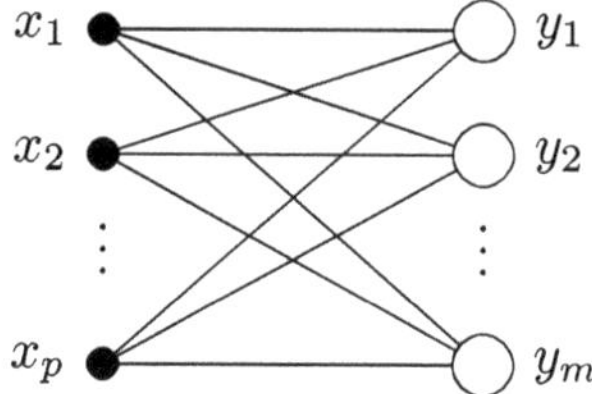

Figure 13.4 A single-layer network with p-dimensional input $\boldsymbol{x}$ and m-dimensional output $\boldsymbol{y}$. To specify the function computed by the network, we specify mp weights, m biases and one activation function. (Neurons in the same layer are assumed to have the same activation function.)

13.3 Single-Layer Networks

Let's now consider fully connected layered networks with a single layer of neurons. That means two layers of *units*: first the input layer, and then – serving as the output layer – the layer of neurons, as shown in Figure 13.4. In the general case, we have p input units (labelled $1, ..., p$) and m output units (labelled $1, ..., m$). Only the latter are neurons. So we have mp connection weights to keep track of, and m biases. For mathematical convenience, we package the biases into an m-dimensional vector $\boldsymbol{b}$. The ith component of this vector, b_i, is the ith neuron's bias. We package the weights into an $m \times p$ matrix $\underline{\boldsymbol{W}}$. The entry at row i and column j, W_{ij}, is the weight of the connection to neuron i from input j. The pre-activation of neuron i, therefore, is

$$z_i = b_i + \sum_{j=1}^{p} W_{ij} x_j.$$

Packaging the pre-activations into an m-dimensional vector $\boldsymbol{z}$ in the obvious way, we have

$$\boldsymbol{z} = \boldsymbol{b} + \underline{\boldsymbol{W}}\boldsymbol{x}. \tag{13.3}$$

13.3.1 Linear Activation: Multivariate Linear Regression

If the neural layer is linear (i.e. if the neurons in the layer have linear activation), then the layer's output vector $\boldsymbol{y}$ is identical to its pre-activation vector $\boldsymbol{z}$. We then have $\boldsymbol{y} = \boldsymbol{b} + \underline{\boldsymbol{W}}\boldsymbol{x}$. That means our single-layer neural network (with linear activation) can express any linear function from $\mathbb{R}^p$ to $\mathbb{R}^m$. If we want to fit a dataset $D = \{(\boldsymbol{x}_k, \boldsymbol{y}_k)\}_{k=1}^n$, we could do that by picking weights $\underline{\boldsymbol{W}}$ and biases $\boldsymbol{b}$ to minimise the MSE,

$$\mathrm{MSE}(\boldsymbol{b}, \underline{\boldsymbol{W}}; D) = \frac{1}{n} \sum_{k=1}^{n} \|\boldsymbol{y}_k - \boldsymbol{b} - \underline{\boldsymbol{W}}\boldsymbol{x}_k\|^2, \tag{13.4}$$

where here the MSE is a mean of squared error vector magnitudes. This is *multivariate* linear regression – linear regression with a vector target.

■ ***Exercise*** 13.2 The RSS when the prediction target is a vector is just n times the MSE defined in equation (13.4). Show that this RSS can be written as a sum of m terms, each an RSS for one component of the prediction target.

In practice, neural networks used for regression tasks will not be single-layer networks. They will, however, usually have linear output layers, and they will often be trained by minimising total squared error (or a regularised variant). As a thought experiment, imagine 'freezing' all the weights and biases of such a network except those associated with the output layer, and then trying to optimise the output layer weights and biases. In that scenario you are effectively doing linear regression. If the rest of the network is already well adapted to the task, then the input to the output layer will be a transformed representation of the network's original input from which it is easy to predict $\boldsymbol{y}$ by applying a final linear transformation.

13.3.2 One Softmax Layer: Multinomial Logistic Regression

We introduced multinomial logistic regression in Section 11.5.1. We'll begin this section with a brisk review using revised notation; the new notation will make it easier to think of multinomial logistic regression in neural network terms.

A multinomial logistic regression model maps a p-dimensional predictor vector $\boldsymbol{x}$ to a discrete probability distribution over m classes in two steps. First, it maps $\boldsymbol{x}$ to a m-dimensional score vector z, defined by

$$z = \boldsymbol{b} + \underline{\boldsymbol{W}}\,\boldsymbol{x},$$

where m-dimensional vector b and $m \times p$ matrix $\underline{\boldsymbol{W}}$ are the parameters of the model. (Notice that this equation is identical to equation (13.3).) Then it maps the score vector to the m-dimensional probability vector $\mathrm{softmax}(z)$, defined component-wise by[6]

$$\mathrm{softmax}(z)_i = \frac{e^{z_i}}{\sum_{j=1}^{m} e^{z_j}}. \tag{13.5}$$

The first step can be implemented by a single layer of linear neurons receiving input $\boldsymbol{x}$. As our notation hints, $\boldsymbol{b}$ is the vector of biases and $\underline{\boldsymbol{W}}$ is the matrix of connection weights. The second step, applying the softmax function, does not translate into neural terms so easily. But notice that softmax is a *fixed* nonlinear function: it has no tunable parameters. In that respect, softmax is a bit like ReLU or the standard sigmoid. And indeed, it is common in the neural network literature to refer to softmax as an activation function, and to think of the map $\boldsymbol{x} \mapsto \mathrm{softmax}(\boldsymbol{b} + \underline{\boldsymbol{W}}\,\boldsymbol{x})$ as something implemented by a single *softmax layer*: a layer of m neurons with softmax activation. The layer's pre-activation vector z is the score vector; passing this through the softmax function yields a probability vector as the layer's output. There is potential for confusion here. The output of the ith neuron in a softmax layer, $\mathrm{softmax}(z)_i$, is not a function solely of the ith neuron's pre-activation. It depends, by way of

[6] The range of summation in this equation differs slightly from that in equation (11.17) because we are counting neurons in a layer from 1 rather than from 0. One advantage of the present convention (standard in the neural network literature) is that it leaves open the possibility of prepending a 'dummy' 1 to any vector as a zeroth component. This notational trick lets us absorb bias vectors into weight matrices. We won't be doing that in this chapter, however.

the normalising denominator in equation (13.5), on the pre-activations of *every* neuron in the layer. Softmax therefore does not conform to the pattern we laid down for activation functions in Section 13.1. It is a special case. Readers who dislike messy exceptions have a way out, however. They can regard talk of 'neural networks with softmax output layers' as shorthand for neural networks with *linear* output layers *whose outputs will be interpreted as multiclass classification score vectors*. Effectively this amounts to absorbing the softmax computation into the loss function. The stakes here are purely terminological. We will continue to speak of softmax layers, though the other perspective will come in handy when we present the backpropagation algorithm in Section 13.10.

■ ***Exercise*** 13.3 Compare a neural network consisting of a single softmax layer to a neural network consisting of a single layer of sigmoid neurons (i.e. neurons with sigmoid activation). For what sort of task might the latter network be suitable? (Hint: *not* multiclass classification.)

In practice, neural networks used for multiclass classification tasks will not be single-layer networks. They will, however, usually have softmax output layers, and they will often be trained by minimising cross-entropy loss. If you freeze all the weights and biases of such a network except those associated with the output layer and then try to optimise the latter, you are effectively doing multinomial logistic regression. If the rest of the network is already well adapted to the task at hand, then the input to the output layer will be a transformed representation of the network's original input from which it is easy to compute a good probabilistic class prediction by applying a linear transformation followed by softmax.

13.4 Hidden Layers and the Universal Approximation Theorem

Neurons in a neural network that do not serve as output units are known as *hidden neurons*. (The idea is that a network user who interacts with the input layer and the output layer can regard the rest of the network as a black box.) Neural networks without hidden neurons can compute only a limited range of functions, but adding even one hidden layer of ReLU neurons changes the situation drastically. In fact, the network architecture depicted in Figure 13.5a – a layered network with p input units, a single hidden layer of ReLUs and m linear output units – can approximate *any* continuous function from a closed, bounded region of $\mathbb{R}^p$ to $\mathbb{R}^m$ to any desired degree of precision, so long as the hidden layer is big enough.[7] This is a rough statement of one form of the *universal approximation theorem*, henceforth UAT. Figure 13.6 should convince you that the $p = m = 1$ special case of the theorem is true. For the general case, see Section 13.13.

■ ***Exercise*** 13.4 Consider the three-piece piecewise linear function from $[0, 1]$ to $\mathbb{R}$

[7] A closed and bounded region of $\mathbb{R}^p$ is the multidimensional analogue of a finite closed interval on the real line. A region of $\mathbb{R}^p$ is *bounded* if it lies within some finite sphere (i.e. if it doesn't stretch off to infinity in any direction), and it is *closed* if it includes its boundary. So, for example, the region $\{\boldsymbol{x} \in \mathbb{R}^p : \|x\| < 1\}$ is bounded but not closed. It would become closed if we replaced '$<$' with '$\leq$'.

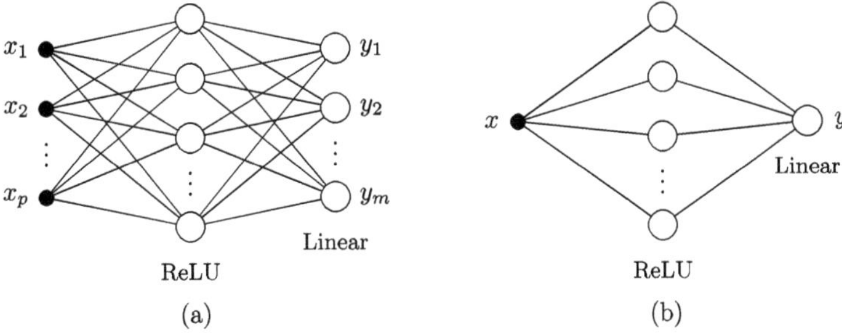

Figure 13.5 (a) A layered neural network with p input units, a single hidden layer of ReLUs and m linear output units. Any continuous function from a closed, bounded region of $\mathbb{R}^p$ to $\mathbb{R}^m$ can be well approximated by such a network. (b) A special case of the architecture in (a): $p = 1$, $m = 1$. Any real continuous function defined on any interval $[a, b]$ can be well approximated by such a network.

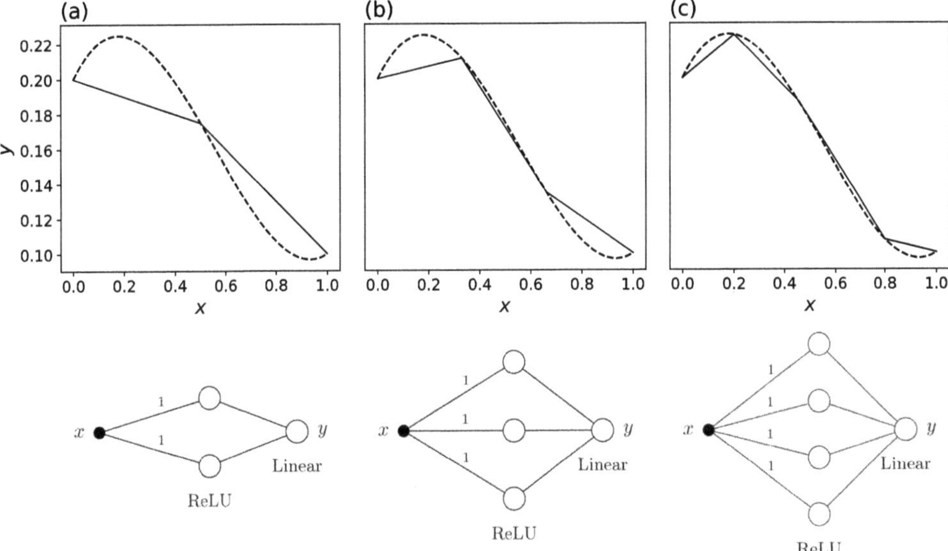

Figure 13.6 (a–c) Piecewise linear approximations (solid) to a smooth target function (dashed). Below each plot, a neural network capable of computing the piecewise linear approximation is shown. Each network is an instance of the architecture in Figure 13.5b. By considering instances with larger and larger hidden layers, we could obtain arbitrarily precise piecewise linear approximations to the target.

defined by

$$f(x) = \begin{cases} x & \text{if } x \le 0.25, \\ 0.25 & \text{if } x \in [0.25, 0.75], \\ 1 - x & \text{if } x > 0.75. \end{cases}$$

Show that this function can be expressed by a neural network that conforms to the pattern shown in Figure 13.6b by finding a set of biases and weights that does the trick. (Notice that the pattern fixes the hidden units' input weights: all are set to 1. You therefore have *four* biases and *three* weights to play with.) Is your solution unique?

■ ***Exercise*** 13.5 Consider a modified version of the architecture shown in Figure 13.5b in which all ReLUs are replaced by linear units. That is, consider networks that have a single input unit, a single hidden layer of *linear* units and a single linear output unit. Show that the family of functions that can be expressed by such networks is the same as the family of functions that can be expressed by a single linear neuron with one input.

■ ***Exercise*** 13.6 Any continuous function from $[a, b]$ to $\mathbb{R}$ can be approximated arbitrarily closely by piecewise linear functions. In this exercise, you will show that this conclusion can fail if the domain is not bounded or not closed. (Here, *bounded* means that the interval has finite length, and *closed* means that it includes its endpoints.) Recall that a piecewise linear function has only finitely many linear pieces.

(a) Give an example of a continuous real-valued function defined on $[0, \infty)$ that cannot be approximated arbitrarily closely by piecewise linear functions.
(b) Give an example of a continuous real-valued function defined on $(0, 1]$ that cannot be approximated arbitrarily closely by piecewise linear functions.

■ ***Exercise*** 13.7 Consider a modified version of the architecture shown in Figure 13.5a in which the linear output layer is replaced by a softmax layer. Show that an arbitrary continuous function from a closed, bounded region of $\mathbb{R}^p$ to the space of probability vectors over m categories can be approximated as closely as we wish by such a network. You may assume the truth of the UAT (in the form stated in the main text).

13.5 Deep Neural Networks

Layered neural networks with more than one hidden layer are known as deep neural networks. (Some writers reserve the term 'deep' for networks with *many* hidden layers.) 'Deep learning' is the interdisciplinary field concerned with designing and training such networks.

Figure 13.7 shows a network with two hidden layers of ReLUs and linear output. The figure introduces a new notational device: parenthetical superscripts are used to indicate layers in the neural network. For example, $\underline{\boldsymbol{W}}^{(1)}$ is the matrix of incoming weights for the first layer of neurons, and $\boldsymbol{b}^{(3)}$ is the vector of biases for the third layer of neurons. We'll denote the vector of neuron activations in the ith hidden layer by $\boldsymbol{h}^{(i)}$ (not shown in the figure). So, stepping through the network in Figure 13.7, we have

$$\boldsymbol{h}^{(1)} = \text{ReLU}\left(\underline{\boldsymbol{W}}^{(1)}\boldsymbol{x} + \boldsymbol{b}^{(1)}\right),$$
$$\boldsymbol{h}^{(2)} = \text{ReLU}\left(\underline{\boldsymbol{W}}^{(2)}\boldsymbol{h}^{(1)} + \boldsymbol{b}^{(2)}\right),$$
$$\boldsymbol{y} = \underline{\boldsymbol{W}}^{(3)}\boldsymbol{h}^{(2)} + \boldsymbol{b}^{(3)},$$

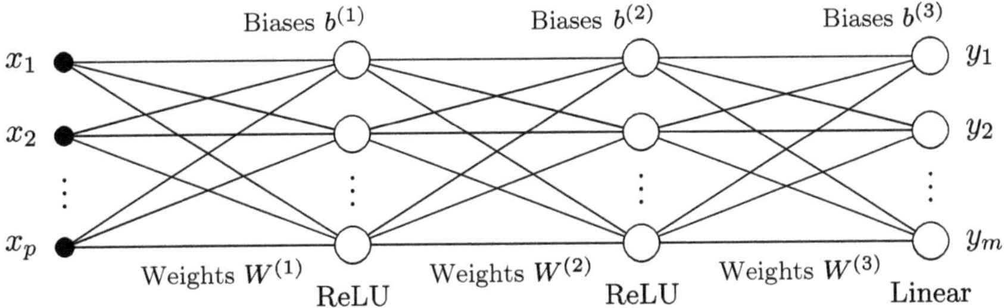

Figure 13.7 A layered neural network with two hidden layers of ReLUs and a linear output layer. Each layer of neurons ($i = 1, 2, 3$) is associated with a matrix of incoming weights, $\underline{\boldsymbol{W}}^{(i)}$, and a vector of biases, $\boldsymbol{b}^{(i)}$.

where ReLU should be understood as applying componentwise. That is, if z is a vector of pre-activations, then $\text{ReLU}(z)$ is the corresponding vector of (ReLU) activations: it is the vector whose ith component is $\text{ReLU}(z_i)$.

Figure 13.7 shows the sizes of the input layer and output layer – p units and m units, respectively – but not the sizes of the two hidden layers. We will denote the number of units in the ith hidden layer by p_i. (A parenthetical superscript is unnecessary here, as there is no risk of confusion with component indices.) So, for example, $\underline{\boldsymbol{W}}^{(1)}$ is a $p_1 \times p$ matrix, and $\boldsymbol{b}^{(1)}$ is a p_1-dimensional vector.

■ ***Exercise*** 13.8 What are the shapes of $\underline{\boldsymbol{W}}^{(2)}$, $\boldsymbol{b}^{(2)}$, $\underline{\boldsymbol{W}}^{(3)}$ and $\boldsymbol{b}^{(3)}$? How many parameters does the network shown in Figure 13.7 have in total? (Each element of a weight matrix or bias vector counts as one parameter.)

The UAT tells us that neural networks with a single hidden layer are universal function approximators. It is not hard to show that the same must be true of deep networks.

■ ***Exercise*** 13.9 In this exercise, you will show that adding extra layers of ReLUs to the network architecture considered in Section 13.4 does not abolish the universal approximation property.

(a) Suppose a function can be computed by a chain of L single-hidden-layer networks of the kind shown in Figure 13.5a chained end-to-end, with the output of the ith network serving as input to the $(i + 1)$th. Show that this function can also be computed by a layered network with L hidden layers of ReLUs and linear output.
(b) Use the result in (a) and the UAT to show that layered networks with L hidden layers of ReLUs are universal function approximators. (Hint: a single-hidden-layer network can approximate the identity function.)

But why bother with deep networks when single-hidden-layer networks are already universal function approximators? One reason is that single-hidden-layer networks capable of approximating even moderately complex real-world functions often require an enormous number of neurons. Adding extra hidden layers typically reduces the total number of neurons required and significantly reduces the number of parameters. Moreover, the resulting slimmer, deeper networks are generally easier to train – not only because of their smaller parameter spaces but also due to their hierarchical structure.

13.6 Training Neural Networks

Knowing that a neural network is capable of approximating the function you need is one thing; getting it to actually do so is quite another. Large modern neural networks have billions of weights and biases. The challenge of finding 'good' parameter vectors in a parameter space with billions of dimensions is a daunting one. As Exercise 13.10 shows, brute-force grid search is certainly not a viable approach.

■ ***Exercise*** 13.10 Suppose we want to do a grid search through the parameter space of a tiny neural network with only 30 parameters. To keep compute costs down, we'll allow each parameter to take only 10 different values. This defines a grid of parameter vectors in a 30-dimensional space. Given a training dataset D and a loss function l, we want to compute $l(\boldsymbol{\theta}; D)$ for each parameter vector $\boldsymbol{\theta}$ on the grid. How many loss function evaluations does that require?

Neural networks are typically trained as follows: start by randomly initialising the weights and biases, then iteratively make small adjustments to reduce the training loss, guided by the gradient of the loss function in the parameter space. Procedures in this ballpark are also used for fitting logistic regression models and even (when datasets are large) linear regression models. So why are we only talking about them now? The reason is that the optimisation problems that arise in logistic and linear regression are relatively straightforward. Your favourite software library's default fitting methods will usually work fine without any tinkering. Neural networks present more challenging optimisation problems. When training large neural networks, you will sometimes need to experiment with different optimisation methods, so it's a good idea to know something about them.

In Section 13.6.1, we introduce the simplest form of gradient-based optimisation: gradient descent. While it serves as an important conceptual foundation, vanilla gradient descent struggles in real-world neural network training. We explain why, and then in Sections 13.6.2 and 13.6.3, we explore modifications that help gradient-based methods work effectively in practice. Of course, to apply any of these methods we need to be able to compute gradients efficiently. The backpropagation algorithm does just that; we will explain how it works in Section 13.10 (after we have presented some concrete examples of network training in Sections 13.7 and 13.8). We will then be able to take a quick look at the problem of vanishing and exploding gradients in deep neural networks (Section 13.11).

Before we get stuck in, we should address a small technical worry. ReLU neurons look like they might cause trouble for gradient-based methods. After all, $z \mapsto \mathrm{ReLU}(z)$ is not differentiable at $z = 0$. However, it *is* differentiable everywhere else, and hitting exactly $z = 0$ is rare in practice. If that does occur we can 'cheat' by defining $\mathrm{ReLU}'(0)$ to be 0 (the left derivative) or 1 (the right derivative). You should assume below that any differentiability issues with ReLU activation functions are handled that way.

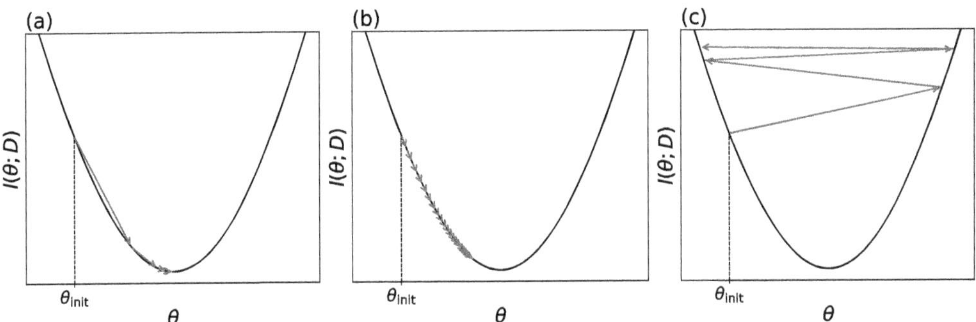

Figure 13.8 (a) Gradient descent in a one-dimensional parameter space with the learning rate hyperparameter η set to an appropriate value. Arrows show gradient descent updates, connecting successive points of the form $(\theta, l(\theta; D))$. Starting from its initial value of θ_{init}, the parameter converges quickly towards the value that minimises the loss. (b) The same process, but with η set to a value that is too small. Convergence is now very slow. (c) The same process, but with η set to a value that is too high. Gradient descent updates now repeatedly overshoot the loss-minimising value, and θ oscillates.

13.6.1 Gradient Descent

Imagine we have a regression or classification problem, an associated training dataset D, a neural network architecture suitable for tackling the problem, and an appropriate loss function l. The loss is a function of both the training data and the network parameters. For notational convenience, we package the parameters into a single P-dimensional parameter vector $\boldsymbol{\theta} = (\theta_1, \theta_2, ..., \theta_P)$, where P is the total number of weights and biases, and we write the loss as $l(\boldsymbol{\theta}; D)$. We control $\boldsymbol{\theta}$ but not D; as usual, our task is to pick a value for $\boldsymbol{\theta}$ that makes $l(\boldsymbol{\theta}; D)$ as small as possible. The intuition behind gradient descent is that we can improve almost any proposed $\boldsymbol{\theta}$ by taking a small step 'downhill' with respect to the loss. By repeatedly taking many such small steps, we hope to arrive at an optimal or close to optimal $\boldsymbol{\theta}$. We'll denote the gradient of the loss in the parameter space by $\nabla l(\boldsymbol{\theta}; D)$. This is the P-dimensional vector whose ith component is $\frac{\partial l(\boldsymbol{\theta};D)}{\partial \theta_i}$. (The partial derivative is taken with all components of $\boldsymbol{\theta}$ other than the ith held constant, and with D held constant too of course.) The direction the gradient vector points in is the direction we would nudge $\boldsymbol{\theta}$ if we wanted to *increase* the loss as quickly as possible. To decrease the loss, we nudge $\boldsymbol{\theta}$ in the opposite direction. So, the basic gradient descent update is

$$\boldsymbol{\theta} \leftarrow \boldsymbol{\theta} - \eta \nabla l(\boldsymbol{\theta}; D), \tag{13.6}$$

where η is a positive hyperparameter known as the *learning rate*. Figure 13.8a is a cartoon illustration of gradient descent working well in a one-dimensional parameter space. The other two panels in the figure show what can go wrong when the learning rate hyperparameter is too small (panel (b)) or too large (panel (c)).

Picking a good value for η becomes challenging in high-dimensional parameter spaces because the loss function may be more strongly curved in some directions than in others. Indeed, this issue can arise in two dimensions, as shown in Figure 13.9a. That is one

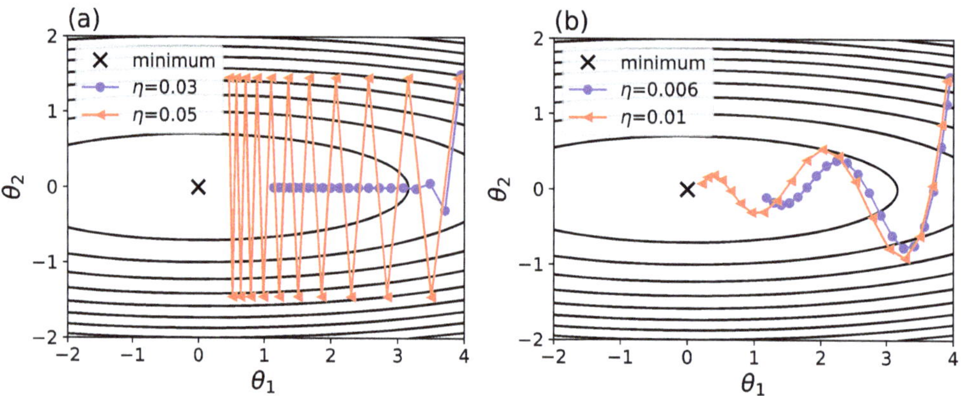

Figure 13.9 Contours of the loss function $l(\boldsymbol{\theta}) = \theta_1^2 + 20\,\theta_2^2$ (black), with parameter vector trajectories for gradient-based optimisation methods superimposed. (a) Basic gradient descent (update rule (13.6)) with two learning rates (20 iterations each). When η is 0.03 progress is slow. Increasing η to 0.05 yields quicker progress in the θ_1 direction, but at the cost of wild oscillations in the θ_2 direction. (b) Gradient descent with momentum (update rule (13.7), $\beta = 0.8$) with two learning rates (20 iterations each). The learning rates shown are the rescaled counterparts of those in panel (a) (see main text). Once again, the smaller learning rate leads to slow progress in the θ_1 direction; but thanks to the smoothing effect of momentum, it is now safe to increase the learning rate.

motivation for tweaking the basic gradient descent recipe. Two other motivations are at least as important. First, neural networks are often trained with very large datasets, for example, millions of examples. The loss $l(\boldsymbol{\theta}; D)$ will typically decompose into an average over terms associated with individual training examples, plus (possibly) a regularisation penalty. The gradient $\nabla l(\boldsymbol{\theta}; D)$ will then have a similar form. Even with an efficient algorithm for computing gradients, computing $\nabla l(\boldsymbol{\theta}; D)$ when D is very large will be expensive. Since gradient descent is an iterative algorithm in which the gradient is recomputed at each step, this is a major issue. Second, neural network loss functions tend to have *multiple minima*, with some minima being better (i.e. achieving smaller loss) than others. If you begin with a randomly initialised parameter vector and perform gradient descent, you may end up at a bad local minimum. This never happens when fitting linear or logistic regression models, because the loss functions in those cases are *convex* with respect to their parameter vectors. A convex function is upward-curving (see Appendix C), which implies that any local minimum will also be a global minimum. Figure 13.10 illustrates the general idea. The problem is that neural network loss functions are almost always *non-convex* in their parameter vectors.

So, we've identified three obstacles to training neural networks by (simply) iterating on update rule (13.6): there may be no learning rate that works well for all directions in parameter space; training datasets may be so large that repeatedly computing $\nabla l(\boldsymbol{\theta}; D)$ is prohibitively expensive; and with non-convex loss functions, training may get stuck at bad local minima. Let's now look at some workarounds.

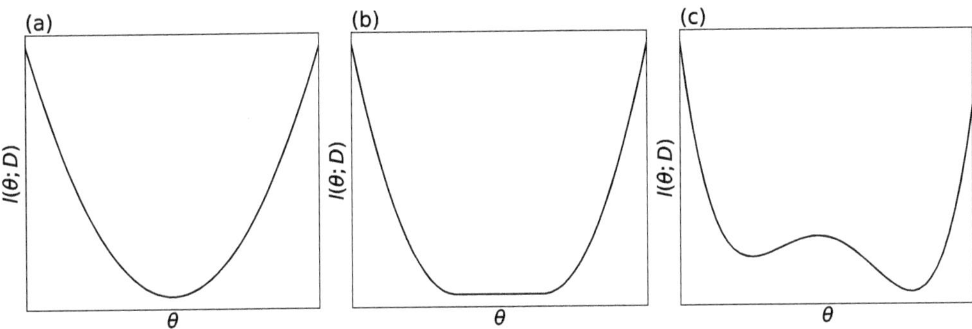

Figure 13.10 Three loss functions defined on one-dimensional parameter spaces. (a) A convex loss function with a single global minimum. Gradient descent begun from a random initial value for θ will converge on the global minimum. (b) A convex loss function with multiple global minima. Gradient descent begun from a random initial value for θ will converge on *a* global minimum. (c) A non-convex loss function with a bad local minimum. Gradient descent begun from a random initial value for θ may get stuck at the inferior minimum.

13.6.2 Momentum and Adaptive Learning Rates

In basic gradient descent, the change to $\boldsymbol{\theta}$ at each iteration depends only on the current gradient $\nabla l(\boldsymbol{\theta}; D)$ and the learning rate. In *gradient descent with momentum*, the change to $\boldsymbol{\theta}$ depends on the recent history of the gradient. To define the update rule, we introduce a new vector $\boldsymbol{v}$ called the *velocity*; it has the same shape (i.e. same number of components) as $\boldsymbol{\theta}$. The update at each iteration is

$$\begin{aligned} \boldsymbol{v} &\leftarrow \beta \boldsymbol{v} + \nabla l(\boldsymbol{\theta}; D), \\ \boldsymbol{\theta} &\leftarrow \boldsymbol{\theta} - \eta \boldsymbol{v}, \end{aligned} \tag{13.7}$$

where η is a positive hyperparameter (still) referred to as the learning rate, and $\beta \in (0, 1)$ is an additional hyperparameter known as the momentum coefficient. Whatever initialisation scheme is used for the parameter vector, $\boldsymbol{v}$ is always initialised to zero at the start of training. The idea is that over many iterations, the velocity will build up in directions with consistent gradients, but not in directions in which gradients often flip sign. So, for example, if $\frac{\partial l(\boldsymbol{\theta};D)}{\partial \theta_1}$ is consistently positive for many consecutive iterations, then the corresponding component of the velocity, v_1, will grow to be much larger than the typical magnitude of $\frac{\partial l(\boldsymbol{\theta};D)}{\partial \theta_1}$. By contrast, if $\frac{\partial l(\boldsymbol{\theta};D)}{\partial \theta_2}$ often flips sign from one iteration to the next, successive updates to v_2 will partially cancel, and the magnitude of v_2 will remain modest relative to the typical magnitude of $\frac{\partial l(\boldsymbol{\theta};D)}{\partial \theta_2}$. This is the situation illustrated in Figure 13.9b.

Momentum makes it easier to find a learning rate that works well for all directions in parameter space. Compare panels (a) and (b) of Figure 13.9. Panel (a) shows basic gradient descent. No learning rate works well for both the θ_1 and θ_2 directions, because any learning rate high enough to yield good progress in the θ_1 direction will lead to wild oscillations in the θ_2 direction. Panel (b) shows gradient descent with momentum. Now learning rates that yield good progress in the θ_1 direction (at least once the velocity has built up) are also well behaved in the θ_2 direction, thanks to the oscillation-damping effects of the partial cancellation dynamic mentioned in the previous paragraph. You may wonder why the learning rates shown in panels (a) and (b) of Figure 13.9 are different. Shouldn't we have used the same two learning rates in each subplot, for an 'apples-to-apples' comparison of gradient descent with and without momentum? In fact matching η values in this way would not make sense, because the η hyperparameter plays subtly different roles in the two update rules. Exercise 13.11 shows one way of converting a learning rate for gradient descent with momentum to a roughly comparable learning rate for basic gradient descent. (We used this conversion when picking the learning rates shown in Figure 13.9.)

■ ***Exercise*** 13.11 Suppose that the gradient $\nabla l(\boldsymbol{\theta}; D)$ has remained constant for many iterations. Show that gradient descent with momentum with learning rate η and momentum coefficient β then matches the parameter vector updates of vanilla gradient descent with learning rate $\eta(1-\beta)^{-1}$.

■ ***Exercise*** 13.12 We've presented momentum as a trick that makes it easier to find a learning rate that works for all directions in parameter space. But momentum can also help prevent training from getting stuck at bad local minima. How?

An alternative response to the challenge of picking a learning rate that works well for all directions in parameter space is to give up and use different learning rates for different directions. When gradient descent is modified in this way, the update rule becomes

$$\boldsymbol{\theta} \leftarrow \boldsymbol{\theta} - \underline{\boldsymbol{\eta}} \nabla l(\boldsymbol{\theta}; D),$$

where $\underline{\boldsymbol{\eta}} = \mathrm{diag}(\eta_1, \eta_2, ..., \eta_P)$ is a diagonal matrix of learning rates – one learning rate for each component of the parameter vector. In practice this idea will usually be combined with an adaptive scheme for adjusting the entries of $\underline{\boldsymbol{\eta}}$ during training. Many strategies are possible, but most of the adaptive schemes in common use rely on statistics (e.g. first and second moments) summarising the recent history of the gradient components. Typically the *overall scale* of the $\underline{\boldsymbol{\eta}}$ matrix will be controlled by a scalar hyperparameter. This hyperparameter may be referred to as 'the learning rate', but you should not read that phrase as implying that each entry in $\underline{\boldsymbol{\eta}} = \mathrm{diag}(\eta_1, \eta_2, ..., \eta_P)$ is the same. The whole point of adaptive schemes is to allow these values to diverge to meet the needs of the current training run.

13.6.3 Stochastic Gradient Descent

Suppose the loss function can be expressed as

$$l(\boldsymbol{\theta}; D) = \frac{1}{n} \sum_{\boldsymbol{\xi} \in D} \boldsymbol{\ell}(\boldsymbol{\theta}; \boldsymbol{\xi}) + R(\boldsymbol{\theta}), \tag{13.8}$$

where $\boldsymbol{\xi}$ ranges over training examples in D, $\boldsymbol{\ell}(\boldsymbol{\theta}; \boldsymbol{\xi})$ is a pointwise loss associated with training example $\boldsymbol{\xi}$, and $R(\boldsymbol{\theta})$ is a regularisation penalty. (If regularisation is not wanted, R is the zero function.) We denote a training example by $\boldsymbol{\xi}$ rather than by $\boldsymbol{x}$ to leave it open whether training examples are labelled (e.g. $\boldsymbol{\xi} = (\boldsymbol{x}, y)$) or unlabelled ($\boldsymbol{\xi} = \boldsymbol{x}$), because the loss function template (13.8) is applicable in both supervised and unsupervised contexts. So, we're supposing the overall loss l can be expressed as a pointwise loss ℓ averaged over the training data plus an optional regularisation penalty. Most loss functions used to train neural networks can indeed be expressed in this way.

■ ***Exercise*** 13.13 Verify that MSE (commonly used for regression) and cross-entropy loss (commonly used for classification) both conform to the pattern laid down in equation (13.8).

The gradient of the loss is

$$\nabla l(\boldsymbol{\theta}; D) = \frac{1}{n} \sum_{\boldsymbol{\xi} \in D} \nabla \ell(\boldsymbol{\theta}; \boldsymbol{\xi}) + \nabla R(\boldsymbol{\theta}). \tag{13.9}$$

The second term on the right-hand side of this equation, $\nabla R(\boldsymbol{\theta})$, is usually easy to compute. It does not involve a sum over training examples, and R will usually have a convenient functional form. For example, if $R(\boldsymbol{\theta}) = \lambda \|\boldsymbol{\theta}\|^2$ (L2 penalty), then $\nabla R(\boldsymbol{\theta}) = 2\lambda\boldsymbol{\theta}$. By contrast, computing $\nabla \ell(\boldsymbol{\theta}; \boldsymbol{\xi}^{(k)})$ is non-trivial. We'll look at the standard algorithm for tackling that job (backpropagation) in Section 13.10. For now, the important point is that computing $\nabla \ell(\boldsymbol{\theta}; \boldsymbol{\xi})$ is, for each $\boldsymbol{\xi}$, relatively expensive. Computing the average $\frac{1}{n} \sum_{\boldsymbol{\xi} \in D} \nabla \ell(\boldsymbol{\theta}; \boldsymbol{\xi})$ once per gradient descent iteration is therefore *very* expensive when the dataset is large. But we can estimate this average using a small random sample – a 'batch' – of the training examples. Let B be a random sample of n_b training examples drawn from D. Then the batch gradient

$$\nabla l(\boldsymbol{\theta}; B) = \frac{1}{n_b} \sum_{\boldsymbol{\xi} \in B} \nabla \ell(\boldsymbol{\theta}; \boldsymbol{\xi}) + \nabla R(\boldsymbol{\theta}) \tag{13.10}$$

provides an unbiased estimate of the true gradient $\nabla l(\boldsymbol{\theta}; D)$. So long as $n_b \ll n$, the batch gradient will be much cheaper to compute than the true gradient $\nabla l(\boldsymbol{\theta}; D)$. Of course, we don't want to base the whole training process on one small and possibly unrepresentative sample of the training data; but that's easy to avoid – we just use a different batch at each iteration. In practice, this idea is implemented as follows. We randomly split our dataset D into a number of disjoint batches $B_1, B_2, ..., B_k$; usually every batch will be the same size, with the possible exception of the final batch. Then we perform the first iteration of gradient descent using batch B_1, the second using batch B_2 and so on. After we have used B_k, we have completed one *pass* over all the data, which we call an *epoch*. We then shuffle the dataset – i.e. form a new random partition into batches – and repeat, continuing for as many epochs as necessary. This is known as *stochastic gradient descent* (SGD). It is stochastic because at each epoch the partitioning into batches is random.

SGD is a cheap approximation to gradient descent. Does that mean we would always prefer the latter to the former if we had infinite computational resources? Not necessarily. The stochasticity of SGD means that it is less likely than vanilla gradient descent to get trapped at a bad local minimum. When we use batches of randomly picked training points in lieu of the full training dataset, we are injecting sampling noise into our computations of the loss function and its gradient. This sampling noise will tend to destabilise 'shallow' minima in the loss function, as shown in Figure 13.11. The batch size – the number of training points per batch – is an important hyperparameter of SGD. The smaller the batch size, the greater the impact of sampling noise. Sampling noise can be beneficial, but we don't want to overdo

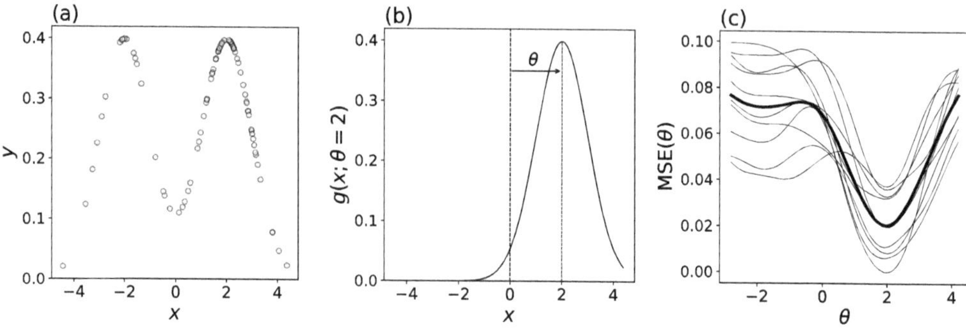

Figure 13.11 'Batching' (using small, random subsets of the training data) can destabilise bad local minima in loss functions. (a) A toy regression dataset where the response (y) depends on the predictor (x) in a double-peaked manner, with most datapoints near the right-hand peak. (b) A family of single-peaked prediction functions parameterised by θ, which controls the peak's position. (c) The MSE of the single-peaked prediction function with parameter θ, plotted against θ. The thick curve shows the MSE computed on the full dataset ($n = 80$). Each of the 10 thin curves shows the MSE computed on one batch of eight datapoints, corresponding to a random partition of the dataset into 10 batches. The bad local minimum appears in some but – crucially – not all of the thin curves. While gradient descent might get trapped at the bad local minimum, stochastic gradient descent is likely to find its way to the global minimum eventually.

it. Large random fluctuations in gradient estimates from one iteration to the next will slow convergence. Also, on modern hardware, SGD will fail fully to leverage parallel processing capabilities if the batch size is set too low. Common default choices include 64 and 128. (Picking a power of two helps streamline memory operations.)

Stochastic gradient descent can be combined with momentum and/or adaptive learning rates. Indeed, one of the optimisation algorithms used most often by people training neural networks, Adam (Adaptive Moment Estimation), exploits all three ideas.

13.7 Classifying Human Speech Sounds with MLPs

For our first concrete examples of neural network training, we return to the human speech sounds classification task introduced in Section 11.7. This task involves distinguishing between 10 vowel sounds (the classes) using a nine-dimensional predictor vector. The dataset contains 2,371 labelled examples. We previously found that multinomial logistic regression with light regularisation achieved a mean cross-validated cross-entropy loss of 0.241. Can simple neural networks do better? Unsurprisingly, yes; but before demonstrating this, we need to discuss our assessment strategy. Training neural networks is computationally expensive. Cross-validation – where training is repeated multiple times – is *very* expensive. Moreover, in deep learning, datasets are usually large. When a large dataset is randomly split into (say) five folds, the folds tend to be statistically similar. It is therefore common for deep learning practitioners to opt for a single train-validation split, allocating their computational budget to exploring a wider range of architectures and hyperparameters instead. That is the approach

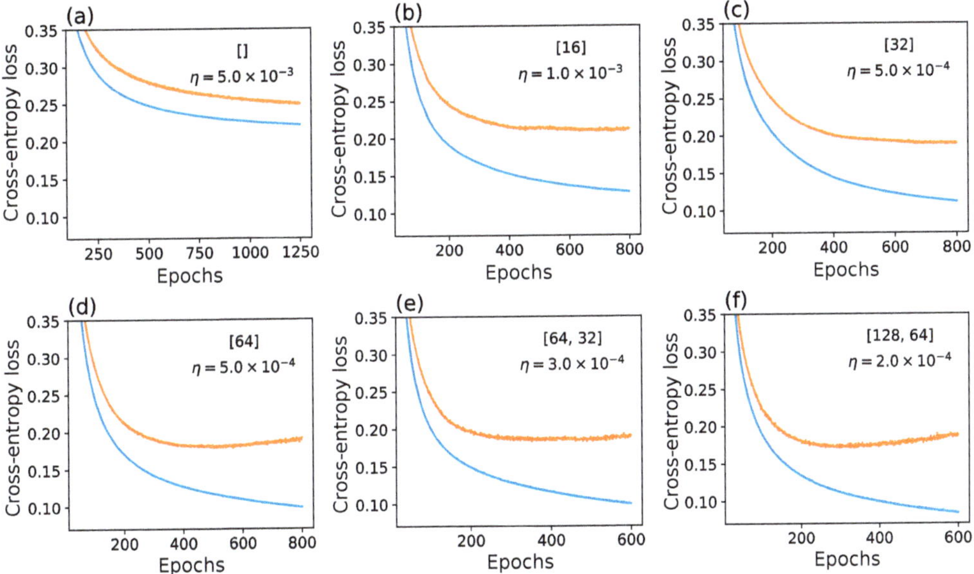

Figure 13.12 (a–f) Training runs for six layered, fully connected neural network architectures on the human speech sounds classification task from Section 11.7. The blue curve represents training loss (lower series in each subplot), and the orange curve represents validation loss (upper series in each subplot). See the main text for further details.

we will take here. We will try a number of different neural network architectures and training procedures and pick the combination that performs best on our validation set. Now, this 'tournament' workflow could trick us into thinking our winning technique is better than it actually is. A *winning* technique probably got at least a bit lucky on the specific validation dataset used in the tournament. So, we will divide our dataset into *three* portions, training (80%), validation (10%) and test (10%), taking care to ensure balanced class statistics in each portion. The test set is reserved for final assessment of the technique that performed best on the validation set (after training on the training set).

Why did we not bother reserving a test set when we were experimenting with multinomial logistic regression in Chapter 11? After all, we used cross-validation to select a 'winning' regularisation strength. Should we not have then used fresh data to reassess the winner? In principle, yes. However, Chapter 11's logistic regression tournament wasn't much of a contest. We had only one hyperparameter to play with. The serious contenders were all variants of 'very light regularisation', and the performance differences between those contenders were tiny. With neural networks, the scope for tinkering with architectures and hyperparameters is much greater, 'winner's curse' is a bigger worry, and the need for a test set is greater. It is worth cutting into the data available for training and validation to form one.

Let's turn now to our neural network tournament. Figure 13.12 shows training runs for six different layered, fully connected neural network architectures. All networks have 9-unit input layers and 10-neuron softmax output layers (to match the structure of the speech sounds classification task); what varies are the number and the size of the hidden layers.

Architectures are therefore represented as sequences of hidden layer sizes. So, for example, '[16]' means a single hidden layer of 16 neurons, while '[128, 64]' means two hidden layers, the first of 128 neurons and the second of 64. '[]' means no hidden layers. In each panel of Figure 13.12, cross-entropy loss is plotted against training epoch for a specific architecture. The optimisation algorithm used is Adam, with a batch size of 64 and the learning rate η displayed on the plot. The blue curve represents cross-entropy loss on the training set; the orange curve represents cross-entropy loss on the validation set.[8] Some trial and error (not shown) went into the choice of learning rates; we tried to pick values that yielded low validation losses.

What do the plots tell us? First, they tell us that simple neural networks can indeed do better on the speech sounds classification task than multinomial logistic regression. In every panel except panel (a), the validation cross-entropy loss dips well below 0.241 (the baseline set in Chapter 11). Panel (a) is an exception because panel (a) *is* multinomial logistic regression. (Recall from Section 13.3.2 that any neural network in which inputs are passed directly to a single softmax layer is mathematically equivalent to a multinomial logistic regression classifier.) If we had trained our no-hidden-layer network for another thousand epochs or so, it would have achieved a validation cross-entropy loss in line with the average cross-validation score we saw in Chapter 11.[9] Second, the plots tell us that letting optimisation run to convergence – that is, run until training loss is no longer significantly declining – can be counterproductive. In every panel, validation loss falls with epoch count initially but in panels (d–f) it eventually starts to increase again. For example, the [128, 64] network trained for 600 epochs performs significantly less well on the validation set than the same network trained for 300 epochs. What is going on? The answer is that over-training can lead to overfitting, precisely because neural networks are such good function approximators. In the early stages of training, a network typically learns to fit the simpler, more obvious patterns in its training data. Further training is then a matter of fine-tuning: the network (if it is large enough) learns to fit more complex, less obvious patterns. Pushed too far, this may lead to the network fitting to noise. In other words, the network may learn pseudo-patterns in the training data that would be unlikely to recur in new data drawn from the same distribution. Validation loss steadily increasing while training loss continues to decrease is a sure sign that this is happening. It is therefore very important to monitor validation loss during the training process. When it starts to increase, we should stop training. (Figure 13.12 breaks that rule in order to demonstrate the phenomenon of over-training.) This simple trick is known as *early stopping*. It is an example of an implicit regularisation strategy – regularisation because it encourages the network to learn simpler functions, and implicit because it does not involve adding a regularisation term to the loss function. Finally, the plots let us crown a victor: the best performance on the validation set was achieved by the [128, 64] architecture trained for 298 epochs with learning rate $\eta = 2 \times 10^{-4}$. The winning validation loss was 0.171. Reassuringly, when we evaluate the winning architecture-plus-training-procedure combination on our *test* (not validation) set, we get an even better result: a cross-entropy test loss of 0.165. This suggests that any 'winner's

[8] These are not batch losses, but losses computed for the whole training set and the whole validation set at the end of each epoch of training.

[9] In practice, general-purpose neural network training algorithms are not used to fit logistic regression classifiers. The loss function in logistic regression is convex, and optimisation proceeds much more quickly when one uses an algorithm that exploits this fact.

curse' effect was small enough to be outweighed by an unrelated statistical fluctuation: the examples in the test set happened to be, on average, a little easier to classify than the examples in the validation set.

If you haven't worked with neural networks before, you might have found this little tournament's outcome surprising. The winning network had many trainable parameters: $128 + 64 + 10 = 202$ biases and $9 \times 128 + 128 \times 64 + 64 \times 10 = 9{,}984$ weights, for a total of 10,186 – *vastly* more than any model we have considered previously. Yet the task was simple, the training set small (1,896 examples) and the loss unregularised. That all sounds like a recipe for overfitting. And indeed, if we had trained to convergence, severe overfitting would have occurred. The implicit regularisation provided by early stopping makes the large network's victory somewhat less surprising. Other forms of implicit regularisation were also at work. For example, the stochasticity introduced by batching reduced the chance of fitting to complex pseudo-patterns in the training data. Still, the fact that gradient-based optimisers so often steer large neural networks towards well-behaved regions of parameter space – regions that lie in the Goldilocks zone between underfitting and overfitting – is striking. The mechanisms are not fully understood; they are a focus of ongoing research.[10]

■ ***Exercise*** 13.14 Explicit regularisation is also an option when fitting neural networks. Ridge penalties are the most widely used. In deep learning contexts, this form of explicit regularisation (applied to weights but not to biases) is known as *weight decay*. Why? (Hint: review the basic gradient descent update rule (13.6) and consider the effect of adding to l a penalty term $\lambda \sum_{w \in W} w^2$, where W is the set of all weights in the network.)

13.8 Convolutional Neural Networks

13.8.1 Designing Networks That Understand Space

In this section, we look at a network architecture often used in computer vision. To keep things simple, we'll focus on one specific task: classifying fashion images. Figure 13.13 shows 10 low-resolution (28×28) greyscale images of fashion products, each with an associated class label. The Fashion MNIST dataset (Xiao et al. (2017)) consists of 70,000 such labelled images: 7,000 per class, across the 10 classes shown. Each image is a 28×28 array of real numbers ranging from -1 (darkest pixels in the image) to $+1$ (brightest pixels in the image).[11] The dataset is split into a training set (60,000 examples) and a validation set (10,000 examples), with class balance maintained.

A neural network for classifying Fashion MNIST images needs an input layer of $28 \times 28 = 784$ units (to handle all the pixels in an image) and a 10-neuron softmax output layer. But what should the guts of the network – the layers between input and output – look like? One option is to stick with the multilayer perceptron (MLP) architecture from Section 13.7. Then the guts of the network will be a sequence of fully connected hidden layers. If we put 128 neurons in the first hidden layer, this layer will have $784 \times 128 = 100{,}352$ input weights (plus 128 biases). That sounds like a lot of parameters for the initial processing of a tiny greyscale image, but big parameter spaces are par for the course in deep learning, so perhaps

[10] See Chapter 20 of Prince (2023) for an introduction to the research literature on this topic.

[11] The raw images are eight-bit integer arrays, but it is standard practice to map values to $[-1,1]$ in pre-processing.

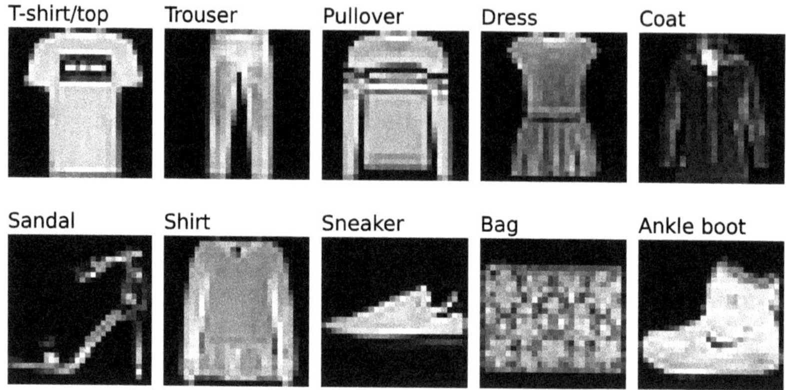

Figure 13.13 Illustrative Fashion MNIST images (with class labels).

we should ignore that qualm. There are better reasons to question the MLP architecture. Connecting every input unit to every neuron in the first hidden layer treats the input image as a mere set of pixels with no geometric structure. For example, prior to training, the network does not 'know' which pixels are neighbours. It might eventually learn proximity relations from the training data (because nearby pixels are correlated). But do we really want to force the network to learn basic facts about space from scratch, using only the data available for the current task? When *you* inspect an image, you immediately perceive how the pixels are arranged in space. Shouldn't an image-processing neural network be able to do the same?

One way to build spatial awareness into the network is to abandon the idea that every neuron in the first hidden layer must process the entire input image. Instead, we can make each neuron responsible for processing just a small portion – say, a 3×3-pixel patch. Also, in the early stages of processing, it makes sense to process each patch in the same way. For instance, if the network is going to look for vertical edges, it should look for them everywhere. (The local features it looks for should be *translation invariant*.) We'll spell out the vertical edges example in detail, as it illustrates the general concept of a *feature map*. Please note that the 'map' in 'feature map' refers to the everyday notion – as in *crime map* or *rainfall map* – and not to an abstract mathematical function. A feature map for vertical edges tells you where (in a specific image) the vertical edges are.

13.8.2 Convolutional Filters and Feature Maps

One sort of vertical edge detector looks at its assigned patch and asks 'is it brighter on the right that on the left? If so, how much brighter?'. A single ReLU neuron receiving inputs from all (and only) pixels in its patch can do that job. For a 3×3-pixel patch, a simple solution is to set the neuron's parameters as follows: zero bias; input weight $+1$ for pixels in the rightmost column; input weight -1 for pixels in the leftmost column; and input weight 0 for pixels in the middle column. This neuron is depicted in Figure 13.14.

■ ***Exercise*** 13.15 Compute the output of the neuron depicted in Figure 13.14 in

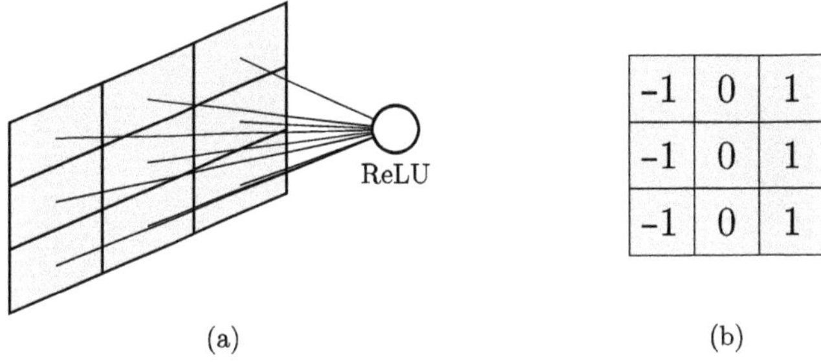

Figure 13.14 (a) A neuron receiving inputs from a 3×3 patch of pixels. (b) One possible set of weights to associate with the inputs. A ReLU with these input weights and zero bias will respond to patches that are brighter on the right than on the left.

response to each of the following patterns of 3×3-pixel-patch input:

$$\text{(a)} \begin{bmatrix} -1 & 0 & 1 \\ -1 & 0 & 1 \\ -1 & 0 & 1 \end{bmatrix}, \qquad \text{(b)} \begin{bmatrix} 1 & 0 & -1 \\ 1 & 0 & -1 \\ 1 & 0 & -1 \end{bmatrix}, \qquad \text{(c)} \begin{bmatrix} 1 & 1 & 1 \\ 1 & 1 & 1 \\ 1 & 1 & 1 \end{bmatrix}.$$

How many copies of the neuron do we need to map the presence (or absence) of vertical edges of the envisaged sort throughout a 28×28-pixel image? That depends on our mapping strategy, but the simplest approach is to use 28×28 of them, associating one copy of the neuron with each pixel position in the input image. Each neuron can then 'see' a 3×3-pixel patch centred on its assigned position.

To spell out the details, we'll use two-dimensional indexing for both the input pixels and the neurons. Let $x_{i,j}$ be the brightness of the pixel in the ith column (counting from left) and jth row (counting from the bottom) of the input image. Similarly, let $h_{i,j}$ be the output of the neuron associated with that position. Then we have

$$h_{i,j} = \text{ReLU}\left(b + \sum_{k=-1}^{1} \sum_{l=-1}^{1} w_{k,l} x_{i+k,j+l}\right), \tag{13.11}$$

where b is the neuron's bias and $w_{k,l}$ is the input weight from the pixel k columns to the right and l rows up from the centre of the neuron's assigned patch. (Notice that neither b nor the weights depend on i or j.) In our vertical-edge-detector example, $b = 0$, $w_{0,0} = 0$, $w_{-1,0} = -1$, and so on. The outputs $h_{i,j}$ form a 28×28-pixel feature map. If i or j takes an extreme value (meaning 1 or 28 in our case), then the right-hand side of equation (13.11) refers to out-of-bounds input pixels. We handle this with *zero padding*: we surround the image with a border of imaginary extra input pixels, all set to zero. Figure 13.15 shows a Fashion MNIST image alongside the feature map produced by the scheme just described. An illustrative pixel in the feature map is highlighted in green, as is the 3×3-pixel patch in the original image from which it was derived.

We will need just a little more jargon. The bias and weights of the neurons used to construct a feature map – one bias and one set of weights, shared by all the neurons – define a *convolutional filter*. (The term originally comes from signal processing.) The weights alone are referred to as the filter's *kernel*. In our example, then, the kernel is a 3×3 array. A

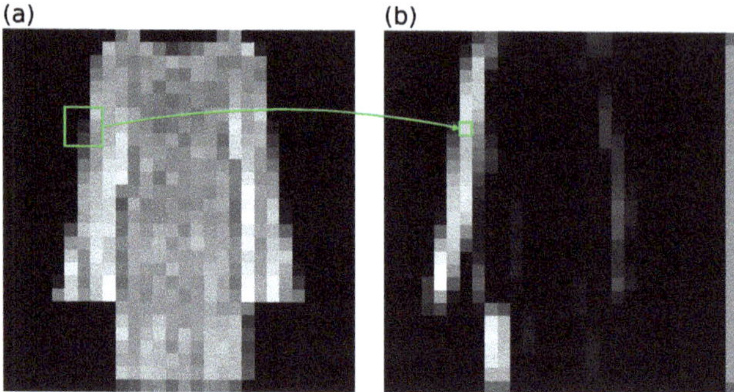

Figure 13.15 (a) An illustrative Fashion MNIST image, with one 3×3-pixel patch highlighted (green box). (b) A feature map of the image, generated by applying the convolutional filter defined in Figure 13.14b, followed by ReLU. The pixel derived from the patch highlighted in (a) is itself highlighted in (b).

convolutional filter is a linear function from local pixel patches to pre-activations that is applied in a translation-invariant way across the input image. When we pass the resulting pre-activations through an activation function (ReLU in our example), we obtain a feature map. A *convolutional layer* is a network layer that computes one or more feature maps, and a *convolutional neural network* is a neural network with one or more convolutional layers. Almost all CNNs include some fully connected layers too.

We now turn to the task of articulating (and motivating) a complete CNN architecture for the Fashion MNIST task. We'll proceed layer by layer, but you may wish to peek ahead at Figure 13.16 to see where we're heading.

13.8.3 A CNN Architecture for the Fashion MNIST Task

As you would expect from the above discussion, we'll use a convolutional layer as our first hidden layer. We might have some hunches about what features it should detect – maybe vertical edges, maybe horizontal edges and maybe certain textures – but we won't try to bake those hunches into the architecture. Instead, we'll let the network learn what features to detect during training. We need only specify the shape of each convolutional filter's kernel (3×3), the activation function to be applied (ReLU) and the number of filters. Let's give our first hidden layer eight filters, so that it will compute eight distinct feature maps. There are $8 \times 28 \times 28 = 6{,}272$ neurons in this layer – by far the biggest layer we have considered so far. But how many trainable parameters are associated with it? Only 80! (Each filter has 10 parameters, and there are eight filters.) Still, the large number of neurons in this layer would be problematic if we opted for a fully connected second hidden layer. We will eventually want to add some fully connected layers, but first we need to compress our representation of the input: 6,272 dimensions is too many. We can do that by using a *max pooling* layer to reduce the spatial resolution of our feature maps (to 'downsample' them). Pooling works by looking at small patches of a feature map, typically 2×2 squares, and replacing each patch with a single value. *Max* pooling means picking the maximum (over the patch) as the single value representing the patch. Patches are usually not allowed to overlap, so when we apply 2×2 max pooling to a 28×28-pixel feature map, we end up with a 14×14-pixel 'thumbnail' representation of the original. Notice that a max pooling layer introduces no new trainable parameters; it just applies a simple, fixed transformation. (We therefore speak of max pooling *units* rather than max pooling neurons.) Now you might wonder: if we want to reduce resolution, why not just downsample the input image from the start? The reason is that we need to first extract useful features from the high-resolution input. If we began by blurring the original image, we would likely have a hard time deciding whether certain features were present or not. After all, features are by definition small, localised patterns. But once we've made our feature maps, we can safely blur *those* a bit. A feature is several pixels wide. The precise location of its centre is unlikely to be crucial information for the Fashion MNIST classification task.

Let's take stock. Our eight feature maps, after max pooling, are each 14×14; so there are $8 \times 14 \times 14 = 1{,}568$ units in the max pooling layer. Max pooling has cut the size of our representation by a factor of 4. Should we now add a fully connected layer and a softmax output layer? There are two reasons to hold off. First, 1,568 units is still rather a lot. Second, there might be more juice to be squeezed from the convolution trick. Just as it made sense for the network to look for the same basic pixel patterns – the same *low-level features* – in every local patch of the original image, so it makes sense for the network to look for the same patterns of low-level features – the same *intermediate-level features* – in every local patch of the first-level feature maps. (An intermediate-level feature could be something like a vertical edge co-occurring with a certain texture, or two vertical edges occurring side by side.) We can force the network to do that by adding a second convolutional layer after the first max pooling layer. What do the convolutional filters look like for this layer? Each receives, as input, eight low-level feature maps. We imagine these as a stack of feature maps eight 'channels' deep. We'll continue to assume our filters process 3×3-pixel patches of *space*, but each such patch now corresponds to a $3 \times 3 \times 8$ block. Thus, filters in the second convolutional layer will be defined by a bias and $3 \times 3 \times 8 = 72$ weights.

Most CNN architectures have more filters in the second convolutional layer than in the first. The idea is that there are lots of potentially interesting ways in which low-level features might be combined to define intermediate-level features. Let's put 16 filters in the second convolutional layer (twice as many as in the first). That means $16 \times 14 \times 14 = 3{,}136$ neurons and $16 \times 73 = 1{,}168$ trainable parameters. After the second convolutional layer, we can avail of max pooling again. The justification is essentially the same as before: the spatial resolution we care about when *using* an intermediate-level feature map for fashion item classification is likely to be coarser than the spatial resolution we needed to compute it in the first place. Another round of 2×2 max pooling compresses the representation back down to 16 7×7-pixel feature maps: 784 units in total. This representation happens to be the same size as the raw input, but its structure is very different. Rather than a single 28×28-pixel greyscale image, it consists of 16 tiny and very low-resolution feature maps. Once the network is trained, the features that are tracked in these maps ought to be highly relevant to the Fashion MNIST classification task. That means there ought to exist – somewhere in mathematical space – a not-too-complicated function that maps the 784-dimensional representations computed by the second max pooling layer to good probability distributions over the 10 class labels. ('Good' here means low cross-entropy loss.) We know from Exercise 13.7 that we can approximate any continuous function from $\mathbb{R}^{784}$ to 10-dimensional probability vectors using one fully connected hidden layer followed by a fully connected 10-neuron softmax output layer. Moreover, if the required function is not too complicated (as we have good reason to hope), we shouldn't need an enormous number of neurons in the penultimate layer. So let's go ahead and complete our CNN architecture: after the second max pooling layer, we'll have a fully connected layer of 128 neurons, followed by the output layer. The architecture is shown in Figure 13.16. Notice that the vast majority of the trainable parameters of this

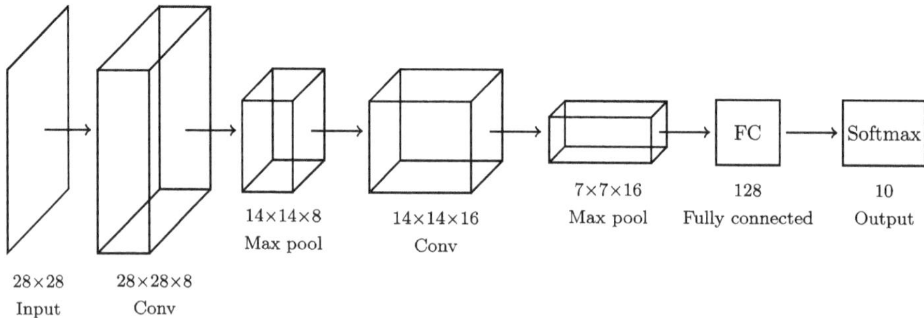

Figure 13.16 The architecture of a CNN suitable for the Fashion MNIST task. The number of units in each layer is shown in the figure. Filters in the first convolutional layer have kernels of shape 3×3; filters in the second convolutional layer have kernels of shape $3 \times 3 \times 8$. Each pooling unit takes the max over a 2×2 window. The convolutional layers and the fully connected layer FC use ReLU activation.

network are associated with the penultimate layer – the fully connected hidden layer just before the softmax output layer, labelled 'FC' in Figure 13.16.

■ ***Exercise*** 13.16 Show that there are 100,480 parameters associated with the penultimate layer, and 103,018 parameters associated with the network as a whole. So the penultimate layer accounts for 97.5% of all the network's parameters.

The earlier layers don't have many parameters associated with them because they are carrying out rather simple local (patch by patch) transformations on the input images. We greatly constrained what these layers could do when we imposed our hunches about how the earlier stages of image processing ought to work. But we have not similarly constrained the final two layers. The job of the final layers is not to perform more localised pattern detection, but to reason about the overall configuration of features across the input. Fully connected layers are well suited to this kind of computation: they treat all input units equally and are free to discover whatever complex combinations of features are useful for the task.

13.8.4 Training the CNN

Figure 13.17a shows a 30-epoch training run for our CNN on the Fashion MNIST dataset. (We used the Adam optimisation algorithm with a batch size of 64 and a learning rate of 3×10^{-4}.) The blue curve represents cross-entropy loss on the training set; the orange curve represents cross-entropy loss on the validation set. For comparison purposes, similar training runs for several different fully connected (MLP) architectures are shown in Figure 13.17b–d. As before, MLP architectures are represented as sequences of hidden layer sizes. Each MLP has roughly the same total number of parameters as the CNN. In this experiment, the CNN

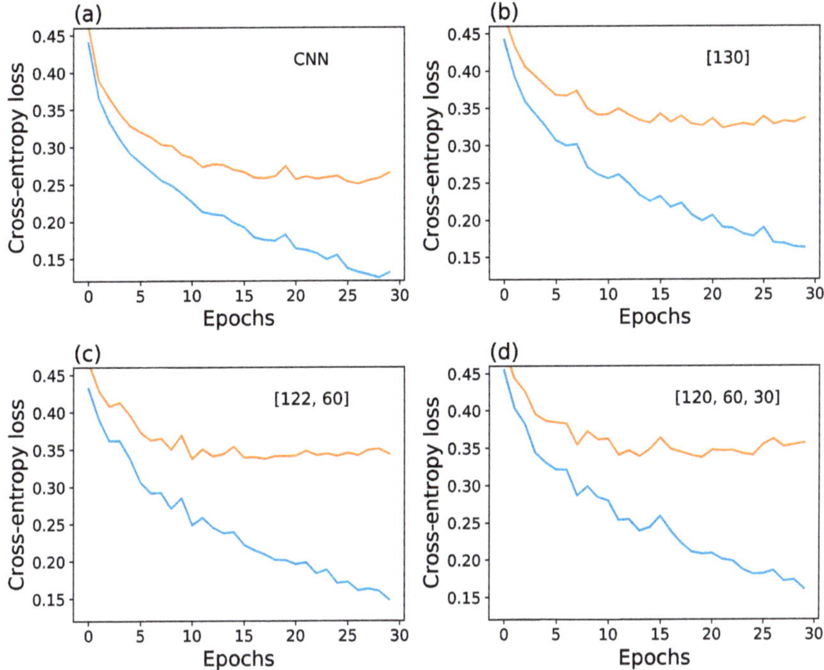

Figure 13.17 Training runs for several neural network architectures on the Fashion MNIST classification task. The blue curve represents training loss (lower series in each subplot), and the orange curve represents validation loss (upper series in each subplot). The architecture in panel (a) is the CNN described in Section 13.8. The architectures in panels (b–d) are MLPs, for comparison; each has a total parameter count similar to the CNN. See main text for further details.

has done only slightly better than the MLPs at fitting the training data. However, its cross-entropy loss on the *validation* data is significantly better than the MLPs. Validation accuracy (not plotted) tells a similar story. After 30 epochs, validation accuracy for the CNN was 91.0%; the corresponding figures for the three MLPs were 88.4%, 88.7% and 88.4%. To put it another way, the MLPs' misclassification rate on validation data, averaged across the three architectures, was about 28% higher than the CNN's (11.5% vs. 9%). In sum, the CNN is generalising more successfully from the training data. This is not surprising. MLPs are extremely flexible function approximators. That means they can learn to fit the training data well, but it also means they are prone to overfitting. For example, if (as is almost bound to be the case) different Fashion MNIST categories are *just by chance* associated with slightly different patterns of background noise in the training images, MLPs can exploit those small differences to drive down the training loss, but this sort of 'learning' won't do anything to reduce the validation loss. The CNN, by contrast, cannot exploit arbitrary pseudo-patterns in the training images. Its architectural constraints bias it towards learning local, translation-invariant features (edges, textures and shapes) that are more likely to be genuinely informative for classification. So although the CNN and the MLPs have similar numbers of trainable parameters, the CNN is using its parameters more efficiently.

■ ***Exercise*** 13.17 Why does training for only 30 epochs suffice for the present task, when we needed hundreds of epochs for the human speech sounds task (Figure 13.12)?

13.8.5 Interpreting the CNN

Now that we've trained our CNN to perform the Fashion MNIST classification task, let's take a look at its convolutional filters. Can we understand what features the network is mapping at the early processing stages, that is, before the fully connected layers? Rather than examining filter weights directly, we'll feed an input image to the network and plot the resulting pre-activations, channel by channel, in each convolutional layer. (Since we're looking at pre-activations rather than activations, we'll speak of pre-activation maps rather than feature maps.) Figure 13.18 shows the input image we used (panel (a)) and the pre-activation maps in the first convolutional layer (panels (b)–(i)). Some of these maps are easy

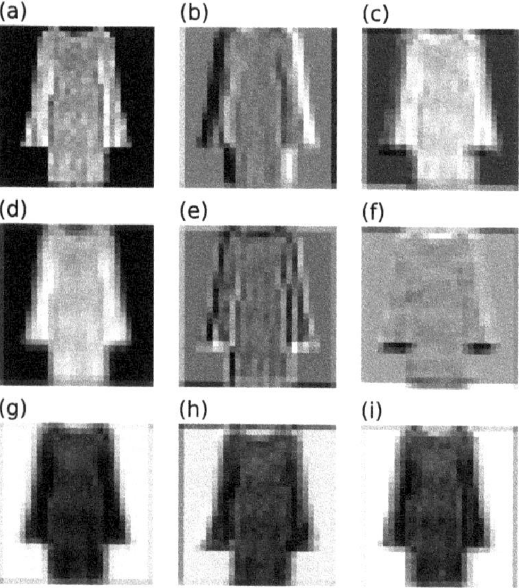

Figure 13.18 (a) A representative Fashion MNIST image. (b)–(i) Pre-activation maps of the image in (a) generated by filters in the first convolutional layer of a CNN trained on the Fashion MNIST dataset.

to interpret. The filter that generated panel (b) is looking for vertical edges of the 'bright left, dark right' type; the right-hand boundary of the dress therefore stands out intensely in panel (b), while the left-hand boundary is dark. The filter that generated panel (f) is looking for horizontal edges of the 'bright bottom, dark top' type. The filter that generated panel (d) is just looking for general brightness (its kernel will be an approximately uniform array of positive weights); effectively it serves as a blurring filter. The filters that generated panels (c) and (e) are not so easy to describe concisely, but it is plausible that they are each doing something useful. What about the filters that generated panels (g)–(i)? These three pre-activation maps look very similar: in each case, pixels outside the dress are bright, while pixels inside the dress are dark. Perhaps it is useful to have one such filter, but why does the network want three of them? It is not clear. Making sense of all the filters in the second convolutional layer is even more challenging. The 16 pre-activation maps (for the same input image) are shown in Figure 13.19. It seems that a couple of the second-convolution-layer filters are serving as

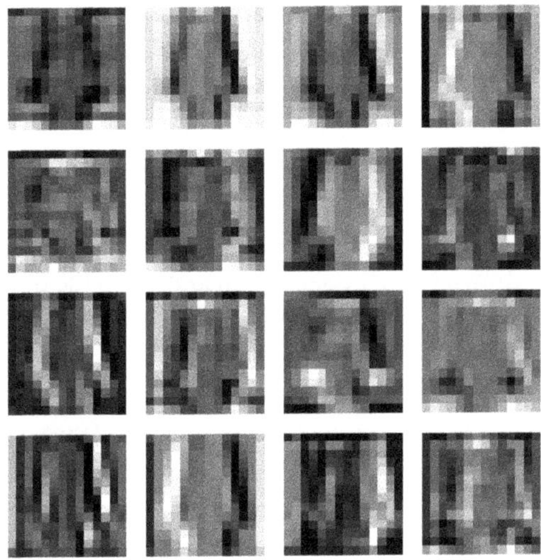

Figure 13.19 Pre-activation maps of the image in Figure 13.18a generated by filters in the second convolutional layer of a CNN trained on the Fashion MNIST dataset.

vertical edge detectors. (Intermediate-level features *can* be more complicated than low-level features, but they do not *have* to be be.) One filter – third row, third column – appears to be detecting sleeve ends. Most of the filters, however, are just hard to describe.

So, peeking under the hood of our CNN has brought up two interesting issues: redundancy (some of the filters are quite similar to each other) and interpretability (some of the filters are hard to describe). Let's take these in turn. Filter redundancy is a well-known phenomenon in deep CNNs. More generally, network architectures in which groups of parallel subnetworks perform subtasks of the same basic kind are prone to redundancy at the subnetwork level. For example, this is true of the transformer models that power the latest generation of generative AI tools. (There the parallel subnetworks are not convolutional filters grouped into convolutional layers but *attention heads* grouped into *multihead attention layers*.) After training, it will often turn out that certain subnetworks closely resemble some of their group-mates. Redundancy is not necessarily a flaw. In fact, it can contribute to trainability, by giving the optimisation process multiple slightly different paths to learning useful representations. It also means the trained network can avail of the 'wisdom of crowds': even if multiple subnetworks compute similar functions, their outputs will usually be slightly different, and the network may benefit from aggregating them. This is known as *ensembling*. Of course, if the goal is to deploy a lightweight model, the benefits of ensembling may not be worth the costs. After training, it is sometimes possible to remove redundant subnetworks – a process known as pruning – without significantly degrading the network's performance. (This is a standard network compression strategy.)

Interpretability is a very general challenge in deep learning. Describing what mathematical function a trained network computes is in principle straightforward: just draw the architecture and specify all the weights and biases. But if you ask how a CNN knows that a given fashion item is a dress, you don't want to be handed a wiring diagram and a 200-page catalogue of parameter values. You want to be told 'it can see a pair of sleeve ends near the bottom of the image and a wide open neck near the top, and it treats both features as strongly predictive of the *dress* category'. Unfortunately, human-friendly stories like that are the exception rather than the rule. With a bit of ingenuity we can find out which intermediate-level features were most important for persuading the CNN that it was seeing a dress (for example, we can temporarily disable certain filters or mask certain feature map patches, and observe the impact on the network's output), but some of those features are likely to be awkward for humans to describe, let alone reason about. This might not matter much when the CNN is used to classify examples drawn from the same distribution as the training data. We know that it works – somehow. But suppose we wanted to apply the trained CNN to examples drawn from a subtly different distribution. In the new distribution, a higher proportion of the dresses are sleeveless, many of the images are cropped to exclude necklines and all the fashion items are lit from the left. Should we still trust the CNN's verdicts? If we can't articulate the CNN's 'reasoning process', it is very hard to know which distributional shifts will tank performance and which will not. Moreover, CNNs lie at the easier end of the interpretability spectrum. At least we know they're looking for local, translation-invariant features of some sort! With large fully connected architectures, interpretation is an even stiffer challenge.

13.9 Other Architectures

Architectural innovation in deep learning is often driven by the desire to build background knowledge – usually of a rather abstract, high-level kind – into network structures. For example, convolutional architectures are motivated by the belief that useful image-processing functions are primarily sensitive to local patterns in the input, and that they should respond similarly to similar patterns regardless of spatial location. Two other influential architectural ideas, each shaped by assumptions about the kinds of functions that are likely to be useful in practice, are residual blocks and transformers.

Residual blocks are motivated by the belief that useful functions in vision and other domains often differ only slightly from the identity function. Concretely, a residual block maps input $\boldsymbol{x}$ to output $f(\boldsymbol{x}) + \boldsymbol{x}$, with f a learned residual function (perhaps computed by an MLP or a CNN). Before training begins, residual functions are initialised to lie close to the zero function; thus, residual blocks are biased towards learning deformations of identity. Deep neural networks are sometimes easier to train when they are organised as chains of such residual blocks.

Transformers, now dominant in natural language processing and increasingly used in vision, make heavy use of residual blocks. However, their core innovation is the self-attention mechanism, which allows networks to attend dynamically to relevant context across their entire input. We will not describe the architecture of self-attention here, but it is inspired by the belief that *pairwise* relationships between input tokens (e.g. words in a text) carry rich structural information. Transformers have proven astonishingly effective at modelling complex, non-local dependencies – provided that training data is very plentiful. For example, models in the GPT family are trained on datasets that include much of the internet.

Many other important architectures, such as mixture of experts, graph neural networks and diffusion models, can also be understood as attempts to build background knowledge into network structure. Deep learning researchers sometimes speak of architectures as having *inductive biases*, or as expressing *priors*. These are not priors in the formal Bayesian sense, that is, they are not explicit probability distributions over parameters. Instead, they are structural constraints that guide learning by limiting the space of functions a network explores. The constraints may be hard (e.g. a convolutional layer cannot detect patterns that are larger than its kernel size), but they may also be soft. For example, a residual block trained by gradient-based optimisation is likely to end up learning a function not too far from identity, even if the block is in principle capable of approximating any $\mathbb{R}^p \to \mathbb{R}^p$ function. Selecting task-appropriate network architectures is essential for getting the most out of the optimisation methods described in Section 13.6; it plays a role roughly analogous to model selection in other inference contexts.

13.10 Backpropagation

Having looked at some examples of neural networks trained using gradient-based optimisation techniques, we now return to the general question of how gradients are actually computed. We noted before that computing $\nabla l(\boldsymbol{\theta}; B)$ (equation (13.10)) was easier than computing $\nabla l(\boldsymbol{\theta}; D)$ (equation (13.9)). To compute, either we have to sum pointwise loss gradients, each of the form $\nabla \boldsymbol{\ell}(\boldsymbol{\theta}; \boldsymbol{\xi})$; but there will be many fewer points in a typical batch than in a typical dataset. This is (one reason) why neural network optimisers are almost always variants of SGD. But what about the individual pointwise loss gradients? How do we compute *those* efficiently? That is the question we address here. To avoid tedious repetition of the word 'pointwise', for the rest of this section and in Section 13.11 we'll just call $\boldsymbol{\ell}$ the loss.

Recall that $\nabla \boldsymbol{\ell}(\boldsymbol{\theta}; \boldsymbol{\xi})$ is the vector whose ith component is $\frac{\partial \ell(\theta;\xi)}{\partial \theta_i}$. Each component represents the sensitivity of the loss to one specific parameter – either the bias of a specific neuron, or the weight of a specific connection. To compute $\nabla \boldsymbol{\ell}(\boldsymbol{\theta}; \boldsymbol{\xi})$, we have to compute the sensitivity of the loss to every weight and every bias in the neural network. If the network is large, this going to be a big job. Suppose for concreteness that $P = 10^9$. One way (*not* a sensible way) in which we might compute $\nabla \boldsymbol{\ell}(\boldsymbol{\theta}; \boldsymbol{\xi})$ is to tackle each component separately. That is, first we compute $\frac{\partial \ell(\theta;\xi)}{\partial \theta_1}$, using whatever calculus tricks we need; we make a note of the answer and throw away any intermediate calculations. Then we do the same for $\frac{\partial \ell(\theta;\xi)}{\partial \theta_2}$, and for $\frac{\partial \ell(\theta;\xi)}{\partial \theta_3}$ and so on, continuing until we write down $\frac{\partial \ell(\theta;\xi)}{\partial \theta_P}$ after completing one billion self-contained calculus exercises. Finally, we package our answers together into a single vector, and that gives us $\nabla \boldsymbol{\ell}(\boldsymbol{\theta}; \boldsymbol{\xi})$. This is a bad strategy because it is wasteful: some of the intermediate calculations we do in order to compute $\frac{\partial \ell(\theta;\xi)}{\partial \theta_1}$ might be useful for computing $\frac{\partial \ell(\theta;\xi)}{\partial \theta_2}$ too. In fact, if the neural network is a deep one, there will inevitably be massive overlap in the intermediate results required to compute different components of the gradient vector.

Consider the bias parameter b of a neuron in an early layer. Through what mechanism is the loss sensitive to changes in this parameter? Well, if we increase the bias by δb, the neuron's pre-activation increases by δb. Unless the neuron is a 'dead' ReLU, its activation (output) will increase too. We would expect that to have an impact on the loss. But the change in the neuron's activation impacts the loss only via its impact on the activations of neurons in the *next* layer, and these activations in turn impact the loss only via their impact on activations of neurons in the layer after that, and so on. So, to compute $\frac{\partial \ell(\theta;\xi)}{\partial b}$, we are going to need to compute the sensitivity of the loss to the activations of every neuron downstream of our early-layer neuron. Then when we go on to compute $\frac{\partial \ell(\theta;\xi)}{\partial b'}$, where b' is the bias of another early-layer neuron, we're going to need all or most of those intermediate results again – because the set of downstream neurons will be very similar.

The rest of this section sketches a sensible way to compute $\nabla \ell(\boldsymbol{\theta};\boldsymbol{\xi})$. We'll begin by assuming that we already know the sensitivity of the loss to every neuron's activation. Given that information, computing $\nabla \ell(\boldsymbol{\theta};\boldsymbol{\xi})$ is easy. We'll show how it's done. Then we'll explain how the sensitivity of the loss to every neuron's activation can be computed efficiently using the backpropagation algorithm.

Things will go more smoothly if we introduce a bit of new notation. We'll give every unit in our network a unique identifier, drawn from some indexing set; subscripts such as i and j will range over this set. Since we are focusing on the network's processing of a single training example $\boldsymbol{\xi}$, it makes sense to speak of the activations of the units. We'll denote the activation of unit i by h_i. If i is a neuron (as opposed to an input node), then it also has a pre-activation, which we'll denote by z_i, and a bias, which we'll denote by b_i. When there is a direct connection from unit j to unit i, we'll denote the weight of that connection by w_{ij}. It will be helpful to have a concise way of referring to the set of units a given unit i receives direct connections from; we'll call that set parents(i). Similarly, we'll call the set of neurons unit i sends direct connections to children(i). If i is an input node, then parents(i) is the empty set. If i is an output-layer neuron, then children(i) is the empty set.

Now consider an arbitrary neuron i, depicted in Figure 13.20a. Notice that it comes with some parameters attached: a single bias b_i, and for each j in parents(i) an incoming connection weight w_{ij}. Applying the usual rule for one neuron (equation (13.1)), we have

$$h_i = \phi_i(z_i), \tag{13.12}$$

where ϕ_i is neuron i's activation function[12] and

$$z_i = b_i + \sum_{j \in \text{parents}(i)} w_{ij} h_j \tag{13.13}$$

is its pre-activation. Now, how does the loss ℓ for the current training example depend on bias b_i and on weight w_{ij} (for each $j \in$ parents(i))? To answer that question, we need to know two things. First, we need to know how h_i depends on the bias and the weights; we can work this out from equations (13.12) and (13.13). Second, we need to know how ℓ depends on h_i.

[12] We assume throughout this section that each neuron's activation is a function of its pre-activation alone. Strictly speaking this excludes softmax activation. We can, however, avail of the trick trailed in Section 13.3.2: when presented with a network whose final layer is described as a softmax layer, we can redefine the final layer as linear (so that its *pre-activation* vector is the final output of the network) and absorb the softmax operation into the loss function.

Per our plan, let's pretend we already know ℓ's sensitivity to h_i. We'll denote this quantity by $\frac{\partial \ell}{\partial h_i}$ (suppressing the explicit dependence on $\boldsymbol{\theta}$ and $\boldsymbol{\xi}$ now, for tidiness). It's important to be absolutely clear about the meaning of this symbol. We're imagining intervening in the neural network and wiggling the value of h_i a little, while holding all activations *that are not downstream of neuron i* constant. That is, we allow the little changes we make to h_i to percolate forward through the network, but we don't make any other changes. The symbol $\frac{\partial \ell}{\partial h_i}$ denotes the gradient of ℓ with respect to h_i in those circumstances. But if we have this gradient, then it is easy to write down expressions for $\frac{\partial \ell}{\partial b_i}$ and $\frac{\partial \ell}{\partial w_{ij}}$ (for any $j \in \text{parents}(i)$). We just need to use the calculus chain rule. We have

$$\frac{\partial \ell}{\partial b_i} = \frac{\partial \ell}{\partial h_i}\frac{\partial h_i}{\partial b_i},$$
$$\frac{\partial \ell}{\partial w_{ij}} = \frac{\partial \ell}{\partial h_i}\frac{\partial h_i}{\partial w_{ij}}.$$

Moreover, from equations (13.12) and (13.13) we have

$$\frac{\partial h_i}{\partial b_i} = \phi_i'(z_i),$$
$$\frac{\partial h_i}{\partial w_{ij}} = h_j\phi_i'(z_i),$$

where $\phi_i'(z)$ is the total derivative of neuron i's activation function, evaluated at its current pre-activation z_i. Thus

$$\begin{aligned}\frac{\partial \ell}{\partial b_i} &= \phi_i'(z)\frac{\partial \ell}{\partial h_i},\\ \frac{\partial \ell}{\partial w_{ij}} &= h_i\phi_i'(z)\frac{\partial \ell}{\partial h_i}.\end{aligned} \tag{13.14}$$

That completes the first stage of our story. We've seen how if we know the sensitivity of the loss to the activation of a particular neuron i, then we can calculate all components of $\nabla\ell$ associated with that neuron's bias and incoming weights. Of course, we also need the gradient of the neuron's activation function, $\phi_i'(z_i)$, and the activations on all the incoming wires, h_j for each $j \in \text{parents}(i)$. But $\phi_i'(z_i)$ is easy, and the incoming activations have to be calculated anyway when processing the current training example.

■ ***Exercise*** 13.18 Suppose ϕ_i is ReLU. Write down an expression for $\phi_i'(z_i)$ as a function of z_i. (If you're worried about the $z_i = 0$ case, refer back to the discussion of this point at the start of Section 13.6.)

Onto the second stage: computing the sensitivity of the loss to every activation in the network. We'll do this recursively, starting with output neurons and then stepping back through the network. Computing $\frac{\partial \ell}{\partial h_i}$ when i is an output neuron will always be straightforward: we just need to pay attention to the structure of the loss function. Exercise 13.19 works through an example.

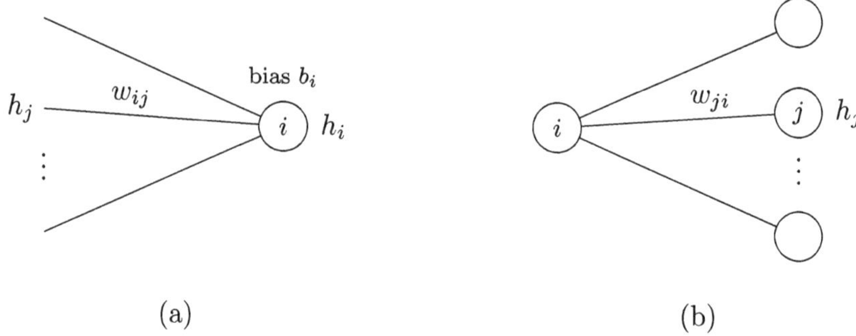

Figure 13.20 (a) Neuron i receiving input signals from its parents. A representative signal h_j is labelled, along with the weight w_{ij} of the corresponding connection. The parents themselves are not shown. The activation of neuron i, h_i, is computed per equations (13.12) and (13.13). Given $\frac{\partial \ell}{\partial h_i}$, we can therefore compute $\frac{\partial \ell}{\partial b_i}$ and $\frac{\partial \ell}{\partial w_{ij}}$ (see equation (13.14)). (b) Neuron i sending output signal h_i (not labelled) to each of its children. A representative child neuron j is labelled, along with the weight w_{ji} of the corresponding connection. The activation h_j of this child also shown. Given $\frac{\partial \ell}{\partial h_j}$ for each $j \in \text{children}(i)$, we can compute $\frac{\partial \ell}{\partial h_i}$ (see equation (13.15)).

■ ***Exercise*** 13.19 You are training a neural network to perform a regression task in which the prediction target is an m-dimensional vector. The network therefore has m output units (one for each component of the prediction target). Using our current notational scheme, every unit in the network gets a unique identifier. Let $i(j)$ be the identifier of the jth output unit, for $j = 1, ..., m$. Suppose you are using MSE as your loss function. Then the pointwise loss on training example $\boldsymbol{\xi} = (\boldsymbol{x}, \boldsymbol{y})$, where $\boldsymbol{y}$ is the true value of the prediction target, is

$$\ell = \sum_{j=1}^{m} (y_j - h_{i(j)})^2.$$

Show that the sensitivity of this loss to the jth output unit's activation is

$$\frac{\partial \ell}{\partial h_{i(j)}} = 2(h_{i(j)} - y_j).$$

Now let i be a neuron that is *not* in the output layer. Neuron i and its children are shown in Figure 13.20b. If you happen to already know $\frac{\partial \ell}{\partial h_j}$ for every $j \in \text{children}(i)$, then you can easily compute $\frac{\partial \ell}{\partial h_i}$. Applying the chain rule again, we have

$$\frac{\partial \ell}{\partial h_i} = \sum_{j \in \text{children}(i)} w_{ji} \phi'_j(z_j) \frac{\partial \ell}{\partial h_j}. \tag{13.15}$$

Equation (13.15) expresses the idea that a non-output-layer neuron impacts the loss only via its children. This simple recursion is the engine of backpropagation.

★ Box 13.1 Backpropagation algorithm

Aim: for a given training example, compute $\frac{\partial \ell}{\partial h_i}$ for every neuron i in a feedforward neural network.

1. Compute $\frac{\partial \ell}{\partial h_i}$ for every neuron i in the output layer, and 'tick off' all output-layer neurons.
2. Identify the set of unticked neurons whose children have all been ticked. Use equation (13.15) to compute $\frac{\partial \ell}{\partial h_i}$ for these neurons, then tick them off.
3. If any neurons in the network remain unticked, repeat step 2. Otherwise stop.

The computational cost of backpropagation is similar whether the network is layered or not, but the layered case is especially tidy. Consider a layered network with an input layer, $L \geq 1$ hidden layers and an output layer. We begin by computing $\frac{\partial \ell}{\partial h_i}$ for every neuron in the output layer – step 1 in Box 13.1. That ticks off all output layer neurons. At this stage, the set of unticked neurons whose children have all been ticked is, precisely, the neurons in the final (Lth) hidden layer. So in our first execution of step 2, we compute $\frac{\partial \ell}{\partial h_i}$ for all those neurons, and tick them off. If $L > 1$, we do the same for all neurons in layer $L - 1$ in our second execution of step 2, and so on, until we have taken care of every hidden layer after precisely L executions of step 2. Textbook presentations of backpropagation often assume the layered case, and use matrix algebra notation specialised for it. We presented a general version of the algorithm because we think the key recursive insight – represented here by equation (13.15) – comes across more clearly that way.

13.11 Vanishing and Exploding Gradients

Equation (13.15) is a recursive relation. Consider any $\frac{\partial \ell}{\partial h_j}$ symbol appearing on the right-hand side. So long as the neuron j in question is not an output layer neuron, we can (by appealing to equation (13.15) itself) replace this symbol with the expression $\sum_{j_1 \in \text{children}(j)} w_{j_1 j} \phi'_{j_1}(z_{j_1}) \frac{\partial \ell}{\partial h_{j_1}}$. We can perform similar substitutions for all the newly appearing $\frac{\partial \ell}{\partial h_{j_1}}$ symbols that don't correspond to output-layer neurons, and so on. Eventually this recursive substitution process will transform the right-hand side of equation (13.15) into a sum of terms that each have the form

$$w_{j_1 i} \phi'_{j_1}(z_{j_1}) w_{j_2 j_1} \phi'_{j_2}(z_{j_2}) \ldots w_{j_s j_{s-1}} \phi'_{j_s}(z_{j_s}) \frac{\partial \ell}{\partial h_{j_s}}, \tag{13.16}$$

where neuron i is a parent of neuron j_1, which is a parent of neuron j_2 and so on, with the chain terminating on an output-layer neuron j_s. Each such term corresponds to one path forward through the neural network from neuron i to the output layer.

If i is an early-layer neuron in a very deep layered network, then all these paths are long. That means each summand in the recursively expanded expression for $\frac{\partial \ell}{\partial h_i}$ is the product of an output-layer $\frac{\partial \ell}{\partial h_{j_s}}$ and *many* factors of the form $w_{j_{r+1} j_r} \phi'_{j_{r+1}}$. If these factors tend to have magnitudes that are significantly less than one, their product will be very small. If they tend to have a magnitudes significantly greater than 1, their product will be very big. Of course, we need to remember that $\frac{\partial \ell}{\partial h_i}$ is a sum over many (13.16)-shaped terms – one for each path forward from neuron i – and that these terms are signed, so partial cancellation can occur. Nevertheless, the potential for gradients to vanish or to explode is clear. Both scenarios are problematic. When gradients associated with neuron i's parameters are tiny, those parameters won't change much during gradient-based training; but since tiny gradients tend to be dominated by noise, simply amplifying them isn't a good fix. When gradients associated with neuron i's parameters are enormous, those parameters will change too quickly during training. Directly editing the gradient vector to dial the magnitudes of exploding components back down to some acceptable maximum ('gradient clipping') can improve stability. However, situations in which the loss is very sensitive to tiny changes in certain parameters are usually situations in which the loss function is very jagged, with sharp peaks and valleys squished tightly together. Optimisation is challenging in those circumstances whether one clips gradients or not.

So, we want to avoid vanishing and exploding gradients; but we also want to be able to train very deep neural networks. This is a tough problem. Historically, one helpful innovation was the widespread adoption of ReLU as a default activation function. Since $\mathrm{ReLU}'(z)$ is either 0 (the $z < 0$ 'dead ReLU' case) or 1 ($z > 0$, 'live ReLU'), the backpropagation recursion relation (13.15) for a network that uses ReLU activation for every non-output layer neuron simplifies to

$$\frac{\partial \ell}{\partial h_i} = \sum_{j \in \mathrm{live\_children}(i)} w_{ji} \frac{\partial \ell}{\partial h_j},$$

where $\mathrm{live\_children}(i)$ is the set of children of neuron i that are live on the current training example. Recursive substitution will transform the right-hand side of the equation above into a sum of terms that each have the form

$$w_{j_1 i} w_{j_2 j_1} \ldots w_{j_s j_{s-1}} \frac{\partial \ell}{\partial h_{j_s}},$$

where as before neuron i is a parent of neuron j_1, which is a parent of neuron j_2 and so on, with the chain terminating on an output-layer neuron j_s. Each such term corresponds to one path forward through the network *via live ReLUs* from neuron i to the output layer. The potential for vanishing or exploding gradients is still there, but now it's all about the weights. If we use a good random initialisation scheme (for example, drawing each weight from a zero-mean normal distribution with a well-chosen variance), then training should at least *begin* in a well-behaved region of parameter space; and with luck it will find its way to a good optimum without running into wilder regions. Needing to be a little lucky is not a deal-breaker, so long as we have the computational resources to run the training process several times from different random initialisations.

There are also architectural strategies for avoiding vanishing and exploding gradients. Organising a deep network around chains of residual blocks (Section 13.9) helps, because it ensures that every neuron is connected to the output layer by at least some short paths. Another approach is to directly control the magnitudes of signals propagating through the network by incorporating normalisation layers, for example, BatchNorm or LayerNorm. The details are beyond the scope of the present chapter.

13.12 Chapter Summary

This chapter introduced neural networks as flexible function approximators constructed by composing layers of simple processing units.

- A network with no hidden layers implements linear regression if its output layer is linear, or logistic regression if its output layer is softmax.
- Hidden layers increase expressivity. A network with one hidden layer and ReLU activations can approximate any continuous function on a closed and bounded region of input space, though approximating complex functions this way may require many hidden units.
- Across a wide range of domains, deep neural networks (with multiple hidden layers) are more efficient and scalable than shallow networks: they can represent complex functions with fewer units and are better suited to learning hierarchical structure.
- Neural networks are trained using gradient-based optimisation methods. Gradients are computed efficiently via the backpropagation algorithm. Training proceeds by iteratively adjusting the weights to reduce a loss function, using small batches of training examples.
- The training workflow includes tuning hyperparameters and comparing architectures using a validation set. Final model performance is assessed on a separate test set.
- Early stopping and the use of small batches act as implicit regularisers. Weight decay (equivalent to a ridge penalty) is sometimes used as an explicit form of regularisation.
- Convolutional neural networks use convolution and pooling layers to exploit spatial structure in image data. More generally, architectural choices often encode high-level domain-specific assumptions.

13.13 Further Exercises

■ ***Exercise*** 13.20 Figure 13.6 amounted to a heuristic proof that any continuous function from an interval $[a, b]$ to $\mathbb{R}$ can be approximated by a neural network with one input unit, a single hidden layer of ReLUs, and one linear output unit. Show how to extend the heuristic proof to show that any continuous function from an interval $[a, b]$ to $\mathbb{R}^m$ can be approximated by a neural network with one input unit, a single hidden layer of ReLUs and m linear output units.

■ ***Exercise*** 13.21 In this guided exercise, you will use Fourier analysis to show that any continuous function from a closed, bounded region of $\mathbb{R}^p$ to $\mathbb{R}^m$ can be approximated by a neural network with p input units, a single hidden layer of ReLUs and m linear output units. In other words, you will prove the universal approximation theorem. Proving the $m = 1$ special case is the main task, because the strategy you used in Exercise 13.20 to generalise to arbitrary m works here too.

You will need to use the following fact: any real continuous function on a closed, bounded region $S \subset \mathbb{R}^p$ can be approximated as closely as we like by a finite Fourier series defined on a p-dimensional box $[a_1, b_1] \times \cdots \times [a_p, b_p]$ that comfortably contains

S, that is, contains S plus some marginal padding. The general form of such a series is

$$\sum_{k \in K} \left(A_k \cos(\boldsymbol{k}^T \boldsymbol{x}) + B_k \sin(\boldsymbol{k}^T \boldsymbol{x})\right),$$

where K is a finite set of frequency vectors that all respect periodic boundary conditions associated with the box. That is, any $\boldsymbol{k} \in K$ will be of the form $(\frac{2\pi n_1}{b_1 - a_1}, ..., \frac{2\pi n_p}{b_p - a_p})^T$, for some integers $n_1, ..., n_p$. The idea is that within S the Fourier series will be a close match to the target function; we don't care about its behaviour outside S.

(a) It turns out to be helpful to imagine that you have two additional hidden layers of *linear* neurons to play with: one between the input and the hidden layer of ReLUs and another between the hidden layer of ReLUs and the linear output layer. (Assume this architecture from now on.) Any network with the expanded architecture can be collapsed back into an instance of the original, single-hidden-layer architecture, without changing the function computed by the network. Explain how.

(b) Show that any expression of the form $\boldsymbol{k}^T \boldsymbol{x}$ can be computed by a single neuron in the first hidden layer (the linear layer you are now imagining comes between the input and the ReLU layer). That is, show that for any given $\boldsymbol{k}$ the network's parameters can be arranged such that the individual neuron's activation will be $\boldsymbol{k}^T \boldsymbol{x}$ when the network's input is $\boldsymbol{x}$ (for any $\boldsymbol{x}$).

(c) Use (b) and the $\mathbb{R} \to \mathbb{R}$ version of the UAT to show that expressions of the form $\cos(\boldsymbol{k}^T \boldsymbol{x})$ or $\sin(\boldsymbol{k}^T \boldsymbol{x})$ can be approximated to any required degree of precision by a single neuron in the third hidden layer (the linear layer that you are imagining comes between the ReLU layer and the output layer).

(d) Use (c) and the Fourier approximation idea to prove the $\mathbb{R}^p \to \mathbb{R}$ version of the UAT.

(e) Finally, prove the full $\mathbb{R}^p \to \mathbb{R}^m$ version of the UAT.

Note: The fact that a function *can* be approximated by a neural network whose structure corresponds to the function's Fourier series does not imply that such a structure will typically emerge from gradient-based training protocols. Large neural networks can approximate target functions in many different ways.

■ ***Exercise*** 13.22 Consider a modified version of the architecture shown in Figure 13.5b in which the ReLU layer is replaced by a layer of neurons with sigmoid activation. Show that any continuous function from $[a, b]$ to $\mathbb{R}$ can be approximated by such a network. (Hint: sharp sigmoids can be used to approximate step functions.) Notice that this result implies that the UAT remains true if the word 'ReLUs' in the theorem statement in Section 13.4 is replaced by 'sigmoids', for the reasoning that extends the $\mathbb{R} \to \mathbb{R}$ special case to the full $\mathbb{R}^p \to \mathbb{R}^m$ theorem remains applicable.

14

Expanding the Toolkit

The problem of inference – making guesses using incomplete information – is ubiquitous in scientific and technical fields, and in life. Our aim in this book has been to introduce the core ideas of inference using a simple toolkit of methods and models. In this final chapter, we briefly explore some ways in which the toolkit can be expanded.

14.1 More Prediction Function Families

A typical regression model takes the form

$$Y = g(\boldsymbol{X}; \boldsymbol{\theta}) + \mathcal{E},$$

where $g(-; \boldsymbol{\theta})$ represents a parametric family of functions. The families we have considered have included single predictor polynomial functions, linear functions of many predictors and some special families based on prior knowledge. In the case of a single predictor, higher-order polynomials are usually not a good choice in most situations. This is because adjusting coefficients to change the value of a polynomial in one location can cause it to whip around violently elsewhere. One way to avoid this problem is to split the domain of the predictor into intervals. We then let g be a different polynomial, often a cubic, within each interval, with coefficients constrained to maintain the continuity of g and its first derivative at the joins or 'knots'. The flexibility of the resulting 'spline' function can be controlled by increasing the number of knots, allowing the function to wiggle around more. An alternative is to penalise the complexity of g via a regulariser in the loss function (or a prior). Typically we penalise the average magnitude of the *curvature* of g, giving a loss

$$l(\boldsymbol{\theta}) = \sum_{i=1}^{n} (y_i - g(x_i; \boldsymbol{\theta}))^2 + \lambda \int \left(\partial_t^2 g(t; \boldsymbol{\theta})\right)^2 dt,$$

where λ is a hyperparameter controlling strength of the penalty. The resulting function is known as a *smoothing spline*. We can also use splines in the classification setting. Consider, for example, the binary model

$$Y|\{X = x\} \sim \text{Bernoulli}(g(x; \boldsymbol{\theta})).$$

To ensure that g is constrained to $[0, 1]$, in logistic regression we let $g(x; \boldsymbol{\theta}) = \sigma(\theta_0 + \theta_1 x)$, where σ is the standard sigmoid. To increase flexibility, we can replace the linear argument $\theta_0 + \theta_1 x$ with a spline.

In the case of multiple predictors, defining function families is more challenging. A natural

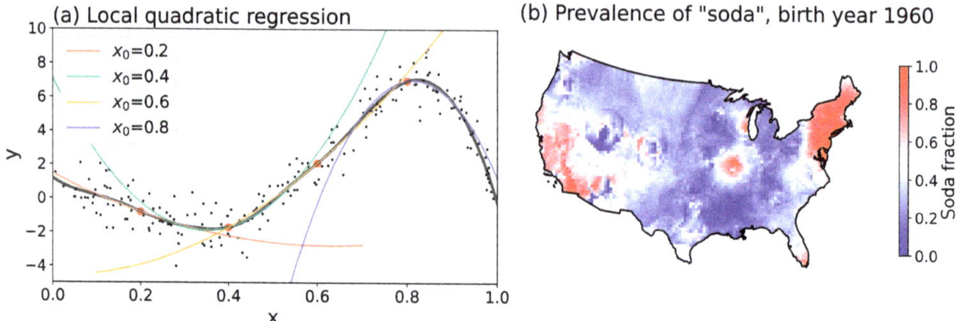

Figure 14.1 (a) The black line shows local quadratic regression function with bandwidth $\lambda = 0.1$. Coloured lines show quadratic fits at selected target points (red dots). (b) Local logistic regression model for fraction of USA population born in 1960 who use the term 'soda' to describe a sweetened carbonated beverage; data extracted from Vaux and Johndal (2025).

extension of the multiple linear model is to allow quadratic dependence on predictors, so that

$$g(\boldsymbol{x}; \boldsymbol{\theta}) = \beta_0 + \sum_{i=1}^{p} \beta_i x_i + \sum_{i=1}^{p} \beta_{ii} x_i^2 + \sum_{i=1}^{n-1} \sum_{j=i+1}^{n} \beta_{ij} x_i x_j.$$

Terms of the form $x_i x_j$ are said to model *interactions* between predictors. If we want to capture more complex dependencies on individual predictors while avoiding the problems of higher-degree polynomials, one possibility is to ignore interactions and write g as a sum of univariate functions, one for each predictor, that is,

$$g(\boldsymbol{x}; \boldsymbol{\theta}_1, \ldots, \boldsymbol{\theta}_p) = \theta_0 + g_1(x_1; \boldsymbol{\theta}_1) + \cdots + g_p(x_p; \boldsymbol{\theta}_p).$$

This is known as a *generalised additive model* (GAM). GAMs allow us to use function families developed for one-dimensional problems, such as splines, in the multidimensional setting.

An elegant alternative to picking out a single member of a parameterised function family to serve as the regression function is to pick a different member for each predictor value $\boldsymbol{x}_0$, using a loss function that weights training points that are closer to $\boldsymbol{x}_0$ more heavily. For example, the locally weighted version of the residual sum of squares loss function is

$$l(\boldsymbol{\theta}; \boldsymbol{x}_0) = \sum_{i=1}^{n} K_\lambda(\boldsymbol{x}_0, \boldsymbol{x}_i)(y_i - g(\boldsymbol{x}_i; \boldsymbol{\theta}))^2,$$

where $K_\lambda(\boldsymbol{x}_0, \boldsymbol{x})$ is a non-negative *kernel* which, viewed as a function of $\boldsymbol{x}$, is concentrated in a region centred on $\boldsymbol{x} = \boldsymbol{x}_0$ with size controlled by the *bandwidth* λ. A common choice is the Gaussian kernel $K_\lambda(\boldsymbol{x}_0, \boldsymbol{x}) = \exp(-\|\boldsymbol{x} - \boldsymbol{x}_0\|^2/\lambda^2)$. The predicted response at $\boldsymbol{x}_0$ is then $\hat{y}_0 = g(\boldsymbol{x}_0; \hat{\boldsymbol{\theta}}(\boldsymbol{x}_0))$, where

$$\hat{\boldsymbol{\theta}}(\boldsymbol{x}_0) = \arg\min_{\theta} l(\boldsymbol{\theta}; \boldsymbol{x}_0).$$

Local regression is computationally expensive because we need to fit a new local model at every point where we want a prediction. However, because local models only need to capture local predictor–response relationships, we can draw them from much simpler function families than would be required for a good global fit. Figure 14.1a shows an example of local regression using a quadratic prediction function. In Figure 14.1b local logistic regression has been used to estimate the fraction of people born in 1960 who use 'soda' as their generic term for a sweetened carbonated beverage, as a function of geographical location in the USA. The data for this model, extracted from the Cambridge Online Survey of World Englishes by Vaux and Johndal (2025), is of the form $D = \{(\boldsymbol{x}_i, y_i)\}_{i=1}^n$, where $n \approx 8 \times 10^4$ and the features of the predictor $\boldsymbol{x}_i = (x_{i1}, x_{i2}, t_i)$ are the latitude, longitude and year of birth of survey respondent i. The response variable $y_i \in \{0, 1\}$ is the indicator that they use the term 'soda' rather than some other term like 'pop', 'coke' or 'fizzy pop'. The map in Figure 14.1b was generated by fitting a model of the form

$$\mathbf{Pr}(Y = 1 | \boldsymbol{X} = \boldsymbol{x}) = g(\boldsymbol{x}; \boldsymbol{\theta}) = \sigma(\theta_0 + \theta_1 x_1 + \theta_2 x_2 + \theta_3 t)$$

at each test point. The data used for training at each test point consisted of the 100 geographically closest respondents out of all those born within 10 years of the target year (1960). Although each local model is very simple, the model as a whole is able to capture complex global usage patterns.

We have highlighted three particularly useful model families – splines, GAMs and local models – but there are many more possibilities, each with different strengths and weaknesses. We recommend Hastie et al. (2001) or Murphy (2022) as comprehensive surveys.

14.2 Differential Equations and Stochastic Models

Scientific models often take the form of differential equations which describe the dynamics of the systems we are interested in. How should we incorporate these models into our inference framework? The most direct approach is to view a differential equation model simply as a way to specify a parametric function family – the family of solutions to the model.

As a concrete example consider a nature reserve, and let $N(t)$ be the size of its elephant population at time t. The population will evolve stochastically, but we can construct a differential equation model of its expected value $n(t) = \mathbb{E}(N(t))$. One simple possibility is the logistic growth model of a population with access to limited resources,

$$\frac{dn}{dt} = rn\left(1 - \frac{n}{K}\right),$$

where r is the intrinsic growth rate and K is the nature reserve's carrying capacity (the maximum sustainable population). This model assumes that when the population is well below the carrying capacity, growth rate is proportional to population size, leading to exponential growth. As the population approaches K, the growth rate tails off because resources are running out. The solution to this equation subject to initial condition $n(0) = n_0$ defines a three-parameter family of models,

$$g(t; \boldsymbol{\theta}) = \frac{K n_0}{n_0 + (K - n_0)e^{-rt}},$$

where $\boldsymbol{\theta} = (n_0, r, K)$. Imagine the population size has been observed at a series of times, yielding the dataset $D = \{(t_i, n_i)\}_{i=1}^{M}$ where n_i is the count at time t_i. One very simple approach is to view these pairs as realisations of the random variables (T, N). If we want to predict the count given the time, we can fit a conditional probabilistic model of N given $T = t$. A sensible choice of model is

$$N|\{T = t\} \sim \text{Poisson}(g(t; \boldsymbol{\theta})),$$

and we are now at a familiar point in the inference pipeline. We could write down the likelihood function and obtain maximum likelihood parameter estimates, or we could take a Bayesian approach. The key point is that differential equation models fit naturally into the inference procedures we have learned, because we can view them as a means to encode scientific knowledge into function families.

In the elephant population example, the process being modelled was inherently stochastic. In such situations, a more sophisticated approach is to directly model the details of the stochastic evolution. If we assume a Markov process model (see Chapter 10) with parameter vector $\boldsymbol{\theta}$ then, assuming ordered observation times $t_0 < t_1 < \cdots < t_M$, the likelihood will be expressible as

$$\mathcal{L}(\boldsymbol{\theta}) = \mathbf{Pr}_{\boldsymbol{\theta}}(N(t_0) = n_0) \prod_{i=1}^{M} \mathbf{Pr}_{\boldsymbol{\theta}}(N(t_{i+1}) = n_{i+1}|N(t_i) = n_i),$$

where $\mathbf{Pr}_{\boldsymbol{\theta}}(N(t_{i+1}) = n_{i+1}|N(t_i) = n_i)$ is the probability of the population changing from size n_i at time t_i to size n_{i+1} at time t_{i+1}, according to our model. The analysis of stochastic processes is a big subject which we have only touched on lightly in this book.[1] The key point is that if we have such a model, it will fit naturally within the inference framework we have presented. As long as we can write down a likelihood, we can proceed.

14.3 Generative Language Models

At the time of writing (2026), machine learning models are emerging with astonishing *generative* capabilities. They are capable of providing long erudite answers to almost any question you care to ask, or of creating high quality original images, or music, given a short text prompt. The output of generative language models feels eerily like the product of an intelligent mind. Many of these models are probabilistic – that is they are models of probability distributions, or stochastic processes. In this final section, we focus on the GPT (generative pre-trained transformer) language model. We aim merely to sketch its high-level modelling strategy; readers keen to dig into architectural details should consult the references provided in Section 13.9.

At heart, GPT defines a probability mass function over sequences of tokens, implemented using a neural network based on the transformer architecture. For example, OpenAI's GPT-4o tokenises the phrase 'Gruber loves Stilton!' as follows:

$$\text{Gr|uber| loves| Stil|ton|!} = (t_1, t_2, t_3, t_4, t_5, t_6).$$

The set of all possible token values is the *vocabulary*, $\mathcal{V}$, of the model. Vocabulary size can be large ($|\mathcal{V}| \approx 10^4$), and each token may be viewed as a one-hot vector in $\mathbb{R}^{|\mathcal{V}|}$. Inside

[1] We recommend Grimmett and Stirzaker (2001) for an in-depth introduction.

the model, each token is represented as a lower-dimensional vector which also encodes the position of the token in the sequence. To keep our explanations simple, we'll describe the operation of GPT in terms of words.

Suppose we have a vocabulary of size K, with every word assigned an integer label from the set $\{1, \ldots, K\}$. Any sequence of N words may then be viewed as a realisation of the sequence of random labels $W_1, \ldots, W_N$. Without loss of generality, the probability distribution over the sequence can be factorised as

$$f_{W_1,\ldots,W_n}(w_1, \ldots, w_N) = f_{W_1}(w_1) \prod_{i=2}^{N} f_{W_i|W_1,\ldots,W_{i-1}}(w_i|w_1, \ldots, w_{i-1}).$$

For a four-word sequence ($N = 4$), this factorisation of the joint distribution has the DAG representation

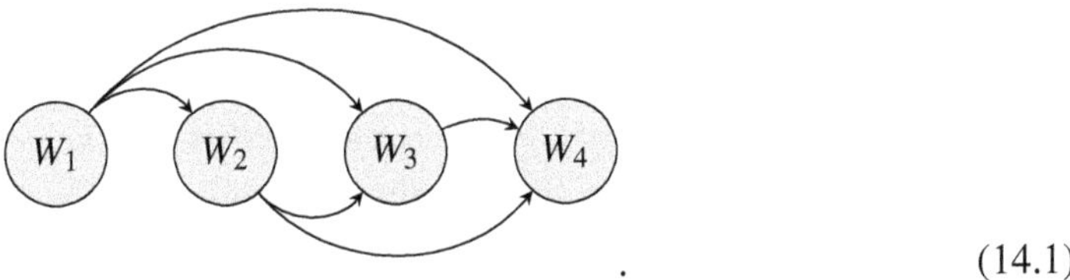

. (14.1)

Any model of the conditional mass functions in this factorisation can be used generatively to produce new text. For example, suppose we start with two words; then the next word can be generated by sampling from $f_{W_3|W_1,W_2}(-|w_1, w_2)$ giving us a three-word sequence. We then sample the fourth word from $f_{W_4|W_1,W_2,W_3}(-|w_1, w_2, w_3)$, and so on. Models that predict the next state of a sequence given a set of previous states are known as *autoregressive*. The maximum sequence length we can condition on is the *context length*. We ignore words that are further back in the sequence than this. Current GPT models have context lengths of hundreds of thousands of tokens. When it comes to estimating conditional probabilities, one possibility is to use empirically observed sequence frequencies. Unfortunately almost all grammatically correct sentences have *never* been written.

■ ***Exercise*** 14.1 A conservative estimate for the size of a typical vocabulary is 10^4 words. If one in a billion random 10-word sentences are grammatically correct, how many grammatically correct 10-word sentences are there?

Instead, we can estimate the conditional probabilities by utilising methods we have learned in this book. Given a sequence of p words, written as a predictor vector $\boldsymbol{x} = (w_1, \ldots, w_p)$, we want to estimate the probability mass function of the next word, $y = w_{p+1}$. Viewing $\boldsymbol{x}$ and y as realisations of the random pair $(\boldsymbol{X}, Y) = ((W_1, \ldots, W_p)^T, W_{P+1})$, we want to learn a vector-valued function $\boldsymbol{g}$ such that

$$Y|\{\boldsymbol{X} = \boldsymbol{x}\} \sim \text{Categorical}(g_1(\boldsymbol{x}), \ldots, g_K(\boldsymbol{x})).$$

We tackled the problem of learning functions of this kind in Chapter 11. Given training data $D = \{(\boldsymbol{x}_i, y_i)\}_{i=1}^n$, if $\boldsymbol{\theta}$ is the parameter vector of $\boldsymbol{g}$, then the negative log-likelihood is

$$l(\boldsymbol{\theta}) = - \sum_{(\boldsymbol{x},y)\in D} \log g_y(\boldsymbol{x}),$$

also known as the cross-entropy loss. To estimate conditional probabilities, we need to define a model for $\boldsymbol{g}$ and minimise this loss.

The approach of GPT is to define a neural network model which defines all conditional probabilities at once. The training data consists of extremely long sequences of text – roughly speaking, all the public-domain documents in the world. The input to the network is a subsequence of M words (tokens, in the real thing). These words are first converted into vector representations that also encode their positions in the sequence. The resulting sequence of vectors is then processed by a neural network with a transformer-based architecture. Self-attention layers *attend* dynamically to relevant context words across their input, allowing them to learn semantic relationships between words. Autoregressive language models use *masked* self-attention layers which attend only to words which appear earlier in the sequence. The output of GPT is a sequence of M probability mass functions over the vocabulary with the mth output viewed as the conditional mass function of the next word, given the first m words in the input. To compute a cross-entropy loss, we need a prediction target. That of course is supplied by the original sequence, shifted by one word. For example, consider the sentence *My uncle Gary loves Stilton cheese*. If we use *My uncle Gary* as the input, then the target is *uncle Gary loves*, and the probabilities relevant to loss computations are

$$\mathbf{Pr}(\text{uncle}|\text{My}), \quad \mathbf{Pr}(\text{Gary}|\text{My uncle}), \quad \mathbf{Pr}(\text{loves}|\text{My uncle Gary}).$$

By repeatedly passing such sequences through the network and minimising cross-entropy loss using gradient-based methods, we eventually arrive at a trained model.

The explanation we have provided here is, of course, highly simplified. We have not described the details of the network, nor the full training process which involves several extra stages and incorporates human feedback. Nevertheless, we hope readers will be excited by the fact that simple principles of probabilistic inference play such an important role in the development of models with human-like abilities.

14.4 Chapter Summary

In this chapter, we have sketched out ways in which the toolkit of inference methods we have covered in Chapters 2–13 can be expanded. We have also explained how probabilistic modelling ideas are exploited in large language models.

- **Splines** are piecewise (local) low-degree polynomial functions. The boundaries between the pieces where the local polynomials join are known as **knots**. Flexibility can be controlled using a **curvature penalty**.
- **Generalised additive models** (GAMs) extend multiple linear regression by replacing each linear term with a non-linear function of a single component of the predictor vector.
- In **local regression**, the predicted response at any given point in predictor space is found by fitting a regression model to the data near that point.
- The solutions to a **differential equation** may be viewed as a function family depending on equation parameters and boundary conditions. This family may then be used to define a probabilistic model.
- **Stochastic process** models define parameterised probability distributions over sequences of observations, allowing inference to be performed using the methods we have learned.

Appendix A

Probability Theory

Probability theory is a mathematical framework for representing uncertainty by assigning numerical values – called probabilities – to possible *events*. We often pick out events by way of statements: we speak of the event that a statement is true. Some example statements are 'my next die roll will be even', 'I will live beyond the age of 80' and 'Dave stole my pencil case'. Whether a statement is true is decided by what we call an *experiment* or *trial*. The set of possible outcomes of an experiment is its *sample space* Ω. Events are *subsets* of Ω. For example, the experiment of rolling a die has sample space $\Omega = \{1, 2, 3, 4, 5, 6\}$ and the event that the roll is even is $A = \{2, 4, 6\}$. In some cases, the experiment will be *repeatable*, but it need not be. Given two events A and B, $A \cap B$ is the event that they both occur and $A \cup B$ is the event that at least one occurs. If $A \cap B = \emptyset$, then A and B are said to be *disjoint* or *mutually exclusive* events. That is, they cannot both occur in the same experiment.

A *probability measure* is a function, **Pr**, which assigns *probabilities* to events, where $\mathbf{Pr}(A) \geq 0$, $\mathbf{Pr}(\Omega) = 1$ and if $A_1, A_2, \ldots$ are disjoint (mutually exclusive) events then $\mathbf{Pr}(A_1 \cup A_2 \cup \ldots) = \mathbf{Pr}(A_1) + \mathbf{Pr}(A_2) + \ldots$. We define the *conditional probability* that A has occurred given that B has occurred as $\mathbf{Pr}(A|B) = \mathbf{Pr}(A \cap B)/\mathbf{Pr}(B)$, where it is assumed that $\mathbf{Pr}(B) > 0$. A double application of this definition gives Bayes' theorem, $\mathbf{Pr}(A|B) = \mathbf{Pr}(B|A)\mathbf{Pr}(A)/\mathbf{Pr}(B)$. Events are said to be *independent*, written $A \perp\!\!\!\perp B$, if $\mathbf{Pr}(A \cap B) = \mathbf{Pr}(A)\mathbf{Pr}(B)$. If $A \perp\!\!\!\perp B$ then $\mathbf{Pr}(A|B) = \mathbf{Pr}(A)$. Suppose that $A_1, \ldots, A_n$ are disjoint events whose union is Ω. They are said to form a *partition* of Ω. The set of events $\{B \cap A_i\}_{i=1}^{n}$ is a partition B. Via the definition of a probability measure, and the definition of conditional probability, we arrive at the *law of total probability* (LOTP), $\mathbf{Pr}(B) = \sum_{i=1}^{n} \mathbf{Pr}(B \cap A_i) = \sum_{i=1}^{n} \mathbf{Pr}(B|A_i)\mathbf{Pr}(A_i)$.

Statistical models describe processes that generate numbers. To identify such processes with probability experiments, we define the concept of a *random variable* that maps elements of the sample space to numbers. Formally, a random variable X is a function $X \colon \Omega \to \mathbb{R}$. For example, when tossing a coin $\Omega = \{H, T\}$, and the function defined by $X(H) = 1$, $X(T) = 0$ is a Bernoulli random variable. Informally, we think of random variables as numerical outcomes of experiments, and events as subsets of the possible values they can take. In fact such events are correctly understood as subsets of Ω. That is, $\{X \in A\}$ is shorthand for $\{\omega \in \Omega | X(\omega) \in A\} \subseteq \Omega$.

The *distribution function* of a random variable X is defined $F_X(x) = \mathbf{Pr}(X \leq x)$. A *discrete random variable* can take a discrete set of values $S = \{x_1, , x_2, \ldots\}$ (finite or countably infinite) and is described by a *probability mass function* $f_X(x) = \mathbf{Pr}(X = x)$, where $\sum_{x \in S} f_X(x) = 1$. The distribution function of a discrete random variable will 'step up' by $f_X(x_i)$ at each $x_i \in S$. Random variables with smooth distribution functions are known as *continuous*. A random variable X is continuous if its distribution function can be written $F_X(x) = \int_{-\infty}^{x} f_X(u)du$, where f_X is a non-negative function called the *probability density* of X, satisfying $f_X(x) = F'_X(x)$ if this derivative exists. Since $\mathbf{Pr}(X \in (a, b]) = F_X(b) - F_X(a)$, if X is continuous $\mathbf{Pr}(X \in (a, b]) = \int_a^b f_X(x)dx$, and $\mathbf{Pr}(X = x) = 0$. Thus, the probability density $f_X(x)$ is *not* the probability that $X = x$, which is zero.

If X is a continuous random variable and $g: \mathbb{R} \to \mathbb{R}$ is a function, then $Y = g(X)$ is also a random variable. If g is strictly increasing then $F_Y(y) = \mathbf{Pr}(Y \le y) = \mathbf{Pr}(g(X) \le y) = \mathbf{Pr}(X \le g^{-1}(y)) = F_X(g^{-1}(y))$. Differentiating the relation $F_Y(y) = F_X(g^{-1}(y))$ yields the *change of variable formula* $f_Y(y) = f_X(g^{-1}(y))\frac{d}{dy}g^{-1}(y)$.

Given two random variables X and Y defined on the same sample space, their joint distribution function is defined $F_{X,Y}(x, y) = \mathbf{Pr}(X \le x, Y \le Y)$, where the argument of $\mathbf{Pr}$ is shorthand for the intersection $\{X \le x\} \cap \{Y \le y\}$. If X and Y are discrete then their joint mass function is $f_{X,Y}(x, y) = \mathbf{Pr}(X = x, Y = y)$. The *marginal* mass function of X is defined $f_X(x) = \mathbf{Pr}(X = x) = \sum_y f_{X,Y}(x, y)$, with a similar expression for the marginal of Y. The conditional mass function of X given Y is $f_{X|Y}(x|y) = \mathbf{Pr}(X = x|Y = y) = \mathbf{Pr}(X = x, Y = y)/\mathbf{Pr}(Y = y) = f_{X,Y}(x, y)/f_Y(y)$. Two random variables X and Y are said to be *independent*, written $X \perp\!\!\!\perp Y$, if $\mathbf{Pr}(X \in A, Y \in B) = \mathbf{Pr}(X \in A)\mathbf{Pr}(Y \in B)$ for all A, B, or (equivalently) if $f_{X,Y}(x, y) = f_X(x)f_Y(y)$ for all x, y. Joint distributions of multiple random variables are often written in vector form. For example, the distribution function of the random vector $\boldsymbol{X} = (X_1, X_2, \ldots, X_n)$ is $F_{\boldsymbol{X}}(\boldsymbol{x}) = \mathbf{Pr}(X_1, \le x_1, \ldots, X_n \le x_n)$. The individual elements of $\boldsymbol{X}$ are independent if and only if $F_{\boldsymbol{X}}(\boldsymbol{x}) = \prod_{i=1}^{n} F_{X_i}(x_i)$.

If X and Y are continuous then their distribution function is expressible in the form $F_{X,Y}(x, y) = \int_{-\infty}^{x}\int_{-\infty}^{y} f_{X,Y}(u, v)dudv$, where $f_{X,Y}(x, y)$ is their joint density function. The *marginal* density of X is $f_X(x) = \int_{-\infty}^{\infty} f_{X,Y}(x, y)dy$. The conditional density of X given Y and the definition of independence take the same form as in the discrete case, replacing all mass functions with the corresponding densities.

The *expectation* or mean of a random variable X is defined as follows:

$$\mathbb{E}(X) = \int xdF_X(x) = \begin{cases} \sum_x xf_X(x), & \text{if } X \text{ is discrete,} \\ \int xf_X(x)dx, & \text{if } X \text{ is continuous.} \end{cases}$$

Here, the notation $\int xdF_X(x)$ is used as shorthand for the discrete and continuous versions of the expectation. Given the distribution of X, we can calculate the expectation of the random variable $Y = g(X)$ using the law of the unconscious statistician (LOTUS), $\mathbb{E}(Y) = \mathbb{E}(g(X)) = \int g(x)dF_X(x)$. The bivariate versions are $\mathbb{E}(g(X,Y)) = \sum_x \sum_y g(x, y)f_{X,Y}(x, y)$ in the discrete case and $\mathbb{E}(g(X,Y)) = \int g(x, y)f_{X,Y}(x, y)dxdy$ in the continuous case. Generalisations to three or more variables are as the reader would expect. Two expectations of special importance are the covariance $\text{Cov}(X, Y) = \mathbb{E}((X - \mathbb{E}(X))(Y - \mathbb{E}(Y)))$ and variance $\text{Var}(X) = \text{Cov}(X, X)$. Conditional expectation is defined as follows:

$$\mathbb{E}(X|Y = y) = \begin{cases} \sum_x xf_{X|Y}(x|y) & \text{if } X \text{ is discrete,} \\ \int xf_{X|Y}(x|y)dx & \text{if } X \text{ is continuous.} \end{cases}$$

If we do not specify the value of Y in the above, then $\mathbb{E}(X|Y)$ may be understood as a random variable depending on the value of Y. It may be shown that $\mathbb{E}(\mathbb{E}(X|Y)) = \mathbb{E}(X)$.

Appendix B

Linear Algebra

A *vector space* is a set of elements called *vectors* equipped with two operations: addition and scalar multiplication. We will be interested in the vector space $\mathbb{R}^n$. Its elements are lists of real numbers of the form

$$\boldsymbol{x} = \begin{pmatrix} x_1 \\ \vdots \\ x_n \end{pmatrix}.$$

By default, vectors are columns. The *transpose* operation flips columns to rows and vice versa. The transpose of the column vector $\boldsymbol{x}$ is the row vector $\boldsymbol{x}^T = (x_1, \ldots, x_n)$. We may write $\boldsymbol{x} = (x_1, \ldots, x_n)^T$ – a space-saving notational trick. A vector of special importance is the *standard unit* or *one-hot* vector $\boldsymbol{e}_k$, every entry of which is zero except for entry k, which has value 1. To add two vectors in $\mathbb{R}^n$, we add their individual components. So if $\boldsymbol{y} = (y_1, \ldots, y_n)^T$, then $\boldsymbol{x} + \boldsymbol{y} = (x_1 + y_1, \ldots, x_n + y_n)^T$. Multiplication by a scalar is defined by $\lambda \boldsymbol{x} = (\lambda x_1, \ldots, \lambda x_n)^T$.

A matrix is a two-dimensional array of numbers,

$$\underline{\boldsymbol{A}} = \begin{pmatrix} a_{11} & a_{12} & \ldots & a_{1n} \\ a_{21} & a_{22} & \ldots & a_{2n} \\ \vdots & \vdots & \ddots & \vdots \\ a_{m1} & a_{m2} & \ldots & a_{mn} \end{pmatrix}.$$

This matrix has m rows and n columns. The set of all such matrices with real number elements is denoted $\mathbb{R}^{m \times n}$. We denote the element in row i, column j of $\underline{\boldsymbol{A}}$ by A_{ij} or $[\underline{\boldsymbol{A}}]_{ij}$. In the example matrix above, $A_{ij} = a_{ij}$. To transpose a matrix we change rows into columns so the ith row of $\underline{\boldsymbol{A}}$ becomes the ith column of $\underline{\boldsymbol{A}}^T$. Elementwise, $[\underline{\boldsymbol{A}}^T]_{ij} = [\underline{\boldsymbol{A}}]_{ji} = A_{ji}$. The product of two matrices $\underline{\boldsymbol{A}} \in \mathbb{R}^{m \times n}$ and $\underline{\boldsymbol{B}} \in \mathbb{R}^{n \times p}$ is the matrix $\underline{\boldsymbol{C}} = \underline{\boldsymbol{A}}\,\underline{\boldsymbol{B}} \in \mathbb{R}^{m \times p}$ defined by

$$C_{ij} = [\underline{\boldsymbol{A}}\,\underline{\boldsymbol{B}}]_{ij} = \sum_{k=1}^{n} A_{ik} B_{kj}. \tag{B.1}$$

This product is defined only when the number of columns of $\underline{\boldsymbol{A}}$ is equal to the number of rows of $\underline{\boldsymbol{B}}$. When $\underline{\boldsymbol{A}} \in \mathbb{R}^{n \times n}$ and $\underline{\boldsymbol{B}} \in \mathbb{R}^{n \times n}$ (they are *square* with the same dimensions), then both $\underline{\boldsymbol{A}}\,\underline{\boldsymbol{B}}$ and $\underline{\boldsymbol{B}}\,\underline{\boldsymbol{A}}$ are defined. However, matrix multiplication is in general not *commutative*, so it can be the case that $\underline{\boldsymbol{A}}\,\underline{\boldsymbol{B}} \neq \underline{\boldsymbol{B}}\,\underline{\boldsymbol{A}}$.

We can view column and row vectors as one-column and one-row matrices, respectively. Applying definition (B.1) then allows us to calculate various different products of vectors and matrices. The matrix product of a row vector with a column vector is a single number $\boldsymbol{x}^T\boldsymbol{y} = \sum_{i=1}^n x_i y_i = \boldsymbol{y}^T\boldsymbol{x}$ known as the *inner product* or *dot product* of $\boldsymbol{x}$ and $\boldsymbol{y}$. The *norm* of a vector is $\|\boldsymbol{x}\| = (\boldsymbol{x}^T\boldsymbol{x})^{1/2}$. The *outer product* of two vectors $\boldsymbol{x} \in \mathbb{R}^m$ and $\boldsymbol{y} \in \mathbb{R}^n$ is the matrix $\boldsymbol{x}\boldsymbol{y}^T \in \mathbb{R}^{m\times n}$ with entries given by $[\boldsymbol{x}\boldsymbol{y}^T]_{ij} = x_i y_j$. The matrix-vector product $\underline{\boldsymbol{A}}\boldsymbol{x}$ is a vector with ith component $[\underline{\boldsymbol{A}}\boldsymbol{x}]_i = \sum_j A_{ij}x_j$. It is sometimes useful to think of a matrix as a row of column vectors or as a column of row vectors, as in

$$\underline{\boldsymbol{A}} = \begin{pmatrix} | & | & & | \\ \boldsymbol{a}_1 & \boldsymbol{a}_2 & \dots & \boldsymbol{a}_n \\ | & | & & | \end{pmatrix} \quad \text{or} \quad \underline{\boldsymbol{B}} = \begin{pmatrix} — & \boldsymbol{b}_1^T & — \\ — & \boldsymbol{b}_2^T & — \\ & \vdots & \\ — & \boldsymbol{b}_m^T & — \end{pmatrix}.$$

Using this notation, we can write the product of two matrices as follows:

$$\underline{\boldsymbol{A}}\,\underline{\boldsymbol{B}} = \begin{pmatrix} — & \boldsymbol{a}_1^T & — \\ — & \boldsymbol{a}_2^T & — \\ & \vdots & \\ — & \boldsymbol{a}_m^T & — \end{pmatrix} \begin{pmatrix} | & | & & | \\ \boldsymbol{b}_1 & \boldsymbol{b}_2 & \dots & \boldsymbol{b}_p \\ | & | & & | \end{pmatrix} = \begin{pmatrix} \boldsymbol{a}_1^T\boldsymbol{b}_1 & \boldsymbol{a}_1^T\boldsymbol{b}_2 & \dots & \boldsymbol{a}_1^T\boldsymbol{b}_p \\ \boldsymbol{a}_2^T\boldsymbol{b}_1 & \boldsymbol{a}_2^T\boldsymbol{b}_2 & \dots & \boldsymbol{a}_2^T\boldsymbol{b}_p \\ \vdots & \vdots & \ddots & \vdots \\ \boldsymbol{a}_m^T\boldsymbol{b}_1 & \boldsymbol{a}_m^T\boldsymbol{b}_2 & \dots & \boldsymbol{a}_m^T\boldsymbol{b}_p \end{pmatrix}.$$

That is, $[\underline{\boldsymbol{A}}\,\underline{\boldsymbol{B}}]_{ij} = \boldsymbol{a}_i^T\boldsymbol{b}_j$. A square matrix of special importance is the *identity matrix*, $\underline{\boldsymbol{I}}$, defined elementwise as $[\underline{\boldsymbol{I}}]_{ij} = \delta_{ij}$, where δ_{ij} is the *Kronecker delta* symbol, defined by

$$\delta_{ij} = \begin{cases} 1 & \text{if } i = j, \\ 0 & \text{otherwise.} \end{cases}$$

The 'diagonal' elements of $\underline{\boldsymbol{I}}$ are all equal to 1, with all other elements equal to 0. We express this using the notation $\underline{\boldsymbol{I}} = \text{diag}(1, \dots, 1)$. For any square matrix $\underline{\boldsymbol{A}}$, $\underline{\boldsymbol{A}}\,\underline{\boldsymbol{I}} = \underline{\boldsymbol{I}}\,\underline{\boldsymbol{A}} = \underline{\boldsymbol{A}}$.

The *span* of a set of vectors $\{\boldsymbol{v}_1, \dots, \boldsymbol{v}_n\}$ is the set of all vectors of the form $\boldsymbol{x} = \lambda_1\boldsymbol{v}_1 + \dots + \lambda_n\boldsymbol{v}_n$. That is,

$$\text{span}(\{\boldsymbol{v}_1, \dots, \boldsymbol{v}_n\}) = \left\{\boldsymbol{x}\,\middle|\,\boldsymbol{x} = \sum_{i=1}^n \lambda_i\boldsymbol{v}_i\right\}.$$

If $\boldsymbol{x} = \boldsymbol{0}$ for some choice of the scalars $\lambda_1, \dots, \lambda_n$ other than $\lambda_i = 0$ for all i, then the set of vectors $\{\boldsymbol{v}_1, \dots, \boldsymbol{v}_n\}$ is said to be *linearly dependent*; otherwise it is *linearly independent*. It can be shown that if $\{\boldsymbol{v}_1, \dots, \boldsymbol{v}_n\}$ is any set of n linearly independent vectors, with each $\boldsymbol{v}_i \in \mathbb{R}^n$, then $\text{span}(\{\boldsymbol{v}_1, \dots, \boldsymbol{v}_n\}) = \mathbb{R}^n$. Any set of linearly independent vectors which span a vector space is called a *basis* for that space. The *standard* or *canonical* basis for $\mathbb{R}^n$ is the set of standard unit vectors $\{\boldsymbol{e}_1, \dots, \boldsymbol{e}_n\}$. Geometrically, we can think of $\mathbb{R}^n$ as an n-dimensional coordinate system with the vector $\boldsymbol{x} \in \mathbb{R}^n$ representing the location of a point, with the entries of $\boldsymbol{x}$ being displacements of that point along the coordinate axes. The canonical basis consists of unit vectors aligned with the coordinate axes, and we have $\boldsymbol{x} = \sum_{i=1}^n x_i\boldsymbol{e}_i$. In general, a basis $\mathcal{B} = \{\boldsymbol{b}_1, \dots, \boldsymbol{b}_n\}$ is *orthonormal* if $\boldsymbol{b}_i^T\boldsymbol{b}_j = \delta_{ij}$ for all i, j. The standard basis is orthonormal. Given a basis we can write any vector in the form $\boldsymbol{x} = \sum_{i=1}^n \lambda_i\boldsymbol{b}_i$, where the scalars λ_i are the *components* or *coordinates* of $\boldsymbol{x}$ with respect to the basis. If the basis is orthonormal, then

$$\boldsymbol{b}_j^T\boldsymbol{x} = \sum_{i=1}^n \lambda_i\boldsymbol{b}_j^T\boldsymbol{b}_i = \sum_{i=1}^n \lambda_i\delta_{ij} = \lambda_j,$$

which provides an efficient method for computing coordinates.

A *linear transformation* is a function $\mathcal{A}: \mathbb{R}^n \to \mathbb{R}^m$ which maps any vector $\boldsymbol{x} \in \mathbb{R}^n$ to a vector $\boldsymbol{y} = \mathcal{A}(\boldsymbol{x}) \in \mathbb{R}^m$ with the property that $\mathcal{A}(\lambda_1\boldsymbol{x}_1 + \lambda_2\boldsymbol{x}_2) = \lambda_1\mathcal{A}(\boldsymbol{x}_1) + \lambda_2\mathcal{A}(\boldsymbol{x}_2)$. The function $\mathcal{A}$ will map each standard basis vector $\boldsymbol{e}_i \in \mathbb{R}^n$ to a vector $\boldsymbol{a}_i = \mathcal{A}(\boldsymbol{e}_i) \in \mathbb{R}^m$. Using the linearity of $\mathcal{A}$, we have

$$\boldsymbol{y} = \mathcal{A}(\boldsymbol{x}) = \mathcal{A}\left(\sum_{i=1}^{n} x_i\boldsymbol{e}_i\right) = \sum_{i=1}^{n} x_i\mathcal{A}(\boldsymbol{e}_i) = \sum_{i=1}^{n} x_i\boldsymbol{a}_i = \begin{pmatrix} | & | & & | \\ \boldsymbol{a}_1 & \boldsymbol{a}_2 & \dots & \boldsymbol{a}_n \\ | & | & & | \end{pmatrix}\begin{pmatrix} x_1 \\ \vdots \\ x_n \end{pmatrix} = \underline{\boldsymbol{A}}\boldsymbol{x}.$$

The action of a linear transformation on $\boldsymbol{x}$ may therefore be represented by a matrix $\underline{\boldsymbol{A}} \in \mathbb{R}^{m\times n}$ with columns given by the transformed basis vectors of $\mathbb{R}^n$. Conversely, every $\underline{\boldsymbol{A}} \in \mathbb{R}^{m\times n}$ represents a linear transformation from $\mathbb{R}^n$ to $\mathbb{R}^m$. The *range* of $\underline{\boldsymbol{A}}$ is the span of the columns of $\boldsymbol{A}$; $\text{range}(\underline{\boldsymbol{A}}) = \{\boldsymbol{v} \in \mathbb{R}^m | \boldsymbol{v} = \underline{\boldsymbol{A}}\boldsymbol{x}, \boldsymbol{x} \in \mathbb{R}^n\}$. This is the set of vectors in $\mathbb{R}^m$ which can be 'reached' by the transformation $\underline{\boldsymbol{A}}$. The *nullspace* of $\underline{\boldsymbol{A}} \in \mathbb{R}^{m\times n}$ is the set of vectors in $\mathbb{R}^n$ which are mapped to the null vector $\boldsymbol{0} \in \mathbb{R}^m$ (*annihilated*) by $\underline{\boldsymbol{A}}$; $\text{nullspace}(\underline{\boldsymbol{A}}) = \{\boldsymbol{x} \in \mathbb{R}^n | \underline{\boldsymbol{A}}\boldsymbol{x} = \boldsymbol{0}\}$. When $m \neq n$ the transformation will not be invertible (injective and surjective). Intuitively, if $n > m$ then the codomain ($\mathbb{R}^m$) is *too small* for an injective (one-one) map to be defined. If $n < m$ then the codomain is *too big* to be surjectively covered (every element in the codomain being mapped to). When $m = n$ the transformation is invertible provided the transformed basis vectors also form a basis for $\mathbb{R}^n$. This can be checked by calculating the *determinant* of $\underline{\boldsymbol{A}}$, equal to the signed volume of the parallelotope spanned by its columns. (The parallelotope is the generalisation of the parallelogram (2D) and parallelepiped (3D) to any number of dimensions.) If $\det(\underline{\boldsymbol{A}}) = 0$ then the columns of $\underline{\boldsymbol{A}}$ are linearly dependent (they span a lower-dimensional space) so do not form a basis for $\mathbb{R}^n$. We say that $\underline{\boldsymbol{A}}$ is *singular*, and since the map is to a proper subspace of the codomain, it cannot be injective. If $\det(\underline{\boldsymbol{A}}) \neq 0$ then the transformation is invertible and the matrix representing its inverse is written $\underline{\boldsymbol{A}}^{-1}$. We have $\underline{\boldsymbol{A}}\,\underline{\boldsymbol{A}}^{-1} = \underline{\boldsymbol{A}}^{-1}\underline{\boldsymbol{A}} = \underline{\boldsymbol{I}}$.

Since $\underline{\boldsymbol{A}}\boldsymbol{x} = \sum_{i=1}^{n} x_i\boldsymbol{a}_i$ where $\boldsymbol{a}_i$ is the ith column of $\underline{\boldsymbol{A}}$, then if $\det(\underline{\boldsymbol{A}}) = 0$, there exists $\boldsymbol{x} \in \mathbb{R}^n$ for which $\underline{\boldsymbol{A}}\boldsymbol{x} = \boldsymbol{0}$. That is, a singular matrix has non-trivial nullspace. Conversely, if $\underline{\boldsymbol{A}}$ is non-singular then $\underline{\boldsymbol{A}}\boldsymbol{x} = \boldsymbol{0}$ has no solutions other than $\boldsymbol{x} = \boldsymbol{0}$, meaning the nullspace is *trivial*. Therefore, $\underline{\boldsymbol{A}}$ has non-trivial nullspace if and only if it is singular.

Given $\underline{\boldsymbol{A}} \in \mathbb{R}^{n\times n}$ we say that λ is an *eigenvalue* of $\underline{\boldsymbol{A}}$ with corresponding (non-null) eigenvector $\boldsymbol{u} \in \mathbb{R}^n$ if $\underline{\boldsymbol{A}}\boldsymbol{u} = \lambda\boldsymbol{u}$. Note that if $\boldsymbol{u}$ is an eigenvector then so is $c\boldsymbol{u}$ for any non-zero $c \in \mathbb{R}$. Rewriting the eigenvalue equation $(\lambda\underline{\boldsymbol{I}} - \underline{\boldsymbol{A}})\boldsymbol{u} = 0$, we see that the matrix $(\lambda\underline{\boldsymbol{I}} - \underline{\boldsymbol{A}})$ has non-trivial nullspace (it annihilates a non-vanishing vector) and so must be singular, yielding the *characteristic equation* $\det(\lambda\underline{\boldsymbol{I}} - \underline{\boldsymbol{A}}) = 0$. This may be solved to find eigenvalues. If $\underline{\boldsymbol{A}}$ is symmetric ($\underline{\boldsymbol{A}}^T = \underline{\boldsymbol{A}}$) then it can be shown that eigenvectors corresponding to different eigenvalues are orthogonal, and that it is possible to construct an orthonormal basis $\{\boldsymbol{u}_1, \dots, \boldsymbol{u}_n\}$ of eigenvectors. Writing $\underline{\boldsymbol{U}}$ for the matrix with columns $\boldsymbol{u}_1, \dots, \boldsymbol{u}_n$ and $\underline{\boldsymbol{\Lambda}}$ for the matrix $\text{diag}(\lambda_1, \dots, \lambda_n)$, where λ_i is the eigenvalue associated with eigenvector $\boldsymbol{u}_i$, we have $\underline{\boldsymbol{U}}\,\underline{\boldsymbol{U}}^T = \underline{\boldsymbol{U}}^T\underline{\boldsymbol{U}} = \underline{\boldsymbol{I}}$, $\underline{\boldsymbol{A}}\,\underline{\boldsymbol{U}} = \underline{\boldsymbol{U}}\,\underline{\boldsymbol{\Lambda}}$, and hence $\underline{\boldsymbol{A}} = \underline{\boldsymbol{U}}\,\underline{\boldsymbol{\Lambda}}\,\underline{\boldsymbol{U}}^T$. Real symmetric matrices are therefore said to be *diagonalisable*.

Appendix C

Jensen's and Gibbs' Inequalities

A function $g: \mathbb{R}^d \to \mathbb{R}$ is said to be *convex* if for any $\boldsymbol{a}, \boldsymbol{b} \in \mathbb{R}^d$ and any $\lambda \in [0, 1]$,

$$g(\lambda \boldsymbol{a} + (1 - \lambda)\boldsymbol{b}) \leq \lambda g(\boldsymbol{a}) + (1 - \lambda)g(\boldsymbol{b}). \tag{C.1}$$

The function g is *strictly convex* if equality holds in equation (C.1) only for $\lambda = 0$ and $\lambda = 1$. To understand this definition geometrically, consider the chord connecting the points $(\boldsymbol{a}, g(\boldsymbol{a}))$ and $(\boldsymbol{b}, g(\boldsymbol{b}))$, in $\mathbb{R}^{d+1}$, given in parameterised form by

$$(\boldsymbol{x}(\lambda), y(\lambda)) = (\lambda \boldsymbol{a} + (1 - \lambda)\boldsymbol{b}, \lambda g(\boldsymbol{a}) + (1 - \lambda)g(\boldsymbol{b})),$$

where $(\boldsymbol{x}, y)$ is shorthand for the vector $(x_1, \ldots, x_n, y)^T \in \mathbb{R}^{d+1}$. If g is convex then for any $\boldsymbol{a}, \boldsymbol{b}$, this chord lies 'above' g in the sense that $y(\lambda) \geq g(\boldsymbol{x}(\lambda))$ for any $\lambda \in [0, 1]$. Figure C.1a shows a function of one variable, with the chord connecting $(a, g(a))$ to $(b, g(b))$ shown in orange. This function is convex because all such chords lie above it. The function in Figure C.1b is not convex because there are chords which lie partially or wholly below the function. Figure C.1c shows an example of a convex function of two variables.

Let g be a convex function. Suppose we distribute unit mass either smoothly or in discrete lumps over the curve, surface or hypersurface given by

$$\mathcal{M} = \left\{ (\boldsymbol{x}, y) \in \mathbb{R}^{d+1} \middle| y = g(\boldsymbol{x}) \right\}.$$

Because the total mass is 1, the mass distribution may be interpreted as a probability mass or probability density function of a random vector $\boldsymbol{X} \in \mathbb{R}^d$. The *centre of mass* of the object may then be written as $(\boldsymbol{x}_{\text{CM}}, y_{\text{CM}}) = (\mathbb{E}(\boldsymbol{X}), \mathbb{E}(g(\boldsymbol{X})))$. Appealing to physical intuition when

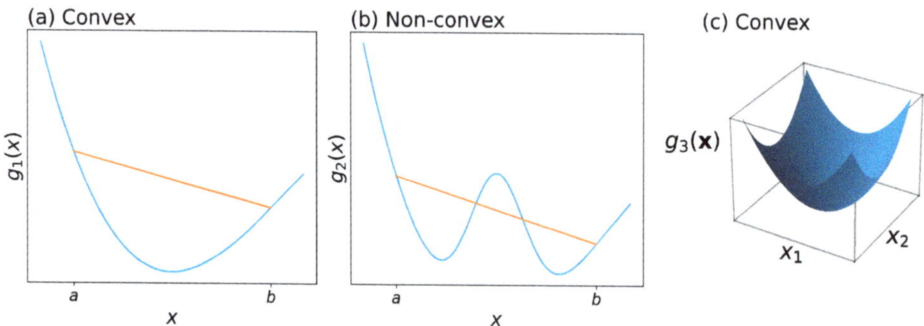

Figure C.1 (a) A convex function of a single variable. (b) A non-convex function of a single variable. (c) A convex function of two variables.

$d \leq 2$, since g is convex then this centre of mass must lie on or above $\mathcal{M}$ in the sense that $y_{\mathrm{CM}} \geq g(\boldsymbol{x}_{\mathrm{CM}})$. Written in terms of expectations, this is *Jensen's inequality*

$$\mathbb{E}(g(\boldsymbol{X})) \geq g(\mathbb{E}(\boldsymbol{X})),$$

which in fact holds for all $d \geq 1$. If g is strictly convex then equality holds only if the random vector $\boldsymbol{X}$ is a constant. Some examples of strictly convex functions $\mathbb{R} \to \mathbb{R}$ are $x^2, -\log x, \exp(x)$ and $x \log x$.

An inequality closely related to Jensen's is Gibbs', which we now introduce and prove. Given two probability density (or mass) functions p and q describing random vectors in $\mathbb{R}^d$, with $q(\boldsymbol{x}) > 0$ for all $\boldsymbol{x}$ such that $p(\boldsymbol{x}) > 0$, the Kullback–Leibler (KL) divergence of p from q is defined as follows:

$$\mathbb{KL}[p\|q] = \mathbb{E}_{\boldsymbol{X}\sim p}\left(\log \frac{p(\boldsymbol{X})}{q(\boldsymbol{X})}\right).$$

To avoid clutter we will focus on the continuous case, where p and q are densities. However, the inequality we prove below holds for the discrete case too, and the proof is essentially the same. Assuming that p and q are indeed densities, we have

$$\mathbb{KL}[p\|q] = \int p(\boldsymbol{x}) \log \frac{p(\boldsymbol{x})}{q(\boldsymbol{x})} d\boldsymbol{x}.$$

(We either restrict the domain of integration to the support of p – that is, to the set of $\boldsymbol{x}$ values where $p(\boldsymbol{x})$ is non-zero – or we adopt the convention that $0 \times \log(0) = 0$.) Often p is understood to be a 'true' distribution and q a model, in which case $\mathbb{KL}[p\|q]$ is a measure of how close q is to p. *Gibbs' inequality* is as follows:

$$\mathbb{KL}[p\|q] \geq 0,$$

with equality if and only if $p = q$. To prove this, let $\boldsymbol{X}$ be the random vector with density p and define the random variable $Z = q(\boldsymbol{X})/p(\boldsymbol{X})$. By the LOTUS,

$$\mathbb{E}(Z) = \int p(\boldsymbol{x}) \frac{q(\boldsymbol{x})}{p(\boldsymbol{x})} d\boldsymbol{x} = 1,$$

and $\mathbb{E}(-\log Z) = \mathbb{KL}[p\|q]$. Since $-\log Z$ is a (strictly) convex function of Z then by Jensen's inequality $\mathbb{E}(-\log Z) \geq -\log(\mathbb{E}(Z)) = 0$. Hence, $\mathbb{KL}[p\|q] \geq 0$. If $p = q$ then $Z = 1$ and $\mathbb{KL}[p\|q] = 0$. Conversely, if $\mathbb{KL}[p\|q] = 0$ then Z is a constant, implying that $p = q$. Therefore, $\mathbb{KL}[p\|q] = 0$ if and only if $p = q$.

The cross-entropy of q relative to p is defined as follows:

$$\mathbb{H}[p, q] = -\int p(\boldsymbol{x}) \log q(\boldsymbol{x}) d\boldsymbol{x}.$$

Using the definition of the KL divergence, we have $\mathbb{H}[p, q] = \mathbb{H}[p] + \mathbb{KL}[p\|q]$ where $\mathbb{H}[p] = \mathbb{H}[p, p]$ is the *entropy* of p. Viewing the cross-entropy as a function of q with p fixed, we have by Gibbs' inequality

$$\mathbb{H}[p, q] \geq \mathbb{H}[p],$$

with equality if and only if $q = p$. Therefore, the cross-entropy achieves its unique global minimum when $q = p$.

References

Alexander, L. V. and Jones, P. D. (2000). Updated precipitation series for the U.K. and discussion of recent extremes. *Atmospheric Science Letters*, 1.

Bishop, C. M. (2006). *Pattern Recognition and Machine Learning (Information Science and Statistics)*. Springer-Verlag.

Bishop, C. M. and Bishop, H. (2024). *Deep Learning: Foundations and Concepts*. Springer.

Clopper, C., Pisoni, D., and de Jong, K. (2005). Acoustic characteristics of the vowel systems of six regional varieties of American English. *The Journal of the Acoustical Society of America*, 118:1661–1676.

Cunningham, G. D., While, G. M., and Wapstra, E. (2017). Climate and sex ratio variation in a viviparous lizard. *Biology Letters*, 13.

Deng, J., Dong, W., Socher, R., Li, L.-J., Li, K., and Fei-Fei, L. (2009). ImageNet: A large-scale hierarchical image database. *2009 IEEE Conference on Computer Vision and Pattern Recognition*, 248–255.

El Kharoua, R. (2024). Diabetes health dataset analysis. Kaggle, www.kaggle.com/datasets/rabieelkharoua/diabetes-health-dataset-analysis.

Gelman, A., Carlin, J. B., Stern, H. S., Dunson, D. B., Vehtari, A., and Rubin, D. B. (2024). *Bayesian Data Analysis*. CRC Press.

Gelman, A. and Rubin, D. B. (1992). Inference from iterative simulation using multiple sequences. *Statistical Science*, 7:457–472.

Grimmett, G. and Stirzaker, D. (2001). *Probability and Random Processes*. Oxford University Press.

Hastie, T., Tibshirani, R., and Friedman, J. (2001). *The Elements of Statistical Learning*. Springer Series in Statistics. Springer New York Inc.

Johnson, R. W. (1996). Fitting percentage of body fat to simple body measurements. *Journal of Statistics Education*, 4(1).

MacKay, D. J. C. (2003). *Information Theory, Inference, and Learning Algorithms*. Cambridge University Press.

Mardia, K., Kent, J., and Taylor, C. (2024). *Multivariate Analysis*. Wiley Series in Probability and Statistics. Wiley.

McShane, B. B., Gal, D., Gelman, A., Robert, C. P., and Tackett, J. L. (2017). Abandon statistical significance. *The American Statistician*, 73:235–245.

Murphy, K. P. (2022). *Probabilistic Machine Learning: An Introduction*. MIT Press.

Press, W. H., Teukolsky, S. A., Vetterling, W. T., and Flannery, B. P. (2007). *Numerical Recipes 3rd Edition: The Art of Scientific Computing*. Cambridge University Press, 3rd ed.

Prince, S. J. (2023). *Understanding Deep Learning*. MIT Press.

Tibshirani, R., Walther, G., and Hastie, T. J. (2000). Estimating the number of clusters in a data set via the gap statistic. *Journal of the Royal Statistical Society: Series B (Statistical Methodology)*, 63.

Tibshirani, R. J. and Tibshirani, R. (2009). A bias correction for the minimum error rate in cross-validation. *The Annals of Applied Statistics*, 3:822–829.

Vaux, B. and Johndal, M. (2025). Cambridge online survey of world Englishes. https://tekstlab.uio.no/cambridge_survey/.

Vehtari, A., Gelman, A., and Gabry, J. (2017). Practical Bayesian model evaluation using leave-one-out cross-validation and WAIC. *Statistics and Computing*, 27:1413–1432.

Xiao, H., Rasul, K., and Vollgraf, R. (2017). Fashion-MNIST: a novel image dataset for benchmarking machine learning algorithms. arXiv preprint arXiv:1708.07747.

Index

accuracy, 199
activation, 256
activation function, 256
  linear, 258
  ReLU, 257
  sigmoid, 258
adaptive moment estimation (Adam), 272
ancestral sampling, 232
argmax, 9
argmin, 9
autocorrelation function, 172
autoregressive model, 297

backpropagation, 266, 285
  algorithm, 289
basis vector, 245
Bayes' theorem, 299
Bayesian
  conjugate prior, 90
  credible interval, 80
  hierarchical model, 188
  highest density interval, 80
  inference, 6, 39
  multiple linear regression, 136
  network, 124
  p-value, 158
  posterior, 40, 75
  prior, 40, 60, 74
  probability, 6
  simple linear regression, 133
Berry–Esseen theorem, 87
bias-variance trade-off, 118
Bienaymé's identity, 101
bootstrap, 19
  nonparametric, 82
  parametric, 81

categorical variable, 192
central limit theorem, 86
change of random variable, 300
classification, 3
classifier, 193
  multiclass, 208
clustering, 5, 222
  hard, 232
  soft, 232
CNN, 278
  interpretation, 282
coefficient of determination, 105
collinearity, 114
conditional
  density, 300
  mass function, 300
conditional correlation, 145, 146
conditional covariance, 113
conditional expectation, 300
conditional independence, 125, 127
conditional probability, 299
confidence interval, 67
continuous random variable, 300
convergence
  in distribution, 86
convolutional
  filter, 277
  neural network, 278
correlation, 98
correlation coefficient, 98
covariance matrix, 146
  estimate, 147
cross-entropy, 179, 234
cross-entropy loss, 195, 210
  categorical, 209
cross-validation, 13, 33, 52, 234
  k fold, 101
  leave one out, 101
curse of dimensionality, 5, 164, 219

decoder, 243
deep learning, 7
density estimation, 4, 229
design matrix, 137
  centred, 244
detailed balance, 166
differential equations, 295
dimensionality reduction, 6, 241
directed acyclic graph, 124
  collider, 128
  constant node, 134
  descendants, 129
  fork, 128
  pipe, 128
  plate, 135
discrete random variable, 299

disjoint, 299
distribution
  Bernoulli, 61
  beta, 79
  beta-binomial, 161
  exponential, 87
  exponential family, 89
  gamma, 90, 141
  inverse gamma, 141
  multivariate normal, 138
  normal, 43
  Poisson, 89
  Rayleigh, 181
distribution function, 299
d-separation, 128, 129

effective sample size, 172
empirical Bayes, 187
empirical covariance, 144
empirical distribution, 82
encoder, 243
epoch, 271
equilibrium, 166
estimator, 64
  biased, 66
  consistent, 66
  maximum a posteriori (MAP), 76
  maximum likelihood, 62
event, 299
evidence, 163
expectation, 300
expectation maximisation, 235
  Gaussian mixtures, 240
  K-means, 225
expected log pointwise predictive density, 178
experiment, 299
explained sum of squares, 106

feature map, 276
Fisher information, 85
frequentist
  inference, 6
  probability, 6
  p-value, 72
  uncertainty, 65

Gaussian mixture model, 29, 230
generalised additive model, 294
Gibbs sampling, 168
gradient descent, 266, 267
  momentum, 269

Hamiltonian Monte Carlo, 168
heteroscedastic, 50
homoscedastic, 50, 104
hypothesis
  alternative, 70
  null, 70
hypothesis test
  power, 71
  size, 71

independent, 299
inductive bias, 285
inertia, 223
irreducible error, 120

joint density, 300
joint distribution, 300

k nearest neighbours, 215
K-means, 224
  algorithm, 225
  elbow method, 226
  gap statistic, 226
Kullback–Leibler divergence, 235

large language model, 296
lasso regression, 151
latent variable, 74, 134, 232, 235, 243
  model, 235
law of the unconscious statistician (LOTUS), 299
learning rate, 269, 270, 274
  adaptive, 270
likelihood, 37
  conditional, 48
  function, 47, 62
  and least squares, 50
  principle, 37, 48, 54
  unconditional, 54
link function, 196
local regression, 295
log odds, 196
logistic regression
  binary, 203
  multinomial, 212
  multiple, 202
  regularised, 202
  simple, 197
logit function, 196

marginal, 112
  density, 300
  distribution, 300
  mass function, 300
marginal likelihood, 163
Markov chain, 165
Markov Chain Monte Carlo, 91, 164
max pooling, 278
Metropolis algorithm, 167
mixing time, 169
mixture model, 27
momentum coefficient, 269
multilayer perceptron, 257, 272
mutually exclusive, 299

neural network, 7, 256
  bias, 256
  deep, 264
  feedforward, 257
  fully connected, 257
  hidden layer, 262
  layer, 257

recurrent, 256
single layer, 260
weight, 256
neuron
artificial, 256
biological, 255
nominal variable, 192

Occam's razor, 226
one hot encoding, 210
ordinal variable, 192
orthogonal projection, 245, 246
overdispersion, 158
overfitting, 3, 14

parents, 124
partition, 299
patch, 276
posterior, 232
posterior dataset predictive, 157
posterior mean, 77
posterior predictive, 159
pre-activation, 256
precision matrix, 144
principal component analysis, 246
algorithm, 250
principal components, 250
prior, 232
choosing, 173
flat, 140
improper, 174
informative, 173
Jeffreys, 174
non-informative, 140
uninformative, 174
weakly informative, 174
probabilistic model
conditional, 42
unconditional, 54
probability density function, 300
probability mass function, 299
probability measure, 299
projection matrix, 246
proposal distribution, 167

R-hat, 171
R-squared, 105
random variable, 299
random walk, 164
reconstruction loss, 243
rectified linear unit (ReLU), 257
reducible error, 120
regression, 3
function, 97
multiple linear, 97, 110
simple linear, 103
regularisation, 144
residual, 46, 104
residual neural network, 284
residual sum of squares, 9, 105
ridge regression, 149
risk, 119

sample space, 299
sample standard deviation, 108
sample variance, 109
sampling distribution, 60, 64
scientific hypothesis, 41
score function, 85
self-attention, 284
shrinkage prior, 147
significance level, 71
smoothing spline, 293
softmax, 211
layer, 261
span, 245
sparse model, 151
spline, 293
standard error, 64, 65
plug-in estimate, 65
standard sigmoid, 196
standardised predictor, 109
stationary process, 165
stochastic gradient descent (SGD), 271
stochastic process, 165, 296
subset selection
best, 116
forward, 116
sufficient statistic, 90
supervised learning, 3

test set, 273
total sum of squares, 105
total variance, 244
trace plot, 171
training set, 273
transformer, 285
trial, 299
type I error, 71
type II error, 71

uncertainty
aleatory, 157
epistemic, 157
underfitting, 14
universal approximation theorem (UAT), 262
unsupervised learning, 4, 25, 222

validation loss, 274
validation set, 273
vector
notation, 47
random, 47

zero padding, 277

For EU product safety concerns, contact us at Calle de José Abascal, 56–1°,
28003 Madrid, Spain or eugpsr@cambridge.org.

www.ingramcontent.com/pod-product-compliance
Ingram Content Group UK Ltd.
Pitfield, Milton Keynes, MK11 3LW, UK
UKHW052108150726
473310UK00012B/571

* 9 7 8 1 0 0 9 6 3 0 7 2 6 *